Two-Dimensional Nanostructures for Biomedical Technology

A Bridge between Material Science and Bioengineering

Edited by

Raju Khan

Senior Scientist
Analytical Chemistry Group, Chemical Sciences & Technology
Division, CSIR-NEIST, Jorhat, Assam, India

CSIR-Advanced Materials and Processes Research Institute (AMPRI)
Bhopal, Madhya Pradesh, India

Shaswat Barua

Assistant Professor
Department of Chemistry, School of Basic Sciences,
Assam Kaziranga University, Jorhat, Assam, India

CSIR-Nehru Fellow
Analytical Chemistry Group, Chemical Sciences & Technology
Division, CSIR-NEIST, Jorhat, Assam, India

ELSEVIER

Elsevier
Radarweg 29, PO Box 211, 1000 AE Amsterdam, Netherlands
The Boulevard, Langford Lane, Kidlington, Oxford OX5 1GB, United Kingdom
50 Hampshire Street, 5th Floor, Cambridge, MA 02139, United States

Two-Dimensional Nanostructures for Biomedical Technology

Notices

Practitioners and researchers must always rely on their own experience and knowledge in evaluating and using any information, methods, compounds or experiments described herein. Because of rapid advances in the medical sciences, in particular, independent verification of diagnoses and drug dosages should be made. To the fullest extent of the law, no responsibility is assumed by Elsevier, authors, editors or contributors for any injury and/or damage to persons or property as a matter of products liability, negligence or otherwise, or from any use or operation of any methods, products, instructions, or ideas contained in the material herein.

ISBN: 978-0-12-817650-4

Publisher: Susan Dennis
Acquisition Editor: Kostas Marinakis
Editorial Project Manager: Ruby Smith
Production Project Manager: Sreejith Viswanathan
Cover Designer: Alan Studholme

Contents

CHAPTER 3   Properties of two-dimensional nanomaterials..........73

N.B. Singh and Saroj K. Shukla

**CHAPTER 4   Graphene-based nanostructures for biomedical
                applications.. 101**

*Keisham Radhapyari, PhD, Suparna Datta, PhD,
Snigdha Dutta, MSc, Nimisha Jadon, PhD and
Raju Khan, PhD*

CHAPTER 7 Transition metal dichalcogenides for biomedical applications...211

Rekha Rani Dutta, PhD, Rashmita Devi, Hemant S. Dutta and Satyabrat Gogoi, PhD

CHAPTER 8 Polymer nanocomposites based on two-dimensional nanomaterials 249

Rajarshi Bayan, BSc, MSc and Niranjan Karak, MTech, PhD

Contributors

Shaswat Barua, PhD
Analytical Chemistry Group, Chemical Sciences & Technology Division, CSIR-NEIST, Jorhat, Assam, India; Department of Chemistry, School of Basic Sciences, Assam Kaziranga University, Jorhat, Assam, India

Rajarshi Bayan, BSc, MSc
Advanced Polymer & Nanomaterial Laboratory, Department of Chemical Sciences, Tezpur University, Tezpur, Assam, India

Kartick Chandra Majhi, BSc, MSc
Department of Applied Chemistry, Indian Institute of Technology (Indian School of Mines), Dhanbad, Jharkhand, India

Suparna Datta, PhD
Regional Chemical Laboratory, Central Ground Water Board, Eastern Region, Ministry of Jal Shakti, Department of Water Resources, River Development and Ganga Rejuvenation, Kolkota, West Bengal, India

Rashmita Devi
Analytical Chemistry Group, Material Sciences & Technology Division, Academy of Scientific and Innovative Research, CSIR North-East Institute of Science & Technology, Jorhat, Assam, India

Rekha Rani Dutta, PhD
Department of Chemistry, School of Basic Sciences, Assam Kaziranga University, Jorhat, Assam

Hemant S. Dutta
Analytical Chemistry Group, Material Sciences & Technology Division, Academy of Scientific and Innovative Research, CSIR North-East Institute of Science & Technology, Jorhat, Assam, India

Snigdha Dutta, MSc
Regional Chemical Laboratory, Central Ground Water Board, North Eastern Region, Ministry of Jal Shakti, Department of Water Resources, River Development and Ganga Rejuvenation, Guwahati, Assam, India

Ilknur Erucar
Assistant Professor, Department of Natural and Mathematical Sciences, Faculty of Engineering, Ozyegin University, Cekmekoy, Istanbul, Turkey

Satyabrat Gogoi, PhD
Analytical Chemistry Group, Material Sciences & Technology Division, Academy of Scientific and Innovative Research, CSIR North-East Institute of Science & Technology, Jorhat, Assam, India

Ezgi Gulcay
Master Student, Department of Mechanical Engineering, Faculty of Engineering, Ozyegin University, Cekmekoy, Istanbul, Turkey

Nimisha Jadon, PhD
School of Studies in Environmental Chemistry, Jiwaji University, Gwalior, Madhya Pradesh, India

Niranjan Karak, MTech, PhD
Advanced Polymer & Nanomaterial Laboratory, Department of Chemical Sciences, Tezpur University, Tezpur, Assam, India; Professor. Department of Chemical Sciences, Tezpur University, Tezpur, Assam, India

Paramita Karfa, BSc, MSc
Department of Applied Chemistry, Indian Institute of Technology (Indian School of Mines), Dhanbad, Jharkhand, India

Raju Khan, PhD
Analytical Chemistry Group, Chemical Sciences & Technology Division, CSIR-NEIST, Jorhat, Assam, India; CSIR-Advanced Materials and Processes Research Institute (AMPRI), Bhopal, Madhya Pradesh, India

Rashmi Madhuri, MSc, PhD
Department of Applied Chemistry, Indian Institute of Technology (Indian School of Mines), Dhanbad, Jharkhand, India; Assistant Professor, Applied Chemistry, Indian institute of technology (Indian School of Mines), Dhanbad, Jharkhand, India

Sujata Pramanik, PhD
All India Institute of Medical Sciences, Bhubaneswar, Odisha, India

Keisham Radhapyari, PhD
Regional Chemical Laboratory, Central Ground Water Board, North Eastern Region, Ministry of Jal Shakti, Department of Water Resources, River Development and Ganga Rejuvenation, Guwahati, Assam, India

Debojeet Sahu, PhD
Department of Chemistry, Assam Royal Global University, Guwahati, India

Pallabi Saikia, PhD
Assistant Professor, Department of Chemistry, School of Basic Sciences, Assam Kaziranga University, Jorhat, Assam, India

Nasifa Shahnaz, PhD
Department of Chemistry, University of Science & Technology, Meghalaya, Baridua, Meghalaya, India

Saroj K. Shukla
Department of Polymer Science, Bhaskaryacharya College of Applied Sciences,
University of Delhi, Delhi, India

N.B. Singh
Professor, Department of Chemistry and Biochemistry, Research and Technology
Development Centre, SBSR and RTDC, Sharda University, Greater Noida, Uttar
Pradesh, India

Dhriti Sundar Das, MBBS, MD
All India Institute of Medical Sciences, Bhubaneswar, Odisha, India

Preface

Nanomaterials have been fascinating the scientific community due to their unique properties. Multifaceted attributes of nanostructured materials have opened newer avenues in the domain of biomedical technology. Consequently, newer possibilities are explored, which offer significant contribution to novel diagnostic and therapeutic applications. However, the shape and size accord, as well as the dimension of a nanostructured material, dictates the overall properties. In terms of dimensions, nanomaterials have been categorized as zero-, one-, two-, and three-dimensional nanostructures. In this book, we are interested to discuss the importance of two-dimensional nanomaterials that signify the nanostructures with only one of their dimensions in the nano regime. It is a very important class that mostly includes nanostructures such as nanosheets including graphene, nanoclay, transition metal dichalcogenides, etc. Owing to their very high aspect ratio, this class of nanomaterials has shown tremendous potential in the field of drug delivery, nanomedicine, and biomedical technology. Another vital aspect of such materials is their excellent electronic and optical properties, which are extremely useful for advanced diagnostic techniques. This book would help researchers to understand the various promising aspects of two-dimensional nanomaterials. In this book, we have showcased the biomedical aspects of such nanostructures in terms of their precursors, structures, morphology, and size. Furthermore, the detailed synthetic methods would guide a reader towards the efficient generation of two-dimensional nanostructures, which is expected to be a timely contribution to the scientific fraternity. Thus this book is useful to the research scholars and scientists working in materials science, nanotechnology, biomedical domain, and composite science and technology and the graduate students from science or engineering background with specialization in nanotechnology or material science.

The book comprises of nine chapters. Chapter 1 deals with the chemistry and general aspects of two-dimensional nanomaterials. It also covers some important aspects of physics, which are integral parts of material research. Chapter 2 and Chapter 3, respectively, discuss the various synthetic methods and properties of two-dimensional nanomaterials. Each of the following chapters elaborately explain a typical two-dimensional nanomaterial. Chapters 4–7 attempt to cover the detailed information about the structure, synthesis, properties, and biomedical aspects of graphene, nanoclay, metal-organic frameworks, and metal dichalcogenides, respectively. Polymer nanocomposites are regarded as the new-generation materials with immense potential in the field of biomedical technology. Hence, Chapter 8 is dedicated to polymer nanocomposites based on two-dimensional nanomaterials. Finally, Chapter 9 presents a critical overview on the future prospects and the commercial viability of two-dimensional nanostructures in biomedical technology. The required references are cited in each chapter for further study on the topic.

Raju Khan
Shaswat Barua

Acknowledgments

The editors thankfully acknowledge the contributors for their sincere and dedicated efforts. The publisher and the publishing team are sincerely acknowledged for their kind assistance in publishing this book. Furthermore, the editors sincerely thank all who have directly or indirectly rendered valuable inputs to this book.

I am extremely indebted to those who have helped me in this book. Thanks especially to Dr. Sunil K Sanghi, Chief Scientist & Head, CSIR-AMPRI, Bhopal for always being available for advice and, even more so, valuable guidance. I would like to thank my colleague Dr. Shaswat Barua from the bottom of my heart for his kind help, hard work in the completion of our book, and encouragement. I thank my family (my mother, wife, and daughter) for their everlasting love, enthusiasm for science, and encouragement to pursue whatever it is I want to do.

Raju Khan

I thankfully acknowledge the Assam Kaziranga University for offering me constant support. I further thank the Council of Scientific and Industrial Research (CSIR), India for the CSIR-Nehru fellowship. Dr. Satyabrat Gogoi, Miss Rashmita Devi, and Mr. Hemant Shankar Dutta are thankfully acknowledged for their cooperation. I sincerely acknowledge the suggestions of Prof. Niranjan Karak, Tezpur University, Assam, India. Special thanks to my parents (Late Devi Charan Barua and Mrs. Mina Barua), my wife (Dr. Swagata Baruah), my brothers (Sashi, Jagadish, and Prantik), my sisters-in-law (Jyotshnali, Meghali, and Rupanjali), my nephew (Nirmaan), and my niece (Aradhya) for their blessings, patience, support, and encouragement, respectively.

Shaswat Barua

Chemistry of two-dimensional nanomaterials

Shaswat Barua, PhD [1,2]**, Debojeet Sahu, PhD** [3]**, Nasifa Shahnaz, PhD** [4]**, Raju Khan** [1,5]

[1]*Analytical Chemistry Group, Chemical Sciences & Technology Division, CSIR-NEIST, Jorhat, Assam, India;* [2]*Department of Chemistry, School of Basic Sciences, Assam Kaziranga University, Jorhat, Assam, India;* [3]*Department of Chemistry, Assam Royal Global University, Guwahati, India;* [4]*Department of Chemistry, University of Science & Technology, Meghalaya, Baridua, Meghalaya, India;* [5]*CSIR-Advanced Materials and Processes Research Institute (AMPRI), Bhopal, Madhya Pradesh, India*

1.1 Introduction

Nanomaterials have been fascinating the scientific community because of their unique attributes in the field of material science as well as biomedical technology. Since the past few decades, a good number of research works have been published in the domain of nanoscience and nanotechnology. It is interesting to note that the properties of a nanomaterial are mostly attributed to its unique structural identity. Thus it has been noted that dimensionally different nanomaterials show different properties [1]. In terms of dimensions, nanomaterials have been categorized as zero-, one-, two (2D)-, and three-dimensional nanostructures. Zero-dimensional materials are those nanostructures that have all the three dimensions within nanometer range. Spherical nanoparticles, quantum dots, etc. have been included in this class. The next category includes the nanostructures with two of their dimensions within nanometer range [2]. Nanotubes, nanofibers, nanowires, etc. are included in this category [3,4]. In this chapter, we are interested to discuss the significance of another category of nanomaterials in the field of biomedical technology. This category, viz. 2D nanomaterials, signifies the nanostructures with only one of their dimensions in the nano regime [5]. It is a very important class that mostly includes nanostructures such as nanosheets (e.g., graphene), nanoclay, and transition metal dichalcogenides (TMDs). Due to very high aspect ratio, this class of nanomaterials have shown tremendous potential in the field of drug delivery, nanomedicine, and biomedical technology [6–8]. Another vital aspect of such materials is their excellent electronic and optical properties, which are extremely useful for advanced diagnostic techniques [9,10]. In this chapter the chemistry of 2D nanomaterials has been thoroughly discussed with respect to their use in biomedical technology.

Most of the 2D nanostructures are sheet-type materials of nanometer thickness. Thus it is very convenient to anchor other nanomaterials or biomolecules onto them

Two-Dimensional Nanostructures for Biomedical Technology. https://doi.org/10.1016/B978-0-12-817650-4.00001-2

via covalent or noncovalent interactions [11]. Here comes the critical analysis of the surface properties of such nanostructures. Nanoclays are the most widely studied 2D nanomaterials because of their low cost and excellent biocompatibility [12]. These are mainly layered aluminosilicates with about 1 nm thickness. Physical forces, such as van der Waals interaction, hold the layers together to form galleries [13]. Among various groups of nanoclays, smectite clays are broadly investigated. Nanoclays, owing to their good biocompatibility and porous structure, offer a platform for cell adhesion, which make them promising candidates for tissue engineering applications [14]. Different forms of nanoclays have been endorsed for different biomedical applications by either combining with other nanomaterials or incorporating appropriate biomolecules [15].

After the discovery of graphene, nanostructured graphene materials have been of utmost interest to the scientific community because of their manifold advanced attributes. Graphene has shown excellent competence in biomedical technology, because of its high surface area (theoretically, 2630 m^2g^{-1}), unique optoelectronic properties, and tunable surface functionalities. These properties are greatly dependent on the number of layers per stack in graphene-based nanomaterials. The lateral dimension of graphene also dictates the biomedical properties such as cellular uptake and transport through blood-brain barrier [16]. Thus it is interesting to delve into the surface chemistry of graphene-based nanostructures to develop advanced biomaterials. Graphene oxide (GO) and reduced graphene oxide (RGO) have also been investigated extensively in this regard.

Another very important class of 2D nanostructures has gained importance in the domains of materials science as well as biomedical technology. This class includes TMDs. Among the TMDs, MoS_2 and WS_2 have been studied widely. 2D MoS_2 and WS_2 resemble graphene in terms of structural features [6]. Thus they have properties similar to those of graphene-based nanostructures. Such nanostructures have opened newer avenues for advanced biomedical diagnostics and therapeutics [1].

Metal-organic frameworks (MOFs) have attracted copious attention of the researchers owing to their ultrathin 2D nanostructures, dimension-based physicochemical properties, and intrinsic porosity [17]. MOF-based 2D materials have shown tremendous potential in catalysis, biomedicine, biosensors, etc. [18]. MOFs are metals or metal clusters chemically linked by organic ligands. Owing to their unique chemical and physical attributes, they have been used in bioimaging and sensing via the photoluminescence mechanism [19]. Good biocompatibility makes them competent candidates for sustained-release drug delivery systems [20]. Biomimetic mineralization of important biomolecules such as DNA and antibody has been reported by using different MOFs [21]. Fig. 1.1 pictorially demonstrates the major biomedical aspects of 2D nanomaterials.

Thus it is interesting to study the chemistry of 2D nanomaterials and their dimension-based properties. This chapter thoroughly discusses the chemical structures of such nanomaterials as well as their unique chemical attributes. However,

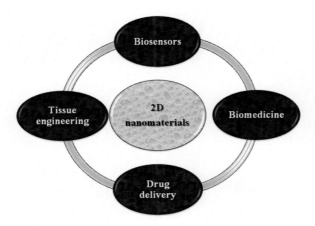

FIGURE 1.1

Some of the major biomedical aspects of two-dimensional (2D) nanomaterials.

the prime concern of this chapter is the biomedical use of such nanostructures, which needs a thorough investigation of their surface functionalities.

1.2 General aspects of two-dimensional nanomaterials

The ideal aspect of a 2D nanomaterial, such as graphene, is its structure, composed of a single layer of atoms. The atoms are strongly bonded by covalent linkages. However, most of the 2D nanostructures are combinations of different layers stacked together by van der Waals forces of attraction [22]. The sheetlike compact structure of 2D nanomaterials imparts excellent stability [23]. The chemical composition dictates the properties of such materials. A 2D nanostructure may be visualized as stacked layers combined together by weak physical forces. This stacking can be disturbed by employing various physical and chemical means. Such modifications result in the rearrangement of layers, which may be either simple delamination or exfoliation [24]. In simple delamination the interlayer spacing increases (Fig. 1.2) that allows other nanomaterials or biomolecules to enter into the galleries of the nanostructure [12,25]. Exfoliation of nanolayers occurs when a strong chemical or physical force is applied. Exfoliation results in the rearrangement of layers in the space (Fig. 1.2). This is a vital criterion for fabricating nanohybrids or nanocomposites [24]. Exfoliated nanostructures can efficiently interact with other nanomaterials, polymers, or biomolecules in the presence of an appropriate medium. Thus surface modification is a major requisite to use a nanomaterial in a particular field of application. Especially, biomedical applications of such nanomaterials demand a biocompatible surface that should also be efficient enough to perform the desired task. For implantable biomaterials, non-immunogenicity is a major requirement [14]. Fig. 1.2 depicts the general steps used in the modification of 2D nanostructures and fabrication of nanohybrids/nanocomposites.

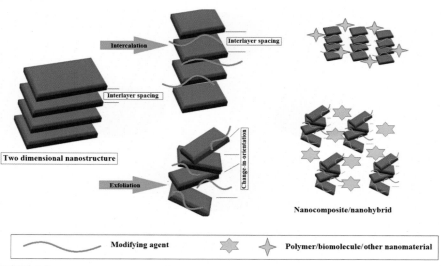

FIGURE 1.2

Modification of two-dimensional nanostructures and the fabrication of nanohybrids/nanocomposites.

1.3 Surface chemistry of two-dimensional nanomaterials

1.3.1 Nanoclay

Clay-based nanomaterials have been widely studied because of their versatile use in the domain of composite science. These naturally occurring layered silicates are composed of tetrahedrally bound Si atoms to $Al(OH)_3$ or $Mg(OH)_2$ edges that are octahedrally shared [26]. The surface-to-volume ratio is very high in case of these 2D nanosilicates. About 1-nm-thick layers are stacked by weaker physical forces of attraction, which allows easy modification [1]. Over the past few decades, nanoclays have attracted considerable research interest mainly because of their easy availability, low cost, and environmentally benign properties. Many works have been carried out exploring the surface properties and stability of nanoclays, fabrication of the surface layer-filled nanocomposites, use of nanoclays as a support for noble metals as catalysts to achieve reusability, etc. Some publications have been reported showing the potential applications of nanoclays in pharmaceutics [27−29], catalysis [30,31], industries [32], and waste water treatment for environmental protection [33]. Moreover, the wider applicability of nanoclays can be attributed to the easy surface modification providing scope for altering the properties such as polarity, surface area, acidity, and pore size, depending on different necessities.

The chemical structure of clay minerals has been fascinating the scientific community for a long time. Pauling first revealed the structure of kaolinite clay, consisting of tetrahedral silica and octahedral alumina in 1:1 ratio by sharing an "O" atom

[34]. Kaolinite clay is composed of silica (46.54%), alumina (39.50%), and water (13.96%). The clay possesses a net negative charge, which imparts reactivity to the clay surface. Another widely explored nanoclay is montmorillonite (MMT) belonging to the smectite group. MMT clays are soft phyllosilicates composed of tetrahedral silica sheets sandwiched around an octahedral alumina core, with 2:1 silica-alumina ratio. The hydrophilic clay galleries expand when they come into contact with water. Due to isomorphic substitution, the MMT clay surface induces a net negative charge, which is balanced by cations, such as Na^+, K^+, Mg^{2+}, and Ca^{2+}, present in the interlamellar spaces [35]. This feature provides the scope of surface modification of such clay structures by cation exchange mechanism [15].

The Al-Mg charge defects in the octahedral layer of MMT clay modify the chemical structure of the clay. Organic surfactants have been widely used to exchange Na^+ ions on the aluminosilicate layers to obtain organically modified MMT (OMMT). This helps to decrease hydrophilicity that makes the clay compatible with different biomolecules and polymers. The molecular model structure of MMT clay has been reported by Drummy et al. [36] (Fig. 1.3A and B).

Furthermore, researchers have revealed that the crystal structure of MMT clay is monoclinic. However, X-ray diffraction study (of 6A OMMT) showed *hkl*-type reflections from MMT lattice [36], which is similar to Cloisite Na^+ (Fig. 1.3C).

1.3.2 Graphene-based nanomaterials

Graphene is an allotrope of carbon, consisting of a single 2D layer of carbon atoms in hexagonal pattern, forming a honeycomb crystal lattice. The carbon atoms are sp^2-hybridized consisting of four bonds, i.e., three sigma bonds and one *pi* bond

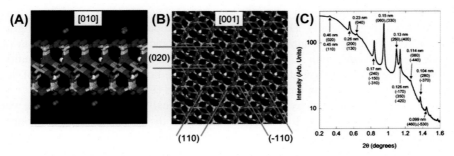

FIGURE 1.3

Molecular structure of (A) montmorillonite (MMT) clay [010] and (B) [001] projections (orange, Si atoms; light purple, Al atoms; red, O atoms; white, H atoms; green, Mg atoms; dark purple, Na atoms). (C) X-ray diffraction patterns of 6A organically modified MMT clay.

Reproduced with permission from Drummy, L. F., Koerner, H., Farmer, K., Tan, A., Farmer, B. L., Vaia, R. A. (2005). High-resolution electron microscopy of montmorillonite and montmorillonite/epoxy nanocomposites. J. Phys. Chem. B, 109(38), 17868–17878.

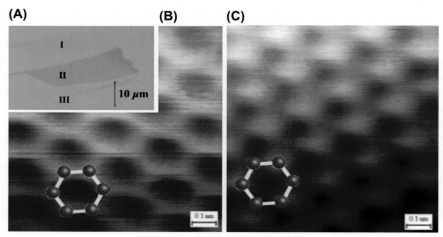

FIGURE 1.4

(A) Optical microscopic images of (I) single-layer graphene, (II) multilayer graphene, and (III) substrate. Scanning tunneling microscopic images of (B) single-layer graphene from the region shown in A-I and (C) multilayer graphene from the region shown in A-II.

Reproduced with permission under Creative Commons License from Stolyarova, E., Rim, K. T., Ryu, S., Maultzsch, J., Kim, P., Brus, L. E., Heinz, T.F., Hybertsen, M.S., Flynn, G. W. (2007). High-resolution scanning tunneling microscopy imaging of mesoscopic graphene sheets on an insulating surface. Proceedings of the National Academy of Sciences, 104(22), 9209–9212.

that is oriented out of the plane. Stolyarova and coworkers [37] observed flake - like morphology of graphene under an optical microscope. Single-layer graphene (Fig. 1.4A–I) appeared with relatively lower optical density than a multilayer graphene flake (Fig. 1.4A-II). They also observed the topography of single-layer graphene with the help of scanning tunneling microscopic (STM) imaging and saw a honeycomb structure with no atomic defect (Fig. 1.4B). However, in case of multilayer graphene, the carbon atoms on the surface are nonequivalent because of the asymmetry in the electron environment of the layers (Fig. 1.4C).

Graphene is produced by mechanical or chemical exfoliation of graphite via chemical vapor deposition. It is hydrophobic due of the absence of oxygen groups. Graphene has a large specific surface area, and the presence of long-range π-conjugation yields extraordinary thermal, mechanical, and electric properties to the molecule. Graphene is a conductive transparent nanomaterial, with low cost and substantial green environmental impact, which make it suitable for catalysis, sensing, drug delivery, and electric and biomedical applications [38–40].

Displacement of the electron density over the plane of the ring imparts geometric strain to the graphene lattice, which generates reactive sites. Stoichiometric functionalization of graphene is, therefore, possible to achieve better compatibility with other nanomaterials, polymers, or biomolecules. Zigzag or armchair conjugation on graphene allows some regioselective reactions such as carbene insertion, cycloaddition, and click reaction [41]. However, the cleavage of the sp^2 bonds is mandatory to form a covalent bond on the basal plane of graphene [42]. Regioselective unzipping of the conjugated tracks on graphene allows the initial site of attack for forming a covalent bond. The detailed functionalization of graphene-based nanomaterials has been discussed in the following sections.

Graphene has mainly four forms: graphene itself, GO, RGO, and graphene quantum dots [43]. In this chapter, only the 2D forms of graphene have been discussed. GO can be obtained by the chemical treatment of graphite followed by sonication. In this context, Hummers' method can be regarded as the most popular technique, which uses concentrated sulfuric acid and potassium permanganate [44]. The surface of GO can be functionalized with various groups such as carbonyl, hydroxyl, and epoxides. Therefore in GO, oxygen atoms are bound to carbon and so it can be considered as a hydrophilic derivative of graphene. In GO, both sp^2- and sp^3-hybridized carbon atoms are present. RGO is generally synthesized by chemical or thermal reduction of GO or graphite oxide; however, some green methods have also been reported in this regard [45]. RGO can be considered as an intermediate between graphene and its oxidized form, GO.

Several methods following either a "top-down" or a "bottom-up" approach have been reported for the synthesis of graphene-based nanomaterials. Graphene-based nanomaterials have gained huge popularity as carrier molecules for therapeutic agents because of their excellent physicochemical properties [46]. The presence of high specific area, π-π stacking, and electrostatic interactions in these nanomaterials facilitates the efficient delivery of partially soluble drugs. Therefore they are mostly used for gene and drug delivery, tissue engineering, biosensing, and anticancer therapy [47]. GO and RGO can be used in nanohybrids to synthesize novel antibacterial agents [48]. The functionalization of GO with various polymers, DNA, proteins, enzymes, etc. to improve solubility, selectivity, and biocompatibility is a vital step to enhance the property of nanomaterials for use in biomedical applications [49]. Although there are widespread applications of graphene-based nanomaterials in the biomedical field, there are limited experimental data on their toxicity. Several studies have revealed that the hydrophobic forms, which accumulate on the surface of cell membranes, are highly toxic compared with the hydrophilic forms, as they can infiltrate the cell membrane. The majority of research on the toxic effects of graphene is at the cellular level rather than the genetic level. However, till date, very few of the GO-based applications have been approved for clinical trials because of issues related to toxicity and biosafety [50,51].

1.3.3 Metal dichalcogenides

The 2D nanostructured sheets have shown great promise in various advanced applications due to their attractive optoelectronic properties as well as reinforcing capacity. TMDs such as molybdenum disulfide (MoS_2), molybdenum diselenide ($MoSe_2$), tungsten disulfide (WS_2), and tungsten diselenide (WSe_2) have been explored in this regard [52,53]. The 2D TMDs are compounds with the general formula, MX_2 (where M is a transition metal typically from groups 4–7 and X is a chalcogen such as S, Se, or Te). Over the past few decades, many 2D TMDs have been broadly studied for their semiconducting properties that may find vast applicability in the development of ultrasmall and low-power transistors [54,55].

The 2D MX_2 compounds are van der Waals solids with strong intralayer and weak interlayer interactions, where each layer consisting of transition metal orbital is sandwiched by two chalcogen orbitals [1]. The charge carriers can be confined in two dimensions (lacking interactions in the z-direction), which makes isolation of monolayers easy that causes dramatic changes in the properties of TMDs [56]. Such distinctive features make TMDs attractive candidates for supercapacitors, batteries, catalysis, electronic and photonic devices, and biosensors [53,56]. Thus interest has been generated by these 2D nanomaterials in numerous technologic fields. Determination of the structure of TMDs dates back to 1923 by Linus Pauling [57,58]. During 1990s the development of fullerene chemistry and different graphite forms (cylindric and polyhedral) further advanced the research interest towards the formation of equivalent stable structures [59]. Later, with the advancement of graphene-based research, the development techniques necessary for studying these layered materials have been explored [60]. Stratified crystals of MoS_2 with hexagonal structure have unit-cell thickness. Exfoliated MoS_2 nanosheets have shown immense potential for biosensing applications because of their interesting optoelectronic attributes [61]. Owing to their large lateral dimension, these nanosheets have very good dispersibility in liquid or gas. Transmission electron micrographs (TEMs) revealed the layered structure of MoS_2 nanosheets, with hexagonal arrangement of atoms. Individual Mo and S has been identified by [62] using a scanning transmission electron microscopy/high-angle annular dark-field fluorescence imaging (Fig. 1.5A–C). Stacking pattern of MoS_2 nanosheets has also been visualized by this technique. Mishra et al. [63] illustrated the flake-like structure of MoS_2 by using field emission scanning electron microscopic and TEM imaging. The lateral dimension of the flakes was found to be ranging from 80 to 120 nm (Fig. 1.6).

1.3.4 Metal-organic frameworks

MOFs have been recognized as an efficient class of 2D nanomaterials, which have immense potential for biomedical applications, especially in the field of advanced diagnostics [64]. These porous crystalline materials are made up of metal ions or clusters linked with organic ligands. Organic ligands with negative charge such as carboxylate and oxalate have been used to fabricate different types of MOFs [65–67]. Tunable structure and functionalities, high surface area, and porosity make this class of 2D materials excellent candidates for myriad applications, including catalysis, biomedicine, sensors, energy storage, etc. [68,69]. Synthesis of 2D MOF nanosheets has been a challenge for researchers due to the complicacy in regulating the vertical dimension within nanometer range, without affecting the lateral directions [17]. Researchers have developed different methods for

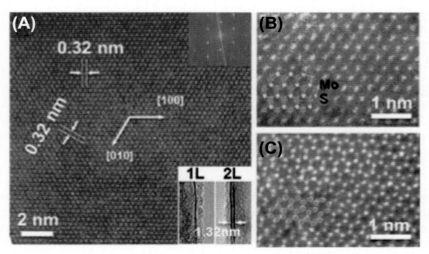

FIGURE 1.5

(A) Scanning transmission electron microscopy/high-angle annular dark-field fluorescence micrograph of two-dimensional MoS$_2$ nanosheets. The upper inset shows the corresponding FFT pattern and the lower inset shows the MoS$_2$ monolayer and bilayer edges, (B,C) Distinction between molybdenum and sulfur atoms in the filtered images of MoS$_2$ layers.

Reproduced with permission under Creative Commons license from Yu, Y., Li, C., Liu, Y., Su, L., Zhang, Y., Cao, L. (2013). Controlled scalable synthesis of uniform, high-quality monolayer and few-layer MoS$_2$ films. Scientific Reports, 3, 1866.

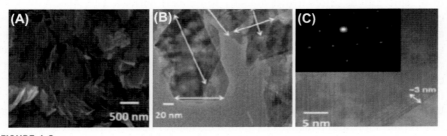

FIGURE 1.6

(A) Flake-like structure of MoS$_2$ nanosheets (field emission scanning electron microscopic image), (B) lateral dimensions of MoS$_2$ nanosheets (transmission electron microscopic image), and (C) electron diffraction pattern of MoS$_2$.

Reproduced with permission under Creative Commons license from Mishra, A. K., Lakshmi, K. V., Huang, L. (2015). Eco-friendly synthesis of metal dichalcogenides nanosheets and their environmental remediation potential driven by visible light. Scientific Reports, 5, 15718.

synthesizing MOFs. Among them, chemical and mechanical exfoliation, ultrasonication, interfacial synthesis, modulated synthesis, surfactant-mediated synthesis, ion intercalation synthesis, etc. have been reported, which are actually either top-down or bottom-up approaches [70–73].

Stacking of 2D layers via weak physical forces along the vertical direction yields layered MOFs. The forces of interaction among the layers are mainly π-π stacking interaction, van der Waals forces, hydrogen bonding, etc. Thus top-down exfoliation helps in overcoming these weak interactive forces to obtain 2D nanosheets [17]. In a typical top-down approach, Wang et al. [74] have synthesized 2D MOF nanosheets by exfoliating mesh-adjustable molecular sieve, $Ni_8(5\text{-bbdc})_6(m\text{-OH})_4$ (MAMS-1), a layered MOF.

In the freeze-thaw process, first the MAMS-1 crystals were dispersed in an appropriate solvent and frozen by using liquid nitrogen. The authors proposed that during the freeze-thaw process, volumetric alteration of hexane imparted shear forces onto the suspended crystals and exfoliated them into discrete nanosheets (Fig. 1.7A). Furthermore, atomic force microscopic images revealed that this kind of exfoliation is efficient enough to yield 2D MOF nanosheets of thickness ~4 nm (Fig. 1.7B). TEM images demonstrated a lattice spacing of 0.24 nm that corresponds to the (420) crystal planes of MAMS-1 (Fig. 1.7C).

1.4 Surface modification of two-dimensional nanomaterials

1.4.1 Nanoclay

Nanoclays act as natural cation exchangers that can exchange inorganic cations Thus, cation exchange is the vital mode of modification of nanoclays. Incorporation of organic cations offers hydrophobicity to the clay surface. This makes the clay compatible with polymers, biomolecules, etc. Quaternary alkylammonium and alkylphosphonium compounds have been broadly explored for the surface modification of nanoclay [75,76]. Modification increases the interlayer spacing, enabling other nanomaterials, polymers, biomolecules, etc. to enter the clay galleries by intercalation mechanism [77]. Furthermore, functionalization of nanoclays by ionic liquids, containing imidazolium, pyridinium, phosphonium cations, has been reported for the preparation of polymeric nanocomposites [78]. It is important to note that isomorphic substitution dictates the cation exchange capacity (CEC) of a nanoclay. The CEC of a typical smectite nanoclay is 0.66% per unit cell [79].

Nanoclays have a strong affinity to agglomerate and form stacks. Thus delamination is required to prevent this shortcoming. Intercalation of polymeric chains in case of polymer nanocomposites can effectively delaminate the clay stackings, imparting stability to the composite system. However, difference in the surface energies between the clay minerals and the polymers significantly affects the stability of the system. Thus surface modification of nanoclay is a prudent step to lower the surface energy [35]. Use of a surfactant has been a widely adopted method for the

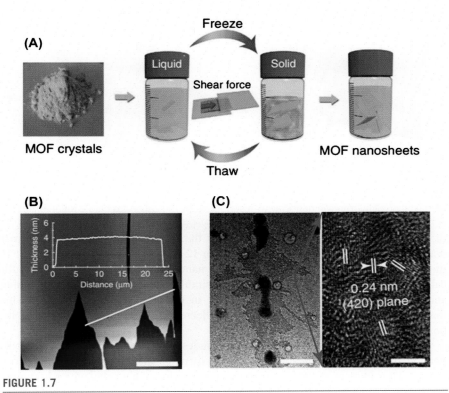

FIGURE 1.7

(A) Synthesis of two-dimensional MAMS-1 nanosheets by the freeze-thaw exfoliation process, (B) atomic force microscopic image of MAMS-1 nanosheets (scale 10 μm), and (C) transmission electron microscopic and high-resolution transmission electron microscopic images of MAMS-1 nanosheets. *MOF*, metal-organic framework.

Reproduced with permission under Creative Commons license from Wang, X., Chi, C., Zhang, K., Qian, Y., Gupta, K. M., Kang, Z.,Jiang, J. Zhao, D. (2017). Reversed thermo-switchable molecular sieving membranes composed of two-dimensional metal-organic nanosheets for gas separation. Nature Communications, 8, 14460.

modification of nanoclays. Surfactants can exchange cations with the clay minerals and expand the interlayer spacing within the clay galleries. The exchangeable alkali cations present in the clay galleries are generally exchanged with the organic "onium" ions of the surfactants, which are attached to the nonpolar hydrocarbon tail (Fig. 1.8).

Surfactants with 1°, 2°, or 3° alkylphosphonium or alkylammonium cations have been explored for clay modification [80]. However, the chemical structure of the organomodifier has a strong role over the extent of intercalation or exfoliation as well as dispersibility of the clay. In case of polymeric nanocomposites the selection criteria of the modifier greatly depend on the structure and nature of the polymer used. It has been observed that the more the number of alkyl chains attached to

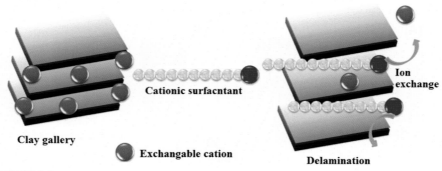

FIGURE 1.8

Modification of hydrophilic nanoclay with surfactant by the cation exchange mechanism.

the surfactant, the better the dispersibility of the modified clay within the polymer matrix. Also, the alkyl chain with more than 12 carbon atoms is more efficient in imparting organophilicity to the hydrophilic nanoclays. However, this is not applicable to the polyamide modifiers, as undesirable interaction between polyamide and alkyl chains is observed with increase in the number of alkyl chains. This directly affects the extent of exfoliation and decreases the stability of the nanocomposite system [15,80].

Although the use of surfactants or polyamides is widely explored for clay modification, this type of modification requires costly reagents and high-temperature conditions. Thus Phua et al. [81] reported the modification of nanoclay with a hormone, dopamine. Furthermore, *Homalomena aromatic* oil has imparted hydrophobicity to hydrophilic bentonite clay [12]. Hydrophilic sodium MMT nanoclay has also been modified by plasma polymerization technique using methyl methacrylate and styrene monomers. This in situ plasma polymerization approach carried out under low-pressure conditions yielded a nanocomposite with enhanced mechanical and thermal properties [82]. Thus functionalization of nanoclays is very important for their biomedical utility. Saponite and MMT clays have been proved to be efficient in cation exchange with some drugs, which makes them competent as drug carriers. In a similar context, smectites and kaolinite group of clays have been studied for controlled drugrelease applications. The highly active surface of nanoclays provides a platform for efficient adsorption of drugs, proteins, etc., which also increases their water solubility.

1.4.2 Graphene-based nanomaterials

The unique properties of 2D graphene and graphene-based nanomaterials opened numerous avenues of applications in material science and biomedical technology. However, graphene as such is very inert and it is challenging to prepare nanohybrids and nanocomposites because of its poor dispersibility in most of the solvents. Hence, surface modification is essential to fabricate graphene-based composites. Different research groups have explored different ways to achieve effective and efficient modification of graphene-based nanomaterials. At first, we can consider the covalent functionalization of such materials, which can impart organic functionalities to the graphene surface. One of the popular methods of covalent functionalization of graphene is the use of free radicals. Reaction of free radicals to the sp^2 carbons of graphene helps in enhancing the bandgap and regulates the dispersibility. Diazonium salts and peroxides are generally used to generate free radicals. Electron transfer occurs between graphene and the aryl diazonium ion, forming a covalent interaction. In this regard, modification of graphene by using nitrophenyl groups has been reported [83]. In a similar context, covalent functionalization of graphene with hydroxylated aryl groups has been carried out via a diazonium addition reaction [84]. Hydroxyl functionalized graphene has been used for preparing polystyrene-based nanocomposites by adopting the atom transfer radical polymerization technique. Also, free radical diazotization reaction has been reported for the covalent functionalization of RGO by using azobenzene [85]. In that report, Feng and coworkers showed the synthesis and chemical structures of RGO-*para*-azobenzene (RGO-*para*-AZO) and RGO-*ortho*-azobenzene (RGO-*ortho*-AZO) hybrids obtained by diazotization reaction (Fig. 1.9A).

Reactions of graphene with organic azides have been widely explored by a number of research groups. Photothermally activated para-substituted perfluorophenylazides have been used to append different chemical functionalities (Fig. 1.9B) to graphene surface via the aziridine ring [86]. Surface functionalization plays a crucial role in the biomedical application of graphene-based nanomaterials. Researchers have witnessed that PEGylation of such materials improves solubility, imparts biocompatibility, and reduces toxicity. Thus functionalization of graphene-based nanomaterials with other materials of nanodimension, such as magnetic iron oxide, gold nanoparticles, and quantum dots, showed excellent potential in cancer diagnosis and treatment. GO and RGO have been studied widely for conjugating appropriate ligand systems to selectively target cancer cells. Such experiments showed promising results in both in vitro and in vivo trials. Thus it is interesting to note that functionalized graphene has tremendous potential in biomedical applications, including drug delivery, cancer diagnosis, multimode bioimaging, biosensing, etc. Yang and coworkers have beautifully summarized (Fig. 1.10) the surface functionalization processes used for GO and RGO, along with their interaction with biocompatible polymers such as polyethylene glycol (PEG) [87].

Besides the aforementioned methods, different other methods can be adopted for the functionalization of GO because of the availability of oxygeneous functionalities on its surface. Amidation, esterification, silanization reactions, etc. have been carried out in this regard [88−90]. Furthermore, enzymes, proteins, drugs, and other biomolecules can be immobilized onto the GO surface via noncovalent interactions. Hydrophobic interactions, π-π stacking, and intrinsic chemical functionalities of GO play vital roles in such immobilization processes [91,92].

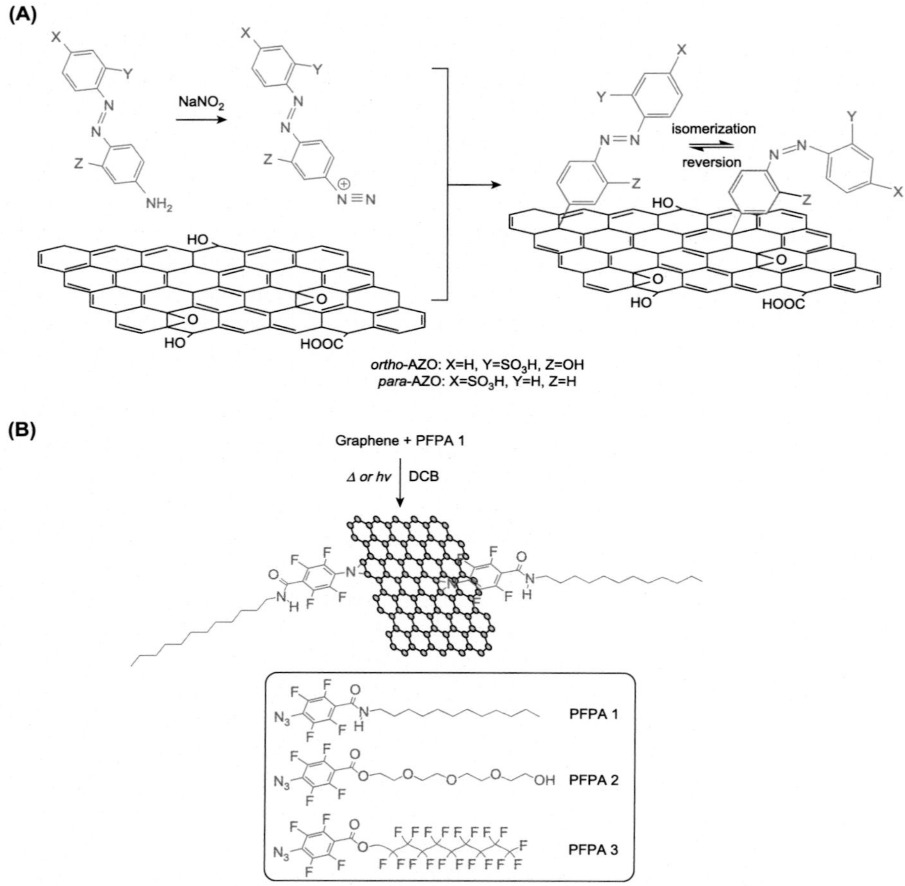

FIGURE 1.9

(A) Covalent functionalization of reduced graphene oxide by using azobenzene. (B) Chemical modification of graphene (prepared by exfoliating graphite in o-dichlorobenzene (DCB)) with para-substituted perfluorophenylazides (PFPAs).

(A) Reproduced with permission from Feng, Y., Liu, H., Luo, W., Liu, E., Zhao, N., Yoshino, K., Feng, W. (2013). Covalent functionalization of graphene by azobenzene with molecular hydrogen bonds for long-term solar thermal storage. Sci. Rep., *3, 3260, Copyright 2013, Nature Publishing Group. (B) Reproduced with permission from Liu, L. H., Lerner, M. M., Yan, M. (2010). Derivitization of pristine graphene with well-defined chemical functionalities.* Nano Lett., *10(9), 3754–3756, Copyright © 2010, American Chemical Society.*

1.4.3 Metal dichalcogenides

As stated earlier, TMDs have attracted widespread attention as 2D analogues of graphene. Surface modification with different functional groups aids advantageous features to TMDs, which endorse them for many important applications. The naturally

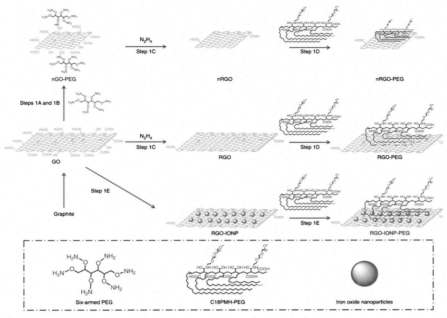

FIGURE 1.10

Preparation, surface functionalization, and bioconjugation of graphene derivatives and nanocomposites. *GO*, Graphene oxide; *IONP*, iron oxide nanoparticle; *PEG*, polyethylene glycol; *RGO*, reduced graphene oxide.

Reproduced with permission from Yang, K., Feng, L., Hong, H., Cai, W., Liu, Z. (2013). Preparation and functionalization of graphene nanocomposites for biomedical applications. Nature Protocols, 8(12), 2392.

occurring dichalcogenides of minerals, molybdenite and tungstenite (MoS_2 and WS_2), are of interest, as these compounds display strong anisotropic optical, electronic, and electrochemical properties due to their layered structure, in addition to excellent friction reduction abilities [93]. The large surface-to-volume ratio of TMDs holds a key role for easy tunable protocols for chemical functionalization to modify their characteristics for different requirements [94]. Thus it becomes very important to understand the approaches for various functionalization methods, which enable TMDs to fulfill different purposes. To date, several surface functionalization strategies have been reported to improve the applicability and chemical reactivity of TMDs. Herein, we aim to highlight the most recent developments in the functionalization of TMDs.

Conjugation of thiolated ligands is a common strategy to make nanomaterials compatible for biological applications. In this context, mercaptoundecanoic acid−conjugated MoS_2 has been used for sensing applications. In a typical modification method, Kim and coworkers [95] have ultrasonicated a dispersion of MoS_2 to create sulfur defects in the peripheral and internal edges. In this study the solution mixing

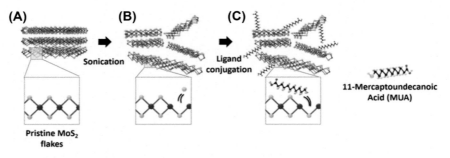

FIGURE 1.11

(A) Pristine MoS$_2$, (B) dispersion of MoS$_2$ to create sulfur defects, and (C) ligand conjugation onto the defects of MoS$_2$.

Reproduced with permission from Kim, J. S., Yoo, H. W., Choi, H. O., Jung, H. T. (2014). Tunable volatile organic compounds sensor by using thiolated ligand conjugation on MoS$_2$. Nano Letters., 14(10), 5941-5947.

method has been employed to conjugate the ligand onto the defect cites of MoS$_2$ (Fig. 1.11).

In another study, covalent functionalization of dibenzothiophene (DBT) on a single layer of MoS$_2$ nanoclusters was studied with the help of STM [96]. The STM images revealed that the MoS$_2$ nanoclusters performed quite well in binding DBT directly. The evidence from thermal stability and the small Mo-−S distance indicated the strong bonding between the organic and the inorganic molecules. The chemical modification can also be achieved based on the formation of Mo-S bonds between exogenous thiols and sulfur vacancies at coordinatively unsaturated Mo sites [97]. However, Voiry and coworkers [98] have developed a simple and efficient method for the covalent functionalization of organic moieties on TMDs irrespective of any conjugation on the defect sites. The authors described that electron transfer between strong electrophiles and negatively charged MoS$_2$ layers of reactants plays a key role in the covalent attachment to the functional groups linked to the chalcogen atoms of the TMDs. The optoelectronic properties of the material can be significantly affected by such attachments.

Chemical exfoliation of MoS$_2$ layers by using lithium intercalation is a popular method to obtain single-layer MoS$_2$. However, crystal deformation is generally observed due to this violent reaction, which makes internal defects visible [99]. Thus chemically exfoliated MoS$_2$, possessing both internal and perimeter edge defects, is more suitable for thiolated ligand conjugation. In this context, Chou and group [100] demonstrated the conjugation between thiol-terminated PEG ligands and chemically exfoliated MoS$_2$. Good colloidal stability of 2D materials bearing organic molecules, such as OH, COO$^-$, and NMe$_3^+$ groups (neutral, anionic, and cationic, respectively), in water has been observed even after 21 days (Fig. 1.12). Thus the ζ-potential and surface functionality of MoS$_2$ layers can be tuned for broad applicability in artificial enzyme receptors.

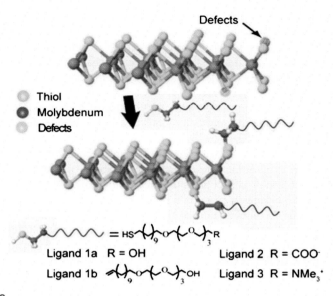

FIGURE 1.12

Ligand conjugation on chemically exfoliated two-dimensional MoS$_2$ sheets.

Reproduced with permission from Chou, S. S., De, M., Kim, J., Byun, S., Dykstra, C., Yu, J.,Huang, J.,Dravid, V.
P. (2013). Ligand conjugation of chemically exfoliated MoS2. Journal of the American Chemical Society,
135(12), 4584–4587.

In a study, 2D MoS$_2$ nanosheets have been functionalized by using M(OAc)$_2$ salts (where M = Ni/Cu/Zn and OAc = acetate). The obtained MoS$_2$-M(OAc)$_2$ showed excellent dispersion in most of the common solvents [101]. The reversal of surface negative charge to positive charge in TMDs effectively prevents agglomeration or restacking. This helps in the functionalization of TMDs with ligands by forming electrostatic guest-host complex. However, the attachments of the ligands on the active cites of the nanosheets decreases the hydrogen evolution reaction activity. This encouraged to use lipoic acid functionalized PEG to modify the surface of chemically exfoliated MoS$_2$. This system has been utilized for the combined photothermal therapy and chemotherapy of cancer by the efficient delivery of doxorubicin [102]. Furthermore, functionalization of 2D MoS$_2$ nanosheets by substitutional doping of transition metal and nonmetal dopants has been explored for advanced applications [103].

1.4.4 Metal-organic frameworks

Chemical versatility, controlled pore size, and high surface area have endorsed MOFs for a range of advanced applications including drug delivery, biosensing, catalysis, etc. Postsynthetic surface functionalization of MOFs with desired chemical functionalities further enhances the scope of applicability for this category of 2D nanostructures. However, retention of the MOF and pore structures has been a challenge for researchers [104]. Sharpless' "click" reaction, i.e., CuI-catalyzed Huisgen cycloaddition of azides, is a model reaction that can overcome these challenges with good efficiency [104]. With such approach, Gadzikwa et al. reported the synthesis of MOFs with silyl protected C≡C bond. Simple organic chemistry tool has been used as the deprotecting agents [105]. Use of metal-chelating agents for grafting of coordinatively unsaturated metal sites is another prudent approach for the functionalization of MOFs.

Encapsulation of MOFs inside the bilayers of lipids can be achieved by means of the physical forces of attraction. Direct coordination of the lipid-ligand, 1,2-dipalmitoyl-*sn*-glycero-3-galloyl, with the metal sites surrounding the MOF surface has been reported by Zhu and coworkers [106]. Low toxicity and nonimmunogenicity of liposomes are the added features obtained by functionalization of lipid layers onto the surface of MOFs [107,108]. Thus liposome-coated MOFs have been explored for controlled drug delivery applications. A phospholipid bilayer—functionalized Zr-MOF has been reported with excellent biocompatibility and efficient cellular uptake [109].

Furthermore, a few research groups have explored phenol-based ligands for functionalization of MOFs. The reversible coordination capacity of phenolic organic ligands helps in constructing MOF-based functional building blocks with excellent potential for advanced biomedical applications [110,111]. In another approach, Clough et al. [112] demonstrated the integration of cobalt dithiolene onto an MOF surface to fabricate a highly efficient electrocatalytic cathode for H_2 generation. For the same purpose, deposition of benzenetricarboxylic acid—functionalized MOFs on carboxyl terminated organic surface has been reported by Shekhah and group [113].

1.5 Physics of two-dimensional nanomaterials

In this chapter, we have discussed the basic chemistry of 2D nanomaterials with recent advancements. Besides the interesting chemistry of such materials, the physics associated with them also attracted copious attention of the scientific community. However, the scope of this chapter allows only a brief discussion on it.

In case of 2D nanoclays, electrostatic forces play a very important role in their interaction with other cations, thereby dictating the clay modification strategies. In this context, during the 1960s, Shainberg and Kemper studied the conductivities of the cations adsorbed in clays saturated with Na^+, Cs^+, and Ca^{2+} [114]. They revealed that due to the presence of Ca^{2+} ions in the tactoid internal surface, the average mobility of adsorbed Ca^{2+} ions is very low. In 1997 the permanent structural charge density of kaolinite clay was studied by determining the surface excess Cs^+ ions selective to charge sites [115]. Nanoclays are well known for imparting many fascinating physical attributes to polymeric nanocomposites. Especially, the dispersibility and rheological properties of many polymers have been augmented by minimal incorporation of nanoclay [116]. Furthermore, antistatic and electromagnetic shielding properties of polyaniline/organoclay nanocomposite have been reported [117]. Researchers observed the dramatic increment of thermomechanical properties of polymers after the formation of nanoclay-based nanocomposites [12]. Moreover, the benign nature of nanoclays is an interesting feature for endorsing them for biomedical applications.

Again, graphene-based 2D nanomaterials not only fascinated the chemist or material scientists but also revolutionized the domain of physics. The structure-dependent properties of graphene and allied materials showed tremendous potential for many advanced applications because of their exceptional physicochemical attributes. Physicists have regarded graphene as a "bridge between quantum electrodynamics and condensed matter physics" [118]. Such materials are efficient semiconductors due to the charge-conjugation symmetry of the charge carriers in graphene. This further forwards the concept of chirality for ultrarelativistic elementary particles. Each "p_z" orbital in graphene contributes one electron to the lattice, forming a half-filled system. The π orbitals are responsible for the unique electronic properties of graphene-based nanostructures. Two energy bands are observed in the energy spectrum originated from the π orbitals in graphene. The higher energy band is the conduction band and the lower energy one is the valence band. Owing to the half-filled nature of graphene the valence band is filled completely. The characteristics of the spectrum near the filled orbitals dictate the electronic attributes of a material in condensed matter physics. Its energy signifies the fermi levels [119]. Thus the character of the energy spectrum determines the physics of graphene nanostructures. The interaction of the hexagonal lattice of graphene with the π electrons imparts exceptional qualities to its energy spectrum. Although it is very interesting to explore the physics of these honeycomb nanostructures, it is very briefly elaborated here.

The optoelectronic properties of 2D TMDs have attracted the attention of researchers. Density state and band structures influence the electronic properties in them. Hence, myriad works have been reported in the past few years on the development of electronic devices and biosensors, using TMDs [1]. The electronic environment near the fermi levels of TMDs can be ascertained by studying the electronic interaction of the "p, s, and d" orbitals. The chalcogen atoms' s orbital possesses a deep energy level in the valence band. Thus its contribution to the band structure is not that significant. The group IVB metals with d^2s^2 electronic configuration, possessing four valence electrons, form 1T phase with chalcogens. The difference in the relative energy of metals and chalcogens is responsible for the change in band structure of TMDs. Even number of electrons in the unit cell imparts semiconducting properties to TMDs [120]. The conduction band minimum depends on the interlayer interaction within the 2D sheet structures of TMDs. Exfoliation of the layered structure of bulk TMDs resulted in the decrease in the number of layers, thereby shifting the conduction minimum to a higher energy Γ point [1]. Thus the bandgap property of TMD nanostructures has shown tremendous potential for nanoelectronics and biosensing applications. In fact, such materials may be regarded superior to analogous graphene, which does not exhibit such bandgap behavior in its native form [121].

Ligand to metal charge transfer (LMCT) transition has been observed in MOFs. Bordiga and coworkers [122] revealed that a Zn_4O_{13} MOF exhibited similar vibrational and electronic properties to those of ZnO quantum dots by studying the LMCT at 350 nm. The charge distribution in MOFs remains unchanged as compared to the building block. Kuc et al. [123] observed that charge distribution alters only in case of the linking atoms in an MOF, $Zn_4O(CO_2)_6$. The group observed very high bandgap for the MOF and classified it to be a semiconductor. They also witnessed that sp^2 states of the carbons in the organic ligands dominated the HOMO-LUMO bands, near the fermi level.

1.6 Biomedical aspects

The following chapters of this book elaborately discuss the biomedical applications of important 2D nanomaterials. Thus in this section, a brief account of the biomedical aspects of a few such materials has been presented for better understanding. The common feature of 2D nanomaterials is their sheetlike structures with very high surface area. Moreover, nanoclays in this category are well known for their biocompatibility and nonimmunogenicity. Hence, implantable biomaterials have been fabricated using modified clay-polymer nanocomposites with immense potential for tissue engineering applications [12]. Cell growth and proliferation is supported by the porous structure and rough surface texture of nanoclays [124]. Nanostructured clays, clay-based hybrid materials, and polymeric nanocomposites exhibited promising performance in drug delivery, microbial fouling resistant coatings, antihemorrhoidal materials, antacids, and regenerative medicines [125]. Clay minerals such as talc, kaolin, and bentonite showed excellent interaction with drug molecules. Gastrointestinal delivery of such drug formulations has been reported by different research groups [126]. Furthermore, certain clays reduce the release rate of drugs and helps in sustained drug delivery. The vital challenges associated with drug delivery formulations are poor solubility and pH sensitivity. MMT clay—based drug formulations have been explored for addressing these issues. In a study, it was observed that MMT clay improved the dispersibility and acts as a sustained-release carrier of the analgesic ibuprofen [127]. Table 1.1 shows some of the clay formulations and their existing and potential biomedical applications.

With the development of graphene research, its successful functioning in biomedical applications paves a new era for the applications of 2D nanomaterials. Recent research showcased the utility of graphene and graphene-based materials in drug delivery, tissue engineering, bioimaging, biosensing, cancer diagnosis and therapy, etc. [136]. Since 2008, graphene-based nanomaterials have been extensively explored to deliver therapeutic agents to cells or tissues both in vitro and in vivo. However, a lot of challenges remain before these drugs qualify for effective

Table 1.1 A few clay formulations and their biomedical applications.

SI no.	Clay used	Hybrid/composite with	Biomedical application	References
1	Organically modified bentonite	Hyperbranched epoxy	Tissue scaffold	[12]
2	OMMT	Silver nanoparticles-hyperbranched epoxy	Antimicrobial tissue scaffold	[14]
3	MMT	Poly(D,L-lactide-co-glycolide)	Oral anticancer drug delivery	[128]
4	MMT	Chitosan-g-lactic acid	Cell proliferation and controlled drug release	[129]
5	MMT	Polyvinyl alcohol-silver nanoparticles	Antibacterial applications	[130]
6	Rectorite	Chitosan	Gene delivery	[131]
7	Sepiolite	Collagen	Biodegradable biomaterials	[132]
8	MMT	Glutathione	Antioxidant delivery	[133]
9	Organically modified MMT	Polyvinyl alcohol	Wound dressing	[134]
10	MMT	Chitosan-hydroxyapatite	Bone tissue regeneration	[135]

MMT, *montmorillonite;* OMMT, *organically modified montmorillonite.*

use on humans. Generally, the interaction of graphene with solutes, proteins, or cellular systems within the body has a significant impact on its behavior or toxicity. Therefore it has become essential to apply suitable approaches, such as modifying graphene or combining it with other molecules, to overcome these problems [137]. Duch et al. explored various strategies to study the biocompatibility of graphene-based nanomaterials in lungs. They directly administered solutions of aggregated graphene, pluronic dispersed graphene, and GO into the lungs of mice. The results demonstrated that the covalent oxidation of graphene is a major contributor to its pulmonary toxicity [138]. Development of efficient artificial enzymes that could combat the disadvantages of natural enzymes is another emerging field in nanobiotechnology. Maji et al. [139] developed a nanohybrid by the immobilization of gold nanoparticles on silica-coated nanosized RGO conjugated with folic acid, a cancer cell–targeting ligand. Multifunctional graphene-based nanomaterials can be utilized as neurodevices to explore their interactions with the signaling machinery of nerve cells and also with the hydrophobic membrane domains. Such interactions may favor graphene translocation, or adhesion to cell membranes, which can potentially interfere with the membrane activities and consequently with the physiologic synaptic transmissions. In this context, Rauti et al. explored the ability of graphene and GO nanosheets to interfere with synaptic signaling neurons. They demonstrated the potential of GO nanosheets to alter different modes of interneuronal communication systems in the central nervous system [140]. In another study, Yang et al. developed a GO-based substrate with hierarchic structures capable of generating synergistic topographic stimulation in order to enhance focal adhesion signaling and neuronal differentiation in human neural stem cells. This type of substrate can be applied for the generation of functional neuronlike cells from human neural stem cells, providing potent therapeutic agents for treating neuronal diseases and disorders [141]. Liu et al. [142] demonstrated the applications of glucose oxidase—immobilized GO electrode as a glucose biosensor, with linear detection range up to 28 mM mm^{-2}. Furthermore, scalable and reproducible biosensors have been fabricated for detecting a range of biomolecules, including DNA [143]. Graphene holds great potential in the field of tissue engineering and biomedical devices and products. Cells are known to interact with nanomaterials differently when presented in the form of a substrate rather than in the form of a suspension. In tissue engineering a stable and supportive substrate or scaffold is necessary that can provide mechanical support, chemical stimuli, and biological signals to cells. Medeiros et al. [144] evaluated the potential of GO/nanohydroxyapatite (nanohydroxyapatite/graphene nanoribbon) in the in vitro analysis of specific genes related to osteogenesis and the in vivo bone regeneration using animal model. Shamekhi et al. synthesized chitosan-based nanocomposite scaffolds by reinforcing with GO nanosheets. They observed that the seeding of the human articular chondrocytes on the nanocomposite scaffolds showed an increased proliferation with augmentation of the GO percentage [145].

Furthermore, TMDs, because of their planar structure and relatively large surface area, allow maximal interaction with the target biomaterial for increased efficiency and easy introduction into biological systems. However, suitable functionalization is required to make TMDs biocompatible and enhance their sensitivity as a diagnostic tool or install them with capabilities such as biosensing, drug delivery, and contrast imaging. Various 2D TMDs have been explored by researchers for advanced electrochemical biosensors, with promising role in bioanalytical applications [146]. Liu and coworkers [147] reported the preparation of WS$_2$ hybrid with multiwalled carbon nanotubes (MWCNTs) by a facile hydrothermal method and demonstrated its application as a biosensing platform for ultrasensitive determination of hepatitis B virus genomic DNA. Huang and group [148] combined MoS$_2$ with MWCNTs to improve the electronic conductivity and electrochemical activity for the fabrication of DNA biosensors with femtomolar sensitivity. Likewise, Cao [149] demonstrated a highly sensitive and selective electrochemical DNA detection platform based on a signal amplification strategy using Au nanoparticles/MoS$_2$ composite. Furthermore, 2D MoS$_2$ nanosheets have shown tremendous potential as a DNA and drug delivery vehicle [6]. PEGylated MoS$_2$ nanosheets have been fabricated for the diagnosis and chemotherapeutic applications of cancer [102].

2D MOFs are regarded as strong candidates for biomedical technology. They have shown tremendous efficacy in drug delivery because of their unique attributes such as drug entrapping capacity, controlled matrix degradation, devoid of "burst out effect" to mediate sustained release of drugs and biomolecules, and real-time monitoring of drug delivery by bioimaging [150]. Fig. 1.13 pictorially depicts the drug delivery application of an MOF and its in vivo release.

Another vital application of MOF is in multimodel bioimaging. A study showed the promising use of MOFs as multimodel imaging probes for contrast agents in magnetic resonance imaging [151]. Furthermore, MOFs also showed good promise as a platform for fabricating ultrasensitive biosensors [152]. The scientific community is expecting many other fascinating biomedical applications of 2D nanostructures in the coming years.

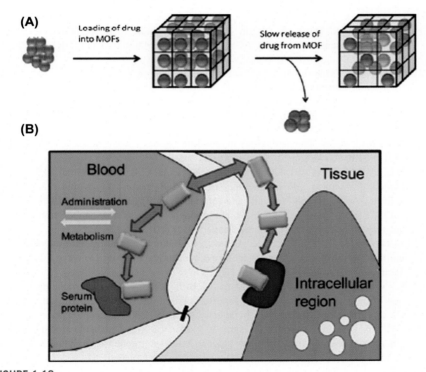

FIGURE 1.13

(A) Metal-organic framework (MOF) in drug delivery application and (B) in vivo release of drug.

Reproduced with permission from Keskin, S., Kızılel, S. (2011). Biomedical applications of metal organic frameworks. Industrial & Engineering Chemistry Research, 50(4), 1799–1812.

1.7 Conclusion

This chapter presented an overview of the general chemistry of 2D nanostructures. The role of surface chemistry of such nanomaterials in their applications has been elaborately explained herein. The importance of surface modification to incorporate the desired chemical functionalities to these nanomaterials has been discussed thoroughly in this chapter. Furthermore, the basic physics and electronics associated with the 2D nanostructures have been included briefly. Finally, a brief account of the biomedical aspects of these nanomaterials has been presented in this chapter. This chapter thus depicts a clear picture of 2D nanomaterials with respect to their structural chemistry and tunable applicability.

References

[1] S. Barua, H.S. Dutta, S. Gogoi, R. Devi, R. Khan, Nanostructured MoS$_2$-based advanced biosensors: a review, ACS Applied Nano Materials 1 (1) (2018) 2−25.

[2] G. Schmid, Nanoparticles, Wiley VCH, 2005.

[3] C.M. Lieber, One-dimensional nanostructures: chemistry, physics & applications, Solid State Communications 107 (11) (1998) 607−616.

[4] Y. Xia, P. Yang, Y. Sun, Y. Wu, B. Mayers, B. Gates, Y. Yin, F. Kim, H. Yan, One-dimensional nanostructures: synthesis, characterization, and applications, Advanced Materials 15 (5) (2003) 353−389.

[5] K.J. Koski, Y. Cui, The new skinny in two-dimensional nanomaterials, ACS Nano 7 (5) (2013) 3739−3743.

[6] B.L. Li, M.I. Setyawati, L. Chen, J. Xie, K. Ariga, C.T. Lim, S. Garaj, D.T. Leong, Directing assembly and disassembly of 2D MoS$_2$ nanosheets with DNA for drug delivery, ACS Applied Materials and Interfaces 9 (18) (2017) 15286−15296.

[7] J. Liu, L. Cui, D. Losic, Graphene and graphene oxide as new nanocarriers for drug delivery applications, Acta Biomaterialia 9 (12) (2013) 9243−9257.

[8] S. Jayrajsinh, G. Shankar, Y.K. Agrawal, L. Bakre, Montmorillonite nanoclay as a multifaceted drug-delivery carrier: a review, Journal of Drug Delivery Science and Technology 39 (2017) 200−209.

[9] R. Devi, S. Gogoi, S. Barua, H.S. Dutta, M. Bordoloi, R. Khan, Electrochemical detection of monosodium glutamate in foodstuffs based on Au@ MoS$_2$/chitosan modified glassy carbon electrode, Food Chemistry 276 (2019) 350−357.

[10] C. Shan, H. Yang, J. Song, D. Han, A. Ivaska, L. Niu, Direct electrochemistry of glucose oxidase and biosensing for glucose based on graphene, Analytical Chemistry 81 (6) (2009) 2378−2382.

[11] S. Barua, P. Chattopadhyay, M.M. Phukan, B.K. Konwar, J. Islam, N. Karak, Biocompatible hyperbranched epoxy/silver−reduced graphene oxide−curcumin nanocomposite as an advanced antimicrobial material, RSC Advances 4 (88) (2014) 47797−47805.

[12] S. Barua, N. Dutta, S. Karmakar, P. Chattopadhyay, L. Aidew, A.K. Buragohain, N. Karak, Biocompatible high performance hyperbranched epoxy/clay nanocomposite as an implantable material, Biomedical Materials 9 (2) (2014) 025006.

[13] D.R. Paul, L.M. Robeson, Polymer nanotechnology: nanocomposites, Polymer 49 (15) (2008) 3187–3204.

[14] S. Barua, P. Chattopadhyay, L. Aidew, A.K. Buragohain, N. Karak, Infection-resistant hyperbranched epoxy nanocomposite as a scaffold for skin tissue regeneration, Polymer International 64 (2) (2015) 303–311.

[15] S. Barua, S. Gogoi, R. Khan, N. Karak, Silicon-based nanomaterials and their polymer nanocomposites, in: Nanomaterials and Polymer Nanocomposites, Elsevier, 2019, pp. 261–305.

[16] M.C.P. Mendonça, E.S. Soares, M.B. de Jesus, H.J. Ceragioli, M.S. Ferreira, R.R. Catharino, M.A. Cruz-Höfling, Reduced graphene oxide induces transient blood–brain barrier opening: an in vivo study, Journal of Nanobiotechnology 13 (1) (2015) 78.

[17] M. Zhao, Y. Huang, Y. Peng, Z. Huang, Q. Ma, H. Zhang, Two-dimensional metal–organic framework nanosheets: synthesis and applications, Chemical Society Reviews 47 (16) (2018) 6267–6295.

[18] P. Horcajada, R. Gref, T. Baati, P.K. Allan, G. Maurin, P. Couvreur, P. Couvreur, G. Ferey, R.E. Morris, C. Serre, Metal–organic frameworks in biomedicine, Chemical Reviews 112 (2) (2011) 1232–1268.

[19] J.C.G. Bünzli, Lanthanide luminescence for biomedical analyses and imaging, Chemical Reviews 110 (5) (2010) 2729–2755.

[20] M.X. Wu, Y.W. Yang, Metal–organic framework (MOF)-based drug/cargo delivery and cancer therapy, Advanced Materials 29 (23) (2017) 1606134.

[21] K. Liang, R. Ricco, C.M. Doherty, M.J. Styles, S. Bell, N. Kirby, S. Mudie, D. Haylock, A.J. Hill, C.J. Doonan, P. Falcaro, Biomimetic mineralization of metal-organic frameworks as protective coatings for biomacromolecules, Nature Communications 6 (2015) 7240.

[22] J.C. Garcia, D.B. de Lima, L.V. Assali, J.F. Justo, Group IV graphene-and graphane-like nanosheets, Journal of Physical Chemistry C 115 (27) (2011) 13242–13246.

[23] X. Zhuang, Y. Mai, D. Wu, F. Zhang, X. Feng, Two-dimensional soft nanomaterials: a fascinating world of materials, Advanced Materials 27 (3) (2015) 403–427.

[24] N. Karak, Fundamentals of Polymers: Raw Materials to Finish Products, PHI Learning Pvt.Ltd, 2009.

[25] M. Batzill, The surface science of graphene: metal interfaces, CVD synthesis, nanoribbons, chemical modifications, and defects, Surface Science Reports 67 (3–4) (2012) 83–115.

[26] M.L. Nehdi, Clay in cement-based materials: critical overview of state-of-the-art, Construction and Building Materials 51 (2014) 372–382.

[27] A.H. Ambre, K.S. Katti, D.R. Katti, Nanoclay based composite scaffolds for bone tissue engineering applications, Journal of Nanotechnology in Engineering and Medicine 1 (3) (2010) 031013.

[28] M.I. Carretero, M. Pozo, Clay and non-clay minerals in the pharmaceutical industry: Part I. Excipients and medical applications, Applied Clay Science 46 (1) (2009) 73–80.

[29] M.I. Carretero, M. Pozo, Clay and non-clay minerals in the pharmaceutical and cosmetic industries Part II. Active ingredients, Applied Clay Science 47 (3–4) (2010) 171–181.

[30] V.A. Vinokurov, A.V. Stavitskaya, E.V. Ivanov, P.A. Gushchin, D.V. Kozlov, A.Y. Kurenkova, P.A. Kolinko, E.A. Kozlova, Y.M. Lvov, Halloysite nanoclay based

CdS formulations with high catalytic activity in hydrogen evolution reaction under visible light irradiation, ACS Sustainable Chemistry & Engineering 5 (12) (2017) 11316−11323.

[31] W. Zhang, M.K. Li, R. Wang, P.L. Yue, P. Gao, Preparation of stable exfoliated Pt−clay nanocatalyst, Langmuir 25 (14) (2009) 8226−8234.

[32] S. Shahidi, M. Ghoranneviss, Effect of plasma pretreatment followed by nanoclay loading on flame retardant properties of cotton fabric, Journal of Fusion Energy 33 (1) (2014) 88−95.

[33] S.M. Lee, D. Tiwari, Organo and inorgano-organo-modified clays in the remediation of aqueous solutions: an overview, Applied Clay Science 59−60 (2012) 84−102.

[34] K.G. Bhattacharyya, S.S. Gupta, Adsorption of a few heavy metals on natural and modified kaolinite and montmorillonite: a review, Advances in Colloid and Interface Science 140 (2) (2008) 114−131.

[35] M.S. Nazir, M.H.M. Kassim, L. Mohapatra, M.A. Gilani, M.R. Raza, K. Majeed, Characteristic properties of nanoclays and characterization of nanoparticulates and nanocomposites, in: Nanoclay Reinforced Polymer Composites, Springer, Singapore, 2016, pp. 35−55.

[36] L.F. Drummy, H. Koerner, K. Farmer, A. Tan, B.L. Farmer, R.A. Vaia, High-resolution electron microscopy of montmorillonite and montmorillonite/epoxy nanocomposites, Journal of Physical Chemistry B 109 (38) (2005) 17868−17878.

[37] E. Stolyarova, K.T. Rim, S. Ryu, J. Maultzsch, P. Kim, L.E. Brus, T.F. Heinz, M.S. Hybertsen, G.W. Flynn, High-resolution scanning tunneling microscopy imaging of mesoscopic graphene sheets on an insulating surface, Proceedings of the National Academy of Sciences 104 (22) (2007) 9209−9212.

[38] M. Hu, Z. Yao, X. Wang, Graphene-based nanomaterials for catalysis, Industrial & Engineering Chemistry Research 56 (13) (2017) 3477−3502.

[39] Q. Zhang, Z. Wu, N. Li, Y. Pu, B. Wang, T. Zhang, J. Tao, Advanced review of graphene-based nanomaterials in drug delivery systems: synthesis, modification, toxicity and application, Materials Science and Engineering: C 77 (2017) 1363−1375.

[40] P.T. Yin, S. Shah, M. Chhowalla, K.B. Lee, Design, synthesis, and characterization of graphene−nanoparticle hybrid materials for bioapplications, Chemical Reviews 115 (7) (2015) 2483−2531.

[41] K.P. Loh, Q. Bao, P.K. Ang, J. Yang, The chemistry of graphene, Journal of Materials Chemistry 20 (12) (2010) 2277−2289.

[42] F.M. Koehler, N.A. Luechinger, D. Ziegler, E.K. Athanassiou, R.N. Grass, A. Rossi, C. Hierold, A. Stemmer, W.J. Stark, Permanent pattern-resolved adjustment of the surface potential of graphene-like carbon through chemical functionalization, Angewandte Chemie International Edition 48 (1) (2009) 224−227.

[43] M.J. Allen, V.C. Tung, R.B. Kaner, Honeycomb carbon: a review of graphene, Chemical Reviews 110 (1) (2009) 132−145.

[44] M. Hirata, T. Gotou, S. Horiuchi, M. Fujiwara, M. Ohba, Thin-film particles of graphite oxide 1:: high-yield synthesis and flexibility of the particles, Carbon 42 (14) (2004) 2929−2937.

[45] S. Thakur, N. Karak, Alternative methods and nature-based reagents for the reduction of graphene oxide: a review, Carbon 94 (2015) 224−242.

[46] D. Bitounis, H. Ali-Boucetta, B.H. Hong, D.H. Min, K. Kostarelos, Prospects and challenges of graphene in biomedical applications, Advanced Materials 25 (16) (2013) 2258−2268.

[47] H. Shen, L. Zhang, M. Liu, Z. Zhang, Biomedical applications of graphene, Theranostics 2 (3) (2012) 283.

[48] S. Barua, S. Thakur, L. Aidew, A.K. Buragohain, P. Chattopadhyay, N. Karak, One step preparation of a biocompatible, antimicrobial reduced graphene oxide–silver nanohybrid as a topical antimicrobial agent, RSC Advances 4 (19) (2014) 9777–9783.

[49] T.P.D. Shareena, D. McShan, A.K. Dasmahapatra, P.B. Tchounwou, A review on graphene-based nanomaterials in biomedical applications and risks in environment and health, Nano-Micro Letters 10 (3) (2018) 53.

[50] A.M. Pinto, I.C. Goncalves, F.D. Magalhaes, Graphene-based materials biocompatibility: a review, Colloids and Surfaces B 111 (2013) 188–202.

[51] F.M. Tonelli, V.A. Goulart, K.N. Gomes, M.S. Ladeira, A.K. Santos, E. Lorençon, L.O. Ladeira, R.R. Resende, Graphene-based nanomaterials: biological and medical applications and toxicity, Nanomedicine 10 (15) (2015) 2423–2450.

[52] S.K. Mahatha, K.D. Patel, K.S. Menon, Electronic structure investigation of MoS_2 and $MoSe_2$ using angle-resolved photoemission spectroscopy and ab initio band structure studies, Journal of Physics: Condensed Matter 24 (47) (2012) 475504.

[53] W. Choi, N. Choudhary, G.H. Han, J. Park, D. Akinwande, Y.H. Lee, Recent development of two-dimensional transition metal dichalcogenides and their applications, Materials Today 20 (3) (2017) 116–130.

[54] W. Xin-Ran, S. Yi, Z. Rong, Field-effect transistors based on two-dimensional materials for logic applications, Chinese Physics B 22 (9) (2013) 098505.

[55] S. Das, J.A. Robinson, M. Dubey, H. Terrones, M. Terrones, Beyond graphene: progress in novel two-dimensional materials and van der Waals solids, Annual Review of Materials Research 45 (2015) 1–27.

[56] M. Chhowalla, Z. Liu, H. Zhang, Two-dimensional transition metal dichalcogenide (TMD) nanosheets, Chemical Society Reviews 44 (9) (2015) 2584–2586.

[57] R.G. Dickinson, L. Pauling, The crystal structure of molybdenite, Journal of the American Chemical Society 45 (6) (1923) 1466–1471.

[58] J.A. Wilson, A.D. Yoffe, The transition metal dichalcogenides discussion and interpretation of the observed optical, electrical and structural properties, Advances in Physics 18 (73) (1969) 193–335.

[59] R. Tenne, L. Margulis, M.E. Genut, G. Hodes, Polyhedral and cylindrical structures of tungsten disulphide, Nature 360 (6403) (1992) 444.

[60] S. Manzeli, D. Ovchinnikov, D. Pasquier, O.V. Yazyev, A. Kis, 2D transition metal dichalcogenides, Nature Rev. Mater. 2 (8) (2017) 17033.

[61] K. Kalantar-zadeh, J.Z. Ou, Biosensors based on two-dimensional MoS_2, ACS Sensors 1 (1) (2015) 5–16.

[62] Y. Yu, C. Li, Y. Liu, L. Su, Y. Zhang, L. Cao, Controlled scalable synthesis of uniform, high-quality monolayer and few-layer MoS_2 films, Scientific Reports 3 (2013) 1866.

[63] A.K. Mishra, K.V. Lakshmi, L. Huang, Eco-friendly synthesis of metal dichalcogenides nanosheets and their environmental remediation potential driven by visible light, Scientific Reports 5 (2015) 15718.

[64] J. Della Rocca, D. Liu, W. Lin, Nanoscale metal–organic frameworks for biomedical imaging and drug delivery, Accounts of Chemical Research 44 (10) (2011) 957–968.

[65] H. Furukawa, K.E. Cordova, M. O'Keeffe, O.M. Yaghi, The chemistry and applications of metal-organic frameworks, Science 341 (6149) (2013) 1230444.

[66] H.C. Zhou, J.R. Long, O.M. Yaghi, Introduction to metal–organic frameworks, Chemical Reviews 112 (2012) 673–674.

[67] M. Anstoetz, T.J. Rose, M.W. Clark, L.H. Yee, C.A. Raymond, T. Vancov, Novel applications for oxalate-phosphate-amine metal-organic-frameworks (OPA-MOFs): can an iron-based OPA-MOF be used as slow-release fertilizer? PLoS One 10 (12) (2015) e0144169.

[68] M. Zhao, K. Yuan, Y. Wang, G. Li, J. Guo, L. Gu, W. Hu, H. Zhao, Z. Tang, Metal—organic frameworks as selectivity regulators for hydrogenation reactions, Nature 539 (7627) (2016) 76.

[69] P.Q. Liao, N.Y. Huang, W.X. Zhang, J.P. Zhang, X.M. Chen, Controlling guest conformation for efficient purification of butadiene, Science 356 (6343) (2017) 1193—1196.

[70] Y. Peng, Y. Li, Y. Ban, H. Jin, W. Jiao, X. Liu, W. Yang, Metal-organic framework nanosheets as building blocks for molecular sieving membranes, Science 346 (6215) (2014) 1356—1359.

[71] M. Zhao, Y. Wang, Q. Ma, Y. Huang, X. Zhang, J. Ping, Z. Zhang, Q. Lu, Y. Yu, H. Xu, Y. Zhao, Ultrathin 2D metal—organic framework nanosheets, Advanced Materials 27 (45) (2015) 7372—7378.

[72] A. Abhervé, S. Manas-Valero, M. Clemente-León, E. Coronado, Graphene related magnetic materials: micromechanical exfoliation of 2D layered magnets based on bimetallic anilate complexes with inserted [Fe III (acac 2-trien)]+ and [Fe III (sal 2-trien)]+ molecules, Chemical Science 6 (8) (2015) 4665—4673.

[73] Y. Ding, Y.P. Chen, X. Zhang, L. Chen, Z. Dong, H.L. Jiang, H. Xu, H.C. Zhou, Controlled intercalation and chemical exfoliation of layered metal—organic frameworks using a chemically labile intercalating agent, Journal of the American Chemical Society 139 (27) (2017) 9136—9139.

[74] X. Wang, C. Chi, K. Zhang, Y. Qian, K.M. Gupta, Z. Kang, J. Jiang, D. Zhao, Reversed thermo-switchable molecular sieving membranes composed of two-dimensional metal-organic nanosheets for gas separation, Nature Communications 8 (2017) 14460.

[75] S. Pavlidou, C.D. Papaspyrides, A review on polymer—layered silicate nanocomposites, Progress in Polymer Science 33 (12) (2008) 1119—1198.

[76] K.B. Yoon, H.D. Sung, Y.Y. Hwang, S.K. Noh, D.H. Lee, Modification of montmorillonite with oligomeric amine derivatives for polymer nanocomposite preparation, Applied Clay Science 38 (1—2) (2007) 1—8.

[77] P. Liu, Polymer modified clay minerals: a review, Applied Clay Science 38 (1—2) (2007) 64—76.

[78] J.U. Ha, M. Xanthos, Functionalization of nanoclays with ionic liquids for polypropylene composites, Polymer Composites 30 (5) (2009) 534—542.

[79] M. Alexandre, P. Dubois, Polymer-layered silicate nanocomposites: preparation, properties and uses of a new class of materials, Materials Science and Engineering: R: Reports 28 (1—2) (2000) 1—63.

[80] K. Majeed, M. Jawaid, A. Hassan, A.A. Bakar, H.A. Khalil, A.A. Salema, I. Inuwa, Potential materials for food packaging from nanoclay/natural fibres filled hybrid composites, Materials and Design 46 (2013) 391—410.

[81] S.L. Phua, L. Yang, C.L. Toh, S. Huang, Z. Tsakadze, S.K. Lau, Y.W. Mai, X. Lu, Reinforcement of polyether polyurethane with dopamine-modified clay: the role of interfacial hydrogen bonding, ACS Applied Materials and Interfaces 4 (9) (2012) 4571—4578.

[82] R.I. Narro-Céspedes, M.G. Neira-Velázquez, L.F. Mora-Cortes, E. Hernández-Hernández, A.O. Castañeda-Facio, M.C. Ibarra-Alonso, Y.K. Reyes-Acosta, G. Soria-Arguello, J.J. Borjas-Ramos, Surface modification of sodium

montmorillonite nanoclay by plasma polymerization and its effect on the properties of polystyrene nanocomposites, Journal of Nanomaterials 2018 (2018).

[83] A. Sinitskii, A. Dimiev, D.A. Corley, A.A. Fursina, D.V. Kosynkin, J.M. Tour, Kinetics of diazonium functionalization of chemically converted graphene nanoribbons, ACS Nano 4 (4) (2010) 1949−1954.

[84] M. Fang, K. Wang, H. Lu, Y. Yang, S. Nutt, Covalent polymer functionalization of graphene nanosheets and mechanical properties of composites, Journal of Materials Chemistry 19 (38) (2009) 7098−7105.

[85] Y. Feng, H. Liu, W. Luo, E. Liu, N. Zhao, K. Yoshino, W. Feng, Covalent functionalization of graphene by azobenzene with molecular hydrogen bonds for long-term solar thermal storage, Scientific Reports 3 (2013) 3260.

[86] L.H. Liu, M.M. Lerner, M. Yan, Derivitization of pristine graphene with well-defined chemical functionalities, Nano Letters 10 (9) (2010) 3754−3756.

[87] K. Yang, L. Feng, H. Hong, W. Cai, Z. Liu, Preparation and functionalization of graphene nanocomposites for biomedical applications, Nature Protocols 8 (12) (2013) 2392.

[88] S. Mallakpour, A. Abdolmaleki, S. Borandeh, Covalently functionalized graphene sheets with biocompatible natural amino acids, Applied Surface Science 307 (2014) 533−542.

[89] D. Yu, Y. Yang, M. Durstock, J.B. Baek, L. Dai, Soluble P3HT-grafted graphene for efficient bilayer− heterojunction photovoltaic devices, ACS Nano 4 (10) (2010) 5633−5640.

[90] Y. Matsuo, T. Fukunaga, T. Fukutsuka, Y. Sugie, Silylation of graphite oxide, Carbon 10 (42) (2004) 2117−2119.

[91] D.Y. Lee, Z. Khatun, J.H. Lee, Y.K. Lee, I. In, Blood compatible graphene/heparin conjugate through noncovalent chemistry, Biomacromolecules 12 (2) (2011) 336−341.

[92] Y. Zhang, J. Zhang, X. Huang, X. Zhou, H. Wu, S. Guo, Assembly of graphene oxide− enzyme conjugates through hydrophobic interaction, Small 8 (1) (2012) 154−159.

[93] S.M. Tan, A. Ambrosi, Z. Sofer, Š. Huber, D. Sedmidubský, M. Pumera, Pristine basal and edge plane oriented molybdenite MoS_2 exhibiting highly anisotropic properties, Chemistry − A European Journal 21 (19) (2015) 7170−7178.

[94] Z. Li, S.L. Wong, Functionalization of 2D transition metal dichalcogenides for biomedical applications, Materials Science and Engineering: C 70 (2017) 1095−1106.

[95] J.S. Kim, H.W. Yoo, H.O. Choi, H.T. Jung, Tunable volatile organic compounds sensor by using thiolated ligand conjugation on MoS_2, Nano Letters 14 (10) (2014) 5941−5947.

[96] A. Tuxen, J. Kibsgaard, H. Gøbel, E. Lægsgaard, H. Topsøe, J.V. Lauritsen, F. Besenbacher, Size threshold in the dibenzothiophene adsorption on MoS_2 nanoclusters, ACS Nano 4 (8) (2010) 4677−4682.

[97] L. Zhou, B. He, Y. Yang, Y. He, Facile approach to surface functionalized MoS_2 nanosheets, RSC Advances 4 (61) (2014) 32570−32578.

[98] D. Voiry, A. Goswami, R. Kappera, C.D.C.C. e Silva, D. Kaplan, T. Fujita, M. Chen, T. Asefa, M. Chhowalla, Covalent functionalization of monolayered transition metal dichalcogenides by phase engineering, Nature Chemistry 7 (1) (2015) 45.

[99] P. Joensen, R.F. Frindt, S.R. Morrison, Single-layer MoS_2, Mate.Res. Bull. 21 (4) (1986) 457−461.

[100] S.S. Chou, M. De, J. Kim, S. Byun, C. Dykstra, J. Yu, J. Huang, V.P. Dravid, Ligand conjugation of chemically exfoliated MoS$_2$, Journal of the American Chemical Society 135 (12) (2013) 4584–4587.

[101] C. Backes, N.C. Berner, X. Chen, P. Lafargue, P. LaPlace, M. Freeley, G.S. Duesberg, J.N. Coleman, A.R. McDonald, Functionalization of liquid-exfoliated two-dimensional 2H-MoS$_2$, Angewandte Chemie International Edition 54 (9) (2015) 2638–2642.

[102] T. Liu, C. Wang, X. Gu, H. Gong, L. Cheng, X. Shi, L. Feng, B. Sun, Z. Liu, Drug delivery with PEGylated MoS$_2$ nano-sheets for combined photothermal and chemotherapy of cancer, Advanced Materials 26 (21) (2014) 3433–3440.

[103] Q. Yue, S. Chang, S. Qin, J. Li, Functionalization of monolayer MoS$_2$ by substitutional doping: a first-principles study, Physics Letters A 377 (19–20) (2013) 1362–1367.

[104] S. Wang, W. Morris, Y. Liu, C.M. McGuirk, Y. Zhou, J.T. Hupp, O.K. Farha, C.A. Mirkin, Surface-specific functionalization of nanoscale metal–organic frameworks, Angewandte Chemie International Edition 54 (49) (2015) 14738–14742.

[105] H.C. Kolb, M.G. Finn, K.B. Sharpless, Click chemistry: diverse chemical function from a few good reactions, Angewandte Chemie International Edition 40 (11) (2001) 2004–2021.

[106] W. Zhu, G. Xiang, J. Shang, J. Guo, B. Motevalli, P. Durfee, J.O. Agola, E.N. Coker, C.J. Brinker, Versatile surface functionalization of metal–organic frameworks through direct metal coordination with a phenolic lipid enables diverse applications, Advanced Functional Materials 28 (16) (2018) 1705274.

[107] S. Wuttke, S. Braig, T. Preiß, A. Zimpel, J. Sicklinger, C. Bellomo, J.O. Rädler, A.M. Vollmar, T. Bein, MOF nanoparticles coated by lipid bilayers and their uptake by cancer cells, Chemical Communications 51 (87) (2015) 15752–15755.

[108] B. Illes, P. Hirschle, S. Barnert, V. Cauda, S. Wuttke, H. Engelke, Exosome-coated metal–organic framework nanoparticles: an efficient drug delivery platform, Chemistry of Materials 29 (19) (2017) 8042–8046.

[109] J. Yang, X. Chen, Y. Li, Q. Zhuang, P. Liu, J. Gu, Zr-based MOFs shielded with phospholipid bilayers: improved biostability and cell uptake for biological applications, Chemistry of Materials 29 (10) (2017) 4580–4589.

[110] H. Ejima, J.J. Richardson, K. Liang, J.P. Best, M.P. van Koeverden, G.K. Such, J. Cui, F. Caruso, One-step assembly of coordination complexes for versatile film and particle engineering, Science 341 (6142) (2013) 154–157.

[111] J. Guo, Y. Ping, H. Ejima, K. Alt, M. Meissner, J.J. Richardson, Y. Yan, K. Peter, D. von Elverfeldt, C.E. Hagemeyer, F. Caruso, Engineering multifunctional capsules through the assembly of metal–phenolic networks, Angewandte Chemie 126 (22) (2014) 5652–5657.

[112] A.J. Clough, J.W. Yoo, M.H. Mecklenburg, S.C. Marinescu, Two-dimensional metal–organic surfaces for efficient hydrogen evolution from water, Journal of the American Chemical Society 137 (1) (2014) 118–121.

[113] O. Shekhah, H. Wang, S. Kowarik, F. Schreiber, M. Paulus, M. Tolan, C. Sternemann, F. Evers, D. Zacher, R.A. Fischer, C. Wöll, Step-by-step route for the synthesis of metal–organic frameworks, Journal of the American Chemical Society 129 (49) (2007) 15118–15119.

[114] I. Shainberg, W.D. Kemper, Electrostatic forces between clay and cations as calculated and inferred from electrical conductivity, Clays and Clay Minerals (1966) 117–132.

[115] B.K. Schroth, G. Sposito, Surface charge properties of kaolinite, Clays and Clay Minerals 45 (1) (1997) 85–91.

[116] Y.T. Lim, O.O. Park, Phase morphology and rheological behavior of polymer/layered silicate nanocomposites, Rheologica Acta 40 (3) (2001) 220–229.

[117] M.A. Soto-Oviedo, O.A. Araújo, R. Faez, M.C. Rezende, M.A. De Paoli, Antistatic coating and electromagnetic shielding properties of a hybrid material based on polyaniline/organoclay nanocomposite and EPDM rubber, Synthetic Metals 156 (18–20) (2006) 1249–1255.

[118] M.I. Katsnelson, K.S. Novoselov, Graphene: new bridge between condensed matter physics and quantum electrodynamics, Solid State Communications 143 (1–2) (2007) 3–13.

[119] N.M.R. Peres, Graphene, new physics in two dimensions, Europhysics News 40 (3) (2009) 17–20.

[120] G.H. Han, D.L. Duong, D.H. Keum, S.J. Yun, Y.H. Lee, Van der Waals metallic transition metal dichalcogenides, Chemical Reviews 118 (13) (2018) 6297–6336.

[121] S.W. Han, H. Kwon, S.K. Kim, S. Ryu, W.S. Yun, D.H. Kim, J.H. Hwang, J.S. Kang, J. Baik, H.J. Shin, S.C. Hong, Band-gap transition induced by interlayer van der Waals interaction in MoS_2, Physical Review B 84 (4) (2011) 045409.

[122] S. Bordiga, C. Lamberti, G. Ricchiardi, L. Regli, F. Bonino, A. Damin, K.P. Lillerud, M. Bjorgen, A. Zecchina, Electronic and vibrational properties of a MOF-5 metal−organic framework: ZnO quantum dot behaviour, Chemical Communications (20) (2004) 2300–2301.

[123] A. Kuc, A. Enyashin, G. Seifert, Metal− organic frameworks: structural, energetic, electronic, and mechanical properties, Journal of Physical Chemistry B 111 (28) (2007) 8179–8186.

[124] M.S. Hosseini, M. Tazzoli-Shadpour, I. Amjadi, A.A. Katbab, E. Jaefargholi-Rangraz, Nanobiocomposites with enhanced cell proliferation and improved mechanical properties based on organomodified-nanoclay and silicone rubber, Engineering Technology 60 (2011) 1159–1162.

[125] W. Chrzanowski, S.Y. Kim, E.A.A. Neel, Biomedical applications of clay, Australian Journal of Chemistry 66 (11) (2013) 1315–1322.

[126] A. López-Galindo, C. Viseras, P. Cerezo, Compositional, technical and safety specifications of clays to be used as pharmaceutical and cosmetic products, Applied Clay Science 36 (1–3) (2007) 51–63.

[127] J.P. Zheng, L. Luan, H.Y. Wang, L.F. Xi, K.D. Yao, Study on ibuprofen/montmorillonite intercalation composites as drug release system, Applied Clay Science 36 (4) (2007) 297–301.

[128] Y. Dong, S.S. Feng, Poly (d, l-lactide-co-glycolide)/montmorillonite nanoparticles for oral delivery of anticancer drugs, Biomaterials 26 (30) (2005) 6068–6076.

[129] D. Depan, A.P. Kumar, R.P. Singh, Cell proliferation and controlled drug release studies of nanohybrids based on chitosan-g-lactic acid and montmorillonite, Acta Biomaterialia 5 (1) (2009) 93–100.

[130] J.H. Park, M.R. Karim, I.K. Kim, I.W. Cheong, J.W. Kim, D.G. Bae, J.W. Cho, J.H. Yeum, Electrospinning fabrication and characterization of poly (vinyl alcohol)/montmorillonite/silver hybrid nanofibers for antibacterial applications, Colloid & Polymer Science 288 (1) (2010) 115.

[131] X. Wang, X. Pei, Y. Du, Y. Li, Quaternized chitosan/rectorite intercalative materials for a gene delivery system, Nanotechnology 19 (37) (2008) 375102.

[132] N. Olmo, M.A. Lizarbe, J.G. Gavilanes, Biocompatibility and degradability of sepiolite-collagen complex, Biomaterials 8 (1) (1987) 67–69.

[133] M. Baek, J.H. Choy, S.J. Choi, Montmorillonite intercalated with glutathione for antioxidant delivery: synthesis, characterization, and bioavailability evaluation, International Journal of Applied Pharmaceutics 425 (1–2) (2012) 29–34.

[134] M. Kokabi, M. Sirousazar, Z.M. Hassan, PVA–clay nanocomposite hydrogels for wound dressing, European Polymer Journal 43 (3) (2007) 773–781.

[135] K.S. Katti, D.R. Katti, R. Dash, Synthesis and characterization of a novel chitosan/montmorillonite/hydroxyapatite nanocomposite for bone tissue engineering, Biomedical Materials 3 (3) (2008) 034122.

[136] D. Chimene, D.L. Alge, A.K. Gaharwar, Two-dimensional nanomaterials for biomedical applications: emerging trends and future prospects, Advanced Materials 27 (45) (2015) 7261–7284.

[137] C. Chung, Y.K. Kim, D. Shin, S.R. Ryoo, B.H. Hong, D.H. Min, Biomedical applications of graphene and graphene oxide, Accounts of Chemical Research 46 (10) (2013) 2211–2224.

[138] M.C. Duch, G.S. Budinger, Y.T. Liang, S. Soberanes, D. Urich, S.E. Chiarella, L.A. Campochiaro, A. Gonzalez, N.S. Chandel, M.C. Hersam, G.M. &Mutlu, Minimizing oxidation and stable nanoscale dispersion improves the biocompatibility of graphene in the lung, Nano Letters 11 (12) (2011) 5201–5207.

[139] S.K. Maji, A.K. Mandal, K.T. Nguyen, P. Borah, Y. Zhao, Cancer cell detection and therapeutics using peroxidase-active nanohybrid of gold nanoparticle-loaded mesoporous silica-coated graphene, ACS Applied Materials & Interfaces 7 (18) (2015) 9807–9816.

[140] R. Rauti, N. Lozano, V. Leon, D. Scaini, M. Musto, I.,I. Rago, F.P. Ulloa Severino, A. Fabbro, L. Casalis, E. Vazquez, K. Kostarelos, Graphene oxide nanosheets reshape synaptic function in cultured brain networks, ACS Nano 10 (4) (2016) 4459–4471.

[141] K. Yang, J. Lee, J.S. Lee, D. Kim, G.E. Chang, J. Seo, E. Cheong, T. Lee, S.W. Cho, Graphene oxide hierarchical patterns for the derivation of electrophysiologically functional neuron-like cells from human neural stem cells, ACS Applied Materials & Interfaces 8 (28) (2016) 17763–17774.

[142] Y. Liu, D. Yu, C. Zeng, Z. Miao, L. Dai, Biocompatible graphene oxide-based glucose biosensors, Langmuir 26 (9) (2010) 6158–6160.

[143] J. Ping, R. Vishnubhotla, A. Vrudhula, A.C. Johnson, Scalable production of high-sensitivity, label-free DNA biosensors based on back-gated graphene field effect transistors, ACS Nano 10 (9) (2016) 8700–8704.

[144] J.S. Medeiros, A.M. Oliveira, J.O.D. Carvalho, R. Ricci, M.D.C. Martins, B.V. Rodrigues, T.J. Webster, B.C. Viana, L.M. Vasconcellos, R.A. Canevari, F.R. Marciano, Nanohydroxyapatite/graphene nanoribbons nanocomposites induce in vitro osteogenesis and promote in vivo bone neoformation, ACS Biomaterials Science & Engineering 4 (5) (2018) 1580–1590.

[145] M.A. Shamekhi, H. Mirzadeh, H. Mahdavi, A. Rabiee, D. Mohebbi-Kalhori, M.B. Eslaminejad, Graphene oxide containing chitosan scaffolds for cartilage tissue engineering, International Journal of Biological Macromolecules 127 (2019) 396–405.

[146] Y.H. Wang, K.J. Huang, X. Wu, Recent advances in transition-metal dichalcogenides based electrochemical biosensors: a review, Biosensors and Bioelectronics 97 (2017) 305–316.

[147] X. Liu, H.L. Shuai, Y.J. Liu, K.J. Huang, An electrochemical biosensor for DNA detection based on tungsten disulfide/multi-walled carbon nanotube composites and hybridization chain reaction amplification, Sensors and Actuators B: Chemical 235 (2016) 603−613.

[148] K.J. Huang, Y.J. Liu, H.B. Wang, Y.Y. Wang, Y.M. Liu, Sub-femtomolar DNA detection based on layered molybdenum disulfide/multi-walled carbon nanotube composites, Au nanoparticle and enzyme multiple signal amplification, Biosensors and Bioelectronics 55 (2014) 195−202.

[149] X. Cao, Ultra-sensitive electrochemical DNA biosensor based on signal amplification using gold nanoparticles modified with molybdenum disulfide, graphene and horseradish peroxidase, Microchimica Acta 181 (9−10) (2014) 1133−1141.

[150] S. Keskin, S. Kızılel, Biomedical applications of metal organic frameworks, Industrial & Engineering Chemistry Research 50 (4) (2011) 1799−1812.

[151] W.J. Rieter, K.M. Taylor, H. An, W. Lin, W. Lin, Nanoscale metal− organic frameworks as potential multimodal contrast enhancing agents, Journal of the American Chemical Society 128 (28) (2006) 9024−9025.

[152] J.W. Zhang, H.T. Zhang, Z.Y. Du, X. Wang, S.H. Yu, H.L. Jiang, Water-stable metal−organic frameworks with intrinsic peroxidase-like catalytic activity as a colorimetric biosensing platform, Chemical Communications 50 (9) (2014) 1092−1094.

Synthesis of two-dimensional nanomaterials

Paramita Karfa, BSc, MSc[1], Kartick Chandra Majhi, BSc, MSc[1], Rashmi Madhuri, MSc, PhD[1,2]

[1]*Department of Applied Chemistry, Indian Institute of Technology (Indian School of Mines), Dhanbad, Jharkhand, India;* [2]*Assistant Professor, Applied Chemistry, Indian institute of technology (Indian School of Mines), Dhanbad, Jharkhand, India*

2.1 Introduction

Two-dimensional (2D) nanostructures have received a lot of attention since the breakthrough of graphene in 2004, owing to their bizarre physicochemical properties, which are mainly attributed to quantum size effect, strange surface morphology, and high aspect ratio [1,2]. They also have large surface area, high stability, cost-effective synthesis ability, exceptional physiochemical/mechanical properties, intercalated morphologic features, and piezoelectric coupling properties [3]. Owing to these extraordinary properties, the 2D materials paved a way toward large number of promising applications such as photocatalysis, electrocatalysis, energy generation, bioimaging, biomedicine, environmental applications, fuel cells, solar cells, battery, and sensing [4−7].

The most popular 2D nanomaterial, i.e., graphene, has a single layer of carbon atoms arranged in a 2D hexagonal structure [8]. With the discovery of graphene and its strange optoelectronic properties, high surface area, outstanding thermal conductivity, great mechanical strength, and high Young's modulus, a new kind of interest has been developed among the researchers and industries working in the field of nanotechnology [9]. Graphene has been used in various applications such as energy storage devices, biological sensing, and high pace electronic and optical devices [10−12]. Therefore after graphene, scientists were involved with tremendous eagerness in the search for 2D nanomaterials including a large variety of materials from metals to insulators, semiconductors, and even to superconductors, which have different properties from their bulk counterparts [13].

After the great efforts of the researchers working in the field of 2D nanomaterials, several new 2D nanomaterials have been fabricated, which show outstanding properties and sometimes better performance than graphene. The single-layered 2D nanomaterials have electron confine effect with outstanding electronic properties. The nanomaterials have high surface area, owing to the exposure of edge sites, and flexibility in atomic level, which made them suitable candidates for

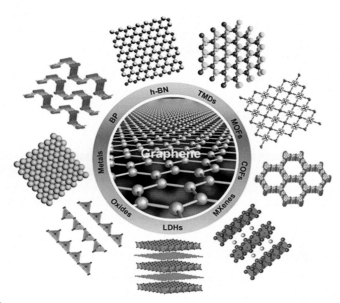

FIGURE 2.1

Two-dimensional graphene analogues such as TMDs, graphitic carbon nitride (g-C₃N₄), hexagonal boron nitride (h-BN), MXenes, and metal-organic frameworks (MOFs). *BP*, black phosphorus; *COFs*, covalent organic frameworks; *LDHs*, layered double hydroxides.

Reproduced with permission from H. Zhang, Ultrathin two-dimensional nanomaterials, ACS Nano 9(10) (2015) 9451–9469.

optoelectronics and transparent electronic devices [14–16]. According to the literature, the edge sites of these nanomaterials are more reactive than the basal planes because of the presence of a large number of active sites on the edges, which allowed their intercalation with ions and electrolytes.

Other than graphene, some popular 2D nanostructures, which are trending nowadays, can be classified into different types or categories. Some of the popular ones are shown in Fig. 2.1, for example, transition metal dichalcogenides (TMDs), graphitic carbon nitride (g-C₃N₄), 2D layered metal hydroxide, 2D metal oxides (Fe₃O₄, NiO, Co₃O₄), phosphorene, germanene, silicene, 2D metal nanoparticles, 2D polymers, and 2D crystals such as metal-organic framework (MOF) [17].

2.2 Synthesis of two-dimensional nanomaterials

The crystal structure, crystal size, and chemicophysical properties of 2D nanomaterials make them an inherent material for the synthesis of new-generation material [18]. In the literature, there are various methods for the synthesis of 2D nanomaterials, but the synthesis methods can be broadly categorized into the top-down and bottom-up approaches [19]. Top-down method is actually a destructive process that is done by forming nanoparticles by removing building blocks from the substrate or cutting or sizing the atomic crystal planes [20]. In contrast, the bottom-up approach or gathering up method is the process in which nanoparticles are synthesized from simpler units or atoms, where atoms are supplemented to the substrate to prepare the nanomaterial. The top-down method is categorized into various types, which are shown in Fig. 2.2. Comparatively, the bottom-up approach is the easier process for the synthesis of 2D nanomaterials and it is more valuable for the researchers because it consumes less energy and results in fewer defects in the nanostructure [20]. The popular process of synthesis used in bottom-up methods is shown in Fig. 2.3. It is true that these two methods are the most popular ways to synthesize 2D nanomaterials. However, it is also true that scientists have been intensely searching for new methods for decades to ease the process of synthesis and to increase the percentage yield of nanomaterials using a small amount of precursor. In the coming sections, we have discussed the bottom-up and top-down processes in detail.

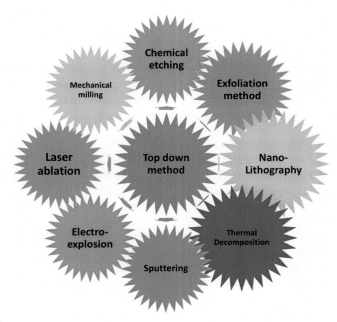

FIGURE 2.2

Top-down approaches popularly used for the synthesis of 2 materials.

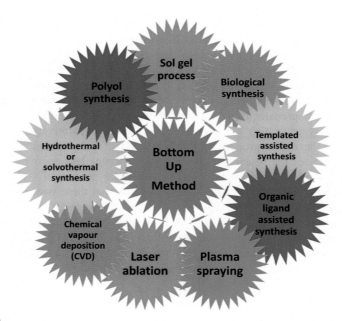

FIGURE 2.3

Bottom-up approaches popularly used for the synthesis of two-dimensional materials.

2.2.1 Bottom-up approach used for the synthesis of two-dimensional nanomaterials

To achieve the desired chemical composition, size, shape, stacking arrangement, crystal structure, and edge or surface defects, different processes have been followed and reported in the literature. Some of them are discussed in the following sections, starting with the bottom-up approach used for the synthesis of 2D nanomaterials.

2.2.1.1 Hydrothermal/solvothermal synthesis

It is now one of the most admired synthetic methods adopted by researchers for the synthesis of 2D nanomaterials [21]. The meaning of the word is within the name itself, hydro means water and thermal means heat; in other words, it is a wet thermal technique, which is operated at a high temperature and pressure. In this process the precursor materials are mixed together and kept in a single reaction vessel, commonly a hydrothermal bomb. The hydrothermal bomb works at high-pressure conditions. In this process the solvent acts as a catalyst for the growth of nanoparticles [22]. This process has several advantages such as low-cost instrumentation, large yield of products, environment-friendly process, and synthesis of different morphologic forms of nanomaterials of controlled size. This process does not require any expensive catalyst or harmful surfactant, thus producing high-quality crystals. This process can be modified by combining with other processes/

techniques or devices such as ultrasound, microwave, electrochemistry, and hot pressing [23,24]. Alam et al. have synthesized Bi-TiO$_2$ nanotube/graphene composites through simple hydrothermal synthesis. Fig. 2.4 shows the process of synthesis of a composite in a hydrothermal bomb at 140°C for 24 h [25].

2.2.1.2 Chemical vapor deposition

Chemical vapor deposition (CVD) is the gas phase method that is popularly used for the synthesis of thin films or crystals by depositing solid materials. In CVD the volatile precursors are placed in the chamber in their gaseous form and allowed to react at an ambient temperature. The surface of the substrate is covered with thin films of the desired product, and the process does not proceed through a particular chemical reaction [26]. By varying the substrate, temperature, pressure, and gas mixture concentration, thin film material of different size varying from nanometer to millimeter, with different chemical and physical properties, can be obtained. Some of the advantages of CVD are production of uniform film with low porosity, high purity, and

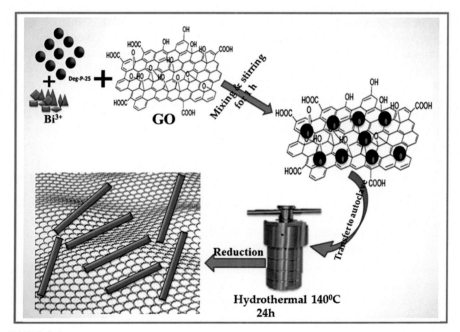

FIGURE 2.4

Graphical abstract showing the synthesis of Bi-TiO$_2$ nanotube/graphene through a simple hydrothermal technique.

Reproduced with permission from U. Alam, M. Fleisch, I. Kretschmer, D. Bahnemann, M. Muneer, One-step hydrothermal synthesis of Bi-TiO$_2$ nanotube/graphene composites: an efficient photocatalyst for spectacular degradation of organic pollutants under visible light irradiation, Applied Catalysis B: Environmental, 218 (2017) 758–769.

stability. However, it has some disadvantages also, as it requires highly expensive instrumentation and emits toxic gaseous by-products during reaction. CVD can be classified into various types, some of the popular ones are plasma-mediated CVD, chemical beam epitaxy, laser-enhanced CVD, and low-pressure CVD [27]. The mechanism of CVD includes the transport of reactant species to the substrate, then the volatile molecule diffuses to the vicinity of the growth site, and the precursor gets adsorbed to the substrate followed by chemical reaction, by-product desorption, and transport of by-products away from the substrate [28].

2.2.1.3 Organic ligand—assisted synthesis

It is also a kind of bottom-up synthesis approach used for obtaining 2D nanomaterials. In this process, at first, crystal sheets of metallates are formed at high temperature, followed by nucleation of nanosheets by bulky intercalating ions [29]. These intercalating ions are long-chain bulky carbon-containing molecules that form metal-ligand complex in a wet colloidal synthesis process. The main endeavor of this process is complex formation (metal-ligand complex), which lowers the total energy of the reaction and provides steric stabilization to bulky ligands. Ligands used in this process can be classified into cationic, anionic, nonionic, and polymeric types. Ionic organic liquid consists of hydrophobic extended carbon chain, and hydrophilic cationic head is commonly used as ionic ligands. Commonly, sodium dodecyl sulfate, cetyltrimethylammonium bromide (CTAB), 1,3-bis(4-carboxybutyl)imidazolium bromide, octadecyl trimethylammonium chloride, tetrabutylammonium ion (TBA+), and 1,2,3-trimethylimidazolium (mmmIm+) methylsulfate are used. Nonionic ligands do not have any cationic head and they mainly precede their work by micelle formation and have weak interaction with metal atoms. For example, trioctylphosphine oxide, trioctylphosphine, polyethylene glycol (PEG), oleylamine, and oleic acid are the popularly used nonionic ligands. These organic ligands are nonvolatile, highly viscous, and soluble in various organic solvents [29,30].

2.2.2 Top-down approach used for the synthesis of two-dimensional nanomaterials

Some of the top-down approaches will be discussed in the following sections. The methods used for the synthesis of 2D nanomaterials, which involve breaking the interlayer spacing to prepare a monolayer nanomaterial from bulk material, will be encompassed.

2.2.2.1 Chemical exfoliation method

Chemical exfoliation is carried out in two steps: first is increasing the gap of interlayer spacing by weakening the van der Waals forces and second is the sonication or fast heating technique applied to exfoliate the layers [31]. The easiest way to overcome the van der Waals forces is dissolving the bulk material in any liquid medium, where London interaction contributes to the potential energy of the layers and helps

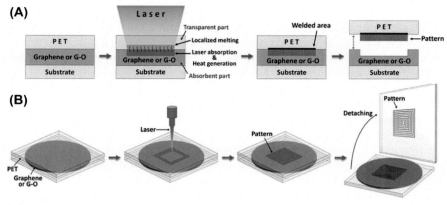

FIGURE 2.5

(A) Cross-sectional and (B) overall views of fabrication of laser-assisted graphene and graphene oxide (GO) film. *PET,* polyethylene terephthalate.

Reproduced with permission from J.S. Oh, S.H. Kim, T. Hwang, H.Y. Kwon, T.H. Lee, A.H. Bae, J.D. Nam, Laser-assisted simultaneous patterning and transferring of graphene, Journal of Physical Chemistry C 117(1) (2012) 663–668.

in cleavage of multilayers into monolayers. Solvents having surface tension around 40 mN m^{-1} are mainly used for exfoliation [32].

2.2.2.2 Laser ablation method

Laser ablation is the technique in which laser removal of a solid substance (precursor of metal) by vaporization is done in a gaseous and wet environment. The laser beam that falls on the material surface has ablation threshold much greater than that of the surface. The laser ablation technique has certain advantages: it is fast, does not need high temperatures, and has an easy synthesis process [33]. To exemplify the method of laser ablation, the report by Oh et al. can be considered, in which they have introduced a facile method for the fabrication of electrically conductive graphene through a laser-assisted synthesis technique [34]. The synthesis procedure involving the laser technique is shown in Fig. 2.5.

2.3 Classification of two-dimensional nanomaterials

In this chapter, we have included and focused on the basic principles and synthesis of 2D nanomaterials. We have classified the materials on the basis of their popularity and related literature.

2.3.1 Graphene

Graphene is one of the novel 2D nanomaterials with a monolayer of carbon atom arranged in a hexagonal honeycomblike structure containing an infinite number of benzene rings [35]. In 1947 the structure of graphene was first brilliantly put forth by Wallace; however, graphene was first synthesized in 2004 in the layered form by Geim et al. through the "scotch tape" mechanical method [36]. Geim and Novoselov won Nobel Prize for this invention, as they put forward a spellbinding innovation to describe whether graphene is a hypothetical structure or not [37]. Graphene possesses sp^2-bonded carbon atom with hexagonal fully bonded structure with alternative double and single bonds. This alternative double and single bond structure provides graphene with a conjugated structure with overlapping p orbital and π electron delocalization. Graphene, being stable and crystalline in nature, holds unique electronic properties such as high thermal conductivity, high specific surface area, optical transparency, and outstanding catalytic properties [38]. Graphene is a well-known nanomaterial and used in various applications such as electronic storage devices, photodetectors, biosensors, light-emitting diodes, drug delivery, and supercapacitor [39,40]. Regardless of having a number of advantages and applications in various fields, graphene has some disadvantages too; for example, it is very much inert toward any chemical reaction, insoluble, and infusible in nature. Therefore to make it more useful, the basal plane, edge, or surface functionalization is needed through covalent or noncovalent interactions.

Graphite is the main precursor of graphene synthesis, which can be synthesized through various top-down and bottom-up approaches, such as physical and chemical exfoliation, CVD, plasma treatment, thermal treatment, seeded growth, epitaxial growth, redox reaction, and liquid-phase stripping [41]. Among all these techniques, CVD produces graphene of high reactivity and high quality and also gives large-scale production. Graphene can also be synthesized through different named reactions such as Brodie method, Staudenmaier method, Hummers method, modified Hummers method, and improved Hummers method [42,43]. Improved Hummers method, which is a chemical exfoliation method, is used widely and is a nontoxic method for the synthesis of graphene, which results in graphene with less defects and nanosized 2D morphologic forms [44].

A few articles cited in the following can give more information on the synthesis of graphene through various top-down and bottom-up methods. Tu et al. have synthesized graphene of large area and high quality on nickel foil as the substrate using laser-assisted chemical vapor deposition (LCVD). By controlling the power of the laser beam, time of growth of layers, and rate of cooling, the layer structure of graphene can be controlled [45]. Jiang et al. have synthesized uniform-layered graphene at an ambient temperature through the LCVD method by using nickel plate as the substrate. The graphene ribbon synthesized in this method is 1.5×16 mm thick and is formed within a few seconds [46].

Hummers method is another popular chemical method used for the synthesis of graphene. Zaaba et al. have synthesized graphene oxide (GO) from graphite flakes through Hummers method in a different solvent to study its structural variance and electronic properties. The reaction was carried out at room temperature without sodium nitrate ($NaNO_3$). The synthesized GO has good solubility in ethanol and acetone owing to the presence of several functional groups on the surface of GO [47]. Yuan et al. have synthesized composites of GO and monohydrated manganese phosphate through the chemical exfoliation Hummers method. The prepared composite had a petal shape. Furthermore, the composite was calcined at 900°C to get the reduced form of GO, i.e., reduced graphene oxide (RGO) composite (RGO-$Mn_2P_2O_7$) [48]. Chang et al. have synthesized low-defect GO nanoribbons by exfoliating multiwalled carbon nanotubes (MWCNTs). The synthesis was done by a two-step process: the first step involves unzipping of stacked layers of MWCNTs through the exfoliation method and the second step includes exfoliation of graphene nanoribbons [49]. Kazemizadeh and Malekfar have synthesized porous GO through the laser ablation top-down approach. The system uses a flow of argon gas at an ultrahigh speed of $2.4 \, L \, min^{-1}$. Here, carbon nanotube is used as the precursor of porous graphene, with Ni as the catalyst [50]. Also, Yusuf et al. have synthesized graphene intercalated with CTAB for the adsorption of acid red 265 and acid orange 7 dyes. Here, they have synthesized GO by reducing graphene with environment-friendly and nontoxic ascorbic acid and noncovalent interface of CTAB as an intercalating agent [51]. Fouda et al. have synthesized high-quality GO nanosheets doped with spherical silicon nanoparticles through the simple and cost-effective hydrothermal method. At first, $C_8H_{20}O_4Si$ (the Si source) is dispersed in GO solution and then autoclaved for the synthesis of the composite [52]. Different other synthesis procedures used for the synthesis of 2D graphene are described in Table 2.1 [53−67].

2.3.2 Transition metal dichalcogenides

Transition metal dichalcogenides (TMDs) are the layered material having the chemical composition or generalized formula MX_2, where M represents the single-layer central transition metal (M = Mo, W, V, and Re) sandwiched between two surrounding layers of chalcogen atoms, X (i.e., O, S, Se) [68]. Similar to graphene, weak van der Waals interaction works between two adjacent layers of chalcogens and each atom in the same layer is covalently bonded together via hexagonal networks. Among the various kinds of TMDs reported in the literature, some, such as MoS_2, $MoSe_2$, WS_2, and WSe_2, are widely studied and more popular [16]. Every year, a number of articles have been published on these nanomaterials, showing their new synthesis procedures and vibrant applications in almost all the areas of material science, especially in the fabrication of energy devices, supercapacitors, electrocatalysts, batteries, etc. [69−71]. The new synthesis approach is tried by several research groups for their morphologic modification, phase engineering, compositional tuning, and heteroatom doping, in order to enhance the performance of 2D TMDs for energy conversions or related applications [72−74]. In general, TMDs hold a vast range of electronic properties, from semimetals to insulators. Studies have demonstrated that TMDs have a very outstanding catalytic property also, which

Table 2.1 Different techniques used for the preparation of two-dimensional graphene and its composites.

S No.	Materials	Methods of synthesis	Precursors of graphene	Application	Ref.
1.	GO and amine-modified GO	Hummers method	Graphite	For tissue engineering	[53]
2.	Gr and functionalized Gr	TCVD	Acetylene	–	[54]
3.	Silicon- and oxygen-codoped Gr	TCVD	Polycarbosilane	Gr/n-type silicon photodetectors	[55]
4.	N-GQDs/MoS$_2$-RGO	Hydrothermal	Graphite powder	ORR	[56]
5.	GeO$_x$-coated RGO balls	Spray pyrolysis	Graphite powder	Lithium-ion batteries	[57]
6.	Sulfur/carboxylated-Gr composite	Microemulsion technique	Carboxylated graphene	Lithium/sulfur batteries	[58]
7.	T-Nb$_2$O$_5$/Gr composite	Polyol-mediated solvothermal method	Natural graphite	Li-ion pseudocapacitor	[59]
8.	TiO$_2$–Gr	Electrospinning	Graphite powder	Lithium-ion batteries	[60]
9.	NiO/Gr	Chemical exfoliation	Graphite powder	p-Type dye-sensitized solar cell	[61]
10.	CNF RGO	Blending	Graphite oxide powder	Biobased flexible devices	[62]
11.	MoS$_2$/Gr composite	Hummers method	Graphite flakes	Sodium-ion batteries	[63]
12.	Cellulose/Gr	Hummers method	Graphite flakes	Triazine pesticide adsorption	[64]
13.	TiO$_2$–Gr composite	Electrospinning	Graphite powder	Photovoltaic and photocatalytic	[65]
14.	BC-RGO	Chemical exfoliation	Graphene oxide	Bacterial device/tissue engineering	[66]
15.	Iron nitride/nitrogen-doped GO	Hummers method	Graphite powder	ORR	[67]

BC = Bacterial cellulose; CNF = Cellulose nanofiber; Gr, Graphene; GO, Graphene oxide; MoS$_2$ = Molybdenum disulfide; Nb$_2$O$_5$ = Niobium oxide; N-GQDs = nitrogen-doped graphene oxide; NiO = Nickel oxide; ORR = Oxygen Reduction Reaction; RGO = Reduced graphene oxide; TCVD = Thermal chemical vapor deposition; TiO$_2$ = Titanium oxide.

make them a suitable and potential candidate as an electrocatalyst in fuel cells as well as in electrocatalysis process. They have gained more interest among researchers because of their magnificent thermal, optical, mechanical, and electric properties, as well as nonzero bandgap, which is one of the benefits of TMDs over graphene. This property made TMDs the next-generation material for optoelectronic and energy conversion devices. Most of the monolayer forms of TMDs have direct bandgap, but some TMDs such as GaSe and ReS_2 have indirect bandgap [75]. TMDs have a broad range of electronic properties and they are mechanically flexible. Suspended layered TMDs nanosheets have a considerable Young's modulus (E) of $\sim 0.33 \pm 0.07$ TPa, which makes them more flexible and durable [76]. The basic salient features and unique properties of TMDs are summarized in Fig. 2.6.

According to the literature, the diverse chemical, physical, optical, and catalytic properties of the TMDs depend mainly on their atomic arrangement [77]. The bulk counterparts have very different properties than the monolayer 2D TMDs and thus the monolayer counterparts are mainly used for sustainable energy applications. Monolayer TMDs have direct bandgap and charge carrier redistribution due to the lack of interlayer interaction. However, having a high surface area and large space between layers, TMDs allow intercalation of ions, small molecules, and chemical functionalization, which is proved to be very useful for reversible oxidation-reduction reaction in electrocatalysis. But the easy intercalation results in a decrease in the total number of active sites of the electrocatalysts, which mainly occurs at the edges [78].

In TMDs the transition metals, mainly from groups IV and VI, are covalently linked to chalcogens, i.e., sulfur, selenium, and tellurium, resulting in different

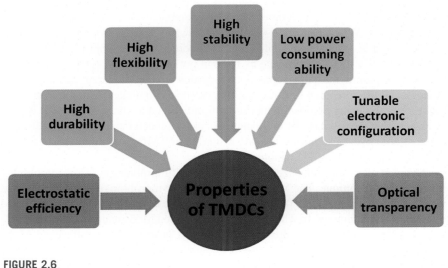

FIGURE 2.6

The basic salient features and unique properties of TMDs.

stacked morphologic forms and types. In the literatures, 1T, 2H, and 3R are the three most important polytypes or identified phase transitions of the TMDs. Unlike the graphene structure, all these polytypes do not hold a single layer of atoms, but they have a sandwich structure with the trigonal prismatic array of transition metals in the middle and hexagonal planes of chalcogens above and below them. In the nomenclature of the polytypes, 1T, 2H, and 3R represent tetragonal (T), hexagonal (H), and rhombohedral (R) unit cells, respectively, with 1, 2, and 3 representing the layers that are in single, double, and triple arrangements, respectively. Among them, the two forms that are naturally abundant and are thermodynamically stable are 2H and 3R. Researchers can easily transform different forms that are 2H to 3R or 1T by partial substitution of a few transition metals with other metal ions or ionic intercalations [79,80]. The most extensively studied polytype is the 2H phase in which each transition metal protrudes its six branches to two tetrahedrons with hexagonal symmetry from the pinnacle view, and thus it can be tuned to different morphologic forms according to its efficiency for different applications. According to the literature, due to the layered structure, large number of defects, easy intercalation ability, and high surface-to-volume ratio of TMDs, they possess and provide a large number of active sites for electrolysis, which might have made them an efficient electrode or electrocatalytic material for electrocatalysis [81].

Nowadays, researchers are inclined toward the use of hydrothermal and solvothermal synthesis procedures due to their fast, scalable, low-cost, and large-scale synthesis of nanomaterials. The similar kind of approach has also been implemented for the synthesis of TMDs. In this method, a transition metal precursor that is a metal salt of Mo, W, or V is used along with chalcogenide sources such as sulfur powder, thiourea, thioacetamide, and L-cysteine. The mixture of both precursors is then transferred to an autoclave at the required temperature and the reaction was performed at the given temperature [82]. For example, Chakravarty and Late [83] have prepared WSe_2 nanorods via the microwave and hydrothermal synthesis approach and explored their application as supercapacitors. Gopi et al. [84] have synthesized WSe_2/RGO composite by the one-pot hydrothermal synthesis approach and also explored the supercapacitor properties of the prepared nanomaterials. Huang et al. [85] prepared layered $MoSe_2$ nanosheets on nickel foam via hydrothermal process by using Se powder and sodium molybdate as the precursors. It is very clear now that day by day, different approaches have been explored and applied for the cost-effective, fast, and large-scale synthesis of TMDs, owing to their vast and versatile applications as 2D nanomaterials.

Zheng et al. have synthesized MoS_2 of different morphologic forms through the hydrothermal technique. The MoS_2 having flowerlike morphologic form was formed via Oswald ripening process [86]. Another composite of MoS_2 has been synthesized by Lejbini and Sangpour, in which they have decorated MoS_2 nanosheets with Fe_2O_3 nanoparticles through the hydrothermal synthesis process. The synthesized MoS_2 nanosheet is hexagonal, with 40-nm-sized hematite nanoparticles embedded on its surface [87]. Upadhyay et al. have synthesized crumpled MoS_2 nanosheets grown on carbon paper via the hydrothermal procedure. In their study, sodium molybdate dihydrate ($Na_2MoO_4 \cdot 2H_2O$) and sodium sulfide (Na_2S) were used as precursors for Mo and S, respectively. The crumpled 2D nanosheet shows good pseudocapacitive behavior [88].

On the other hand, large-scale synthesis of TMDs is done mainly by the CVD process. Wafer-scale TMDs can be controllably fabricated through CVD [89]. The CVD practice shows great assurance to manufacture high-quality TMDs layers with controllable thickness, scalable size, and excellent electronic properties. Other than the basic CVD technique, some modifications have also been done in this technique to develop a new and large-scale synthesis method of TMDs. For example, Zhao et al. [90] have discovered a method for the large-scale synthesis of $MoSe_2$ films on SiO_2/Si substrates by CVD, with uniform thickness and high crystallinity. Hyun et al. have synthesized MoS_2 layered crystals through CVD by using MoO_3 and sulfur powder as the Mo and S source, respectively. They have fully discussed the mechanism of synthesis of MoS_2 layers via CVD [91]. Mahyavanshi and his group have synthesized monolayer MoS_2 nanoribbons by CVD [92]. The group has optimized different concentrations of Mo precursor and sulfur-enriched environment for the synthesis of both MoS_2 ribbons and triangular structures. Han et al. have adopted a CVD technique for the synthesis of MoS_2 nanorods. They varied the MoS_2 morphologic forms by changing the gas flow direction and the concentration of gaseous MoO_3 [93]. Lu et al. have synthesized monolayer $MoSe_2$ films on silica substrate through CVD. The prepared films have high crystallinity and 2D hexagonal structure [94]. Wang et al. have studied the effects of hydrogen and reactive environment on the growth of $MoSe_2$ nanoflakes prepared through CVD in silica substrate [95].

In the recent time, sonication-assisted exfoliation has been developed as a new technique for the synthesis of TMDs using different solvents. For example, Gupta et al. [96] have synthesized layered MoS_2 nanosheets by the liquid exfoliation method with N-methyl-2-pyrrolidone (NMP) as the solvent. Li et al. [97] have also synthesized few-layered MoS_2 nanosheets by the use of shear-exfoliating device, i.e., high-speed dispersive homogenizer, and also used NMP as the solvent. Marqus and group [98] have synthesized MoS_2 quantum dot through the constant multicycle microfluidic process to break the bulk material. Ansari et al. [99] have synthesized M-MoS_2-polyaniline nanocomposite by the mechanical exfoliation method of MoS_2 sheet with polyaniline. Yang et al. [100] have synthesized MoS_2 thin films via the polymer-assisted deposition procedure. Truong et al. [101] have prepared MoS_2 and $MoSe_2$ nanosheets by the solution-phase exfoliation method using the supercritical fluid method. Lin et al. have synthesized MoS_2 and WS_2 nanosheets from their bulk counterparts through the chemical exfoliation method using $NaNO_3$/HCl and the yield of MoS_2 and WS_2 is found to be 58% and 52%, respectively. The prepared nanosheet exhibits great catalytic activity for oxygen evolution reaction [102]. Jawaid et al. have synthesized MoS_2 through the liquid-phase exfoliation technique, in which NMP was used as a solvent. According to the authors, the higher the moisture content, the higher rate of dissolution [103]. Lukowski et al. have synthesized MoS_2 for the application of hydrogen evolution reaction. The

TMDs are exfoliated through lithium intercalation and converted from 2H-MoS$_2$ to 1T-MoS$_2$ [104]. Huang et al. have synthesized TMDs through an easy and environment-friendly technique by exfoliating through a biopolymer silk fibroin in an aqueous medium. The silk fibroin was carboxyl modified and the synthesized nanomaterial was used for antibacterial dressing of wounds [105].

Other than the exfoliation methods, laser ablation technique is also very popular for the synthesis of TMDs. Oztas et al. [106] have synthesized 2D and three-dimensional (3D) colloidal MoS$_2$ nanostructures by the laser ablation technique in nanoseconds using 2H-MoS$_2$ crystals in methanol. Fominski and group have synthesized MoSe$_{x>3}$/Mo nanoparticles by pulsed laser ablation using MoSe$_2$ as the target surface. The prepared nanoparticle was ball shaped and was used for hydrogen evolution reaction [107]. Fominski et al. [108] have synthesized quasi-amorphous thin films of MoS$_x$ through the laser ablation technique and used them for hydrogen evolution reaction.

Some of the examples of TMDs synthesis via the organic ligand—mediated synthesis approach is also discussed here. For example, Chen et al. [109] have synthesized ultrathin nanosheets of MoS$_2$ incorporated with oxygen using PEG400 as a nonionic organic ligand, which plays a major role in structural determination. Li et al. have synthesized MoS$_2$ microspheres using 1-ethyl-3-methylimidazolium bromide ([EMIM]Br) as the organic cationic ligand. This ionic liquid has used the soft templating method by stacking metal-ligand complexes and thus forming the desired shape by altering the concentration of the organic ligand [110]. Zhang et al. have synthesized MoS$_2$ nanospheres with the surfactant-assisted hydrothermal synthesis method. The surfactant or ligand used here is an ionic organic ligand, CTAB. Here, with a change in the concentration of CTAB, the shape and size of MoS$_2$ are changed. The optimized amount of CTAB for the growth of the desired MoS$_2$ nanosphere is found to be 6.0 g per liter [111]. Guo et al. have synthesized flowerlike MoSe$_2$ using nonionic organic ligands, i.e., oleylamine and oleic acid. They have used molybdenum hexacarbonyl [Mo(Co)$_6$] as the molybdenum precursor and selenium powder as the Se precursor [112]. Xu et al. have synthesized sulfur-doped MoSe$_2$ and made their composite with nitrogenated graphene with the help of triethylenetetramine as the cationic ligand. Here, the ligand has played a dual role: it helped in graphene functionalization with the NH$_2$ group and was involved in the exfoliation of MoSe$_2$ layers [113].

Contrary to these, Chen and his group have synthesized edge-terminated MoS$_2$ with considerable interlayer spacing with the help of graphene and CTAB. At first, the GO surface, which is full of oxygen-carrying functional groups, interacts with the cationic organic ligand CTAB for the formation of positively charged GO, which was then attracted by the anionic MoO$_4^{2-}$. Then MoS$_2$ on graphene is formed with the help of nucleation [114]. Kirubasankar et al. have synthesized a competent electrode material using 2D MoSe$_2$-Ni(OH)$_2$ through the one-step hydrothermal synthesis and used it in supercapacitor application. The prepared nanomaterial via hydrothermal synthesis has a large number of electroactive sites and showed higher specific capacitance [115].

2.3.3 Hexagonal boron nitride (h-BN)

Hexagonal boron nitride (h-BN) has the hexagonal Bernal structure with equal number of boron (B) and nitrogen (N) atoms procuring the A and A$'$ sublattices, unlike graphene, and has AB stacked structure, with B and N atoms above each other and vice versa. In h-BN, both boron and nitrogen have unlike electronegativities, giving it a limited ionic character. The h-BN forms have showed potential applicability in different fields such as removal of organic and inorganic pollutants from wastewater, as photocatalysts, electronic devices, and many others. But their large-scale uniform synthesis is a major challenge for researchers [116].

BN was first invented by Balmain in the year 1842 from the reaction of potassium cyanide (KCN) and molten boric acid (H_3BO_3) [117]. h-BN has noteworthy structural resemblance to that of graphene and in its single-layer form it is called as "white graphene." Boron nitride has four crystalline forms. Among them, the most important is h-BN, next is rhombohedral boron nitride (r-BN) with sp^2 hybridized carbon atom, and the other two are cubic boron nitride (c-BN) and wurtzite boron nitride (w-BN), having sp^3 hybridized carbon atom [6,118]. Like carbon, 2D boron nitride also has nanostructures of different dimensions such as zero-dimensional boron nitride fullerenes, one-dimensional boron nitride nanotubes, and 2D boron nitride nanosheets. The 2D h-BN has both armchair and zigzag edge-terminated structures, with each layer attached to another by weak van der Waals forces. The h-BN being isostructural with graphene has a high surface area, good stability, high thermal conductivity, high bandgap, fine mechanical stability, and good carrier mobility [119].

At present, BN is synthesized through various chemical synthesis processes, including heating the powder of boric acid or oxides of boron with nitrogen-containing compounds such as urea, melamine, amine, and ammonia at a high temperature and then annealing in a nitrogen atmosphere. Some of the chemical reactions involved in BN synthesis are [120]

$$2B(OH)_3 \rightarrow B_2O_3 + 3H_2O$$

$$NH_2\text{-}CO\text{-}NH_2 \rightarrow HNCO + NH_3$$

$$B_2O_3 + 2NH_3 \rightarrow 2BN + 3H_2O$$

For the top-down approach the van der Waals forces between the layers have to be broken to form monolayered or few-layered h-BN from boron nitride crystals. Therefore h-BN can also be synthesized by dehydrogenation of NH_3BH_3. However, the well-known synthesis method of h-BN is the CVD method, which provides pure, high-quality, desirable, atomic-layered BN. Several types of precursors such as borazine and hexachloroborazine ($B_3N_3Cl_6$) are used as a precursor for the CVD process. In situ growth technique is used for the synthesis of BN-metal oxide composite, in which BN acts as a template for the growth of metal oxide using metal salt as the precursor [121]. Wang et al. have synthesized boron nitride nanosheets via a two-step process: the first step is the combustion of citrate-nitrate for the synthesis of boron-powder-coated cobalt oxide and the second step is the catalytic CVD of boron precursor. The synthesized nanotubes were of 20−80 nm in size [122]. da Silva et al. have synthesized nanostructures of boron nitride through thermal CVD at 1150°C using iron compounds as the catalyst on alumina nanostructures. The catalyst FeS/Fe_2O_3 was obtained from dithiocarbamate and the solvothermal reaction proceeds through decomposition of $Fe(S_2CNEt_2)_3$ precursor [123].

After TMDs, boron nitride is also very popularly synthesized by the hydrothermal process. Hao et al. [124] have synthesized boron nitride nanocrystals through the hydrothermal autoclave technique at 300°C for 24 h, using boric acid as precursor, trimethylamine as the nitrating agent, and hydrazine hydrate as the solvent. Chen and coworkers have synthesized carbon nitride−modified h-BN using the hydrothermal calcination process and used this as a photocatalyst. The CN-BN composite was prepared by mixing melamine and different concentrations of BN, with ethanol as the hydrothermal solvent [125]. Peng et al. made a novel approach for the synthesis of 2D MXene (i.e., Ti_3C_2, Nb_2C) through the hydrothermal etching process. The etching agent used here is $NaBF_4$ and HCl, which produces Ti_3C_2 with high lattice parameter, considerable interlayer gap, and larger surface area [126].

Laser ablation method is sometimes also used for the synthesis of boron nitride. For example, Nistor et al. [127] have synthesized boron nitride nanotubes by simply changing the pulse of laser under liquid environment. Maleki et al. [128] have synthesized porous boron nitride using two inorganic templates, one is polystyrene latex another one is CTAB. Different other methods applied for the synthesis of 2D hexagonal boron are described in Table 2.2 [129−143].

2.3.4 Graphitic carbon nitride (g-C$_3$N$_4$)

After the proposal of Liu and Cohen [144] that carbon nitrides have towering hardness property, with high chemical and mechanical durability, low density, and water and wear resistivity, outrageous interest has been gained toward this material. Graphitic carbon nitride (g-C_3N_4) has structural similarity with graphite and because of the presence of s-triazine units, more unique physiochemical properties are present in it. Graphitic carbon nitride (g-C_3N_4) consists of C-N bond, so more number of π electrons with sp^2 hybridized hexagonal rings of carbon atom is present in it. Based on the X-ray powder diffraction study, their different crystalline structures, such as α-C_3N_4, β-C_3N_4, γ-C_3N_4, zinc blende, and cubic C_3N_4, are reported in the literature [145]. It is reported that (g-C_3N_4) is the most stable allotrope of carbon nitride. Graphitic carbon nitride consists of loaded surface properties because of the existence of a small amount of hydrogen, which rendered g-C_3N_4 with more surface defects. However, the presence of an extra electron in the nitrogen atom makes graphitic carbon nitride a good catalyst for hydrogen evolution, oxygen evolution, CO_2 activation, and photodegradation of dyes [146].

Table 2.2 Different techniques applied for the synthesis of 2D hexagonal boron nitride (h-BN) and its composites.

S No.	Materials	Methods of synthesis	Precursors of boron	Application	Ref.
1.	BNNSs and BNQDs	High-temperature solvothermal treatment	h-BN powder	Cell imaging	[129]
2.	h-BN nanosheet/polymer composite nanofiber	Vacuum ball milling	h-BN powders	Fast MG absorption and separation of MG and RhB dyes	[130]
3.	BNNTs	Catalytic pyrolysis	H_3BO_3 and ethylenediamine	—	[131]
4.	Urchin-like h-BN	TCVD	B powder and urea	Removal of heavy metal ions	[132]
5.	h-BN nanotubes	TCVD	B powder and NH_3	—	[133]
6.	h-BN nanosheet/MWCNT nanocomposite	Ultrasonication	Powder of bulk BN	Determination of β-agonists	[134]
7.	BNNTs	Annealing	B_2O_3, B powder, MgB_2, NH_3	Preparation of thermoplastic polyurethane-based composite	[135]
8.	Rhombohedral h-BN	Ammonothermal	MgB_2 and supercritical NH_3	—	[136]
9.	BNQDs on GO	Solvothermal	2D h-BN nanosheets	Detection of organophosphate pesticides such as MP, DIA, and CHL	[137]
10.	Graphenelike h-BN-modified BiOBr flower	IL-assisted solvothermal process	H_3BO_3 and urea	Degradation of TC and RhB	[138]
11.	AgNP-decorated BN	Pyrolysis	H_3BO_3 and urea	Removal of TC and RhB	[139]
12	Highly porous h-BN	Template-free synthesis	H_3BO_3 and urea	CO_2 adsorption	[140]
13.	Nanoporous boron nitride	Pyrolysis	H_3BO_3 and dicyandiamide	Adsorbent for light hydrocarbons	[141]
14.	Silicon carbonitride/h-BN	Pyrolysis	h-BN powder	Li-ion battery	[142]
15.	Terbium (III) complexes into a porous BN	Pyrolysis	H_3BO_3 and melamine	Light-emitting diodes	[143]

AgNP= Silver nanoparticle; BN= Boron nitride; BNNSs = boron nitride nanosheets; BNNTs = Boron nitride nanotubes; B powder = Boron powder; BNQDs = boron nitride quantum dots; B_2O_3 = boron oxide; CHL = chlorpyrifos; CO_2= Carbon dioxide; DIA = diazinon; GO = Graphene oxide; h-BN = hexagonal boron nitride; H_3BO_3 = Boric acid; IL = Ionic liquid; MG = malachite green; MgB_2 = magnesium diboride; MP = methyl parathion; MWCNT = multiwalled carbon nanotube; NH_3 = Ammonia; RhB = rhodamine (B); TCVD = Thermal chemical vapor deposition; TC = tetracycline; 2D = two-dimensional.

Synthesis of g-C$_3$N$_4$ is mainly done by using nitrogen-abundant compounds, which are free from any oxygen-containing groups with C-N structure in the core. Derivatives of triazine such as heptazine and melamine are mainly preferred but they are explosive and unstable in nature [147,148]. The use of melamine as one of the precursors for the synthesis of g-C$_3$N$_4$ through pyrolysis and condensation is shown in Fig. 2.7. In recent times, g-C$_3$N$_4$ can be synthesized by pyrolysis or partial carbonization of dicyandiamide. For this, 1,3,5-trichloromelamine (C$_3$N$_3$Cl$_3$) was used as the precursor and heated with sodium amide (NaNH$_2$) at a high temperature (1000°C) and in an inert atmosphere [149]. Tanaka and his group have synthesized carbon nitride by the microwave plasma-assisted CVD method. Si/SiO$_2$ was used as the substrate, CH$_4$ and N$_2$ were used as the reaction gases, and the reaction was carried out at 1.1—4.0 kPa pressure and 400—800 W microwave power [150]. Also, Feng et al. have synthesized g-C$_3$N$_4$ by treating bulk graphitic carbon nitride using Hummers method. The bulk g-C$_3$N$_4$ was synthesized by calcination of dicyanamide. The prepared material is used for photocatalytic activities [151]. Different other techniques used for the synthesis of 2D g-C$_3$N$_4$ are described in Table 2.3 [152—161].

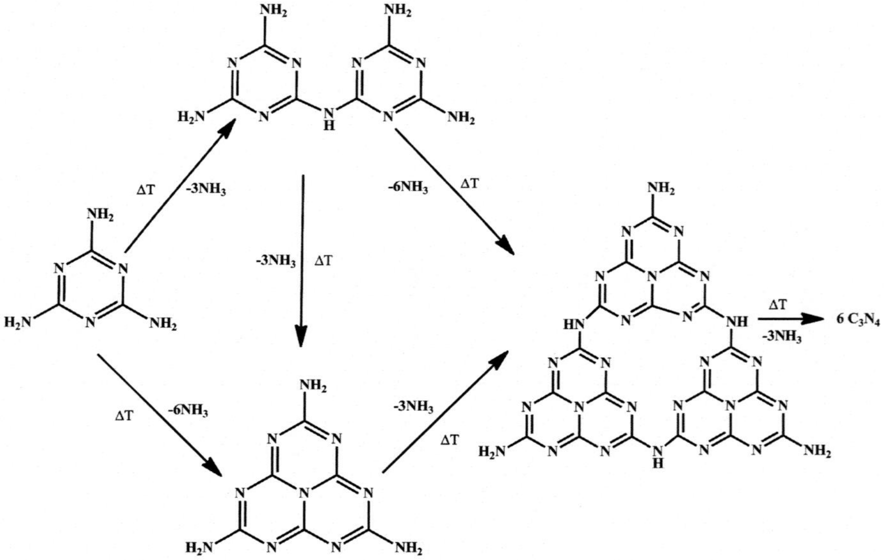

FIGURE 2.7

Postulated chemical reaction for the synthesis of graphitic carbon nitride from melamine.

Reproduced with permission from B. Jürgens, E. Irran, J. Senker, P. Kroll, H. Müller, W. Schnick, Melem (2, 5, 8-triamino-tri-s-triazine), an important intermediate during condensation of melamine rings to graphitic carbon nitride: synthesis, structure determination by X-ray powder diffractometry, solid-state NMR, and theoretical studies, Journal of the American Chemical Society 125(34) (2003) 10288—10300.

Table 2.3 Different techniques used for the synthesis of two-dimensional g-CN and its composites.

S No.	Materials	Methods of synthesis	Precursors used	Application	Ref.
1.	g-CN polymers	Pyrolysis	Melamine	Reduction of CO_2 to CO with visible light	[152]
2.	CN-based hydrogels	Photopolymerization	Melamine and cyanuric acid	Photocatalytic hydrogen production	[153]
3.	Fe_2O_3/R-CN/Co-Pi	Pyrolysis	Melamine	Photoelectrochemical water splitting	[154]
4.	Porous CN@RGO aerogel	Sol gel	Dicyandiamide	CO_2 capture	[155]
5.	Porous carbon-rich g-CN	Calcination	Melamine	Water cleaning	[156]
6.	CN/RGO	CVD	Dicyandiamide	Oxygen chemiresistor sensor	[157]
7.	Iron-doped CN	Hydrothermal	Formamide and citric acid	Visible light–driven hydrogen evolution	[158]
8.	Au/C_3N_4	Thermal decomposition	Cyanamide	CO oxidation	[159]
9.	Se-doped CN	Thermal polycondensation	Dicyandiamide and cyanuric acid	Solar fuel production from CO_2	[160]
10.	P-g-C_3N_4	Thermal decomposition	Melamine	Photocatalytic CO_2 reduction	[161]

Au= gold; CN = carbon nitride; CO_2= Carbon dioxide; Co-Pi = cobalt phosphate; CVD = chemical vapor deposition; g-CN = graphitic carbon nitride; P-g-C_3N_4 = Phosphorus-doped graphitic carbon nitride; RGO = reduced graphene oxide; R-CN = reassembled carbon nitride.

2.3.5 Transition metal oxides and layered hydroxides

The first discovered layered transition metal oxide (TMO) is delaminated perovskite, $K[Ca_2Na_{n-3}Nb_nO_{3n+1}]$, which was in 1990. Layered oxides are mainly oxides of transition metals such as MO_3, WO_3, TiO_2, TaO_3, V_2O_5, and several oxides of perovskite, such as $K_{1.5}Eu_{0.5}Ta_3O_{10}$, $K_{0.45}MnO_2$, $LaNb_2O_7$, and $Bi_4Ti_3O_{12}$ [162]. TMOs are negatively charged and highly polarizable due to the presence of oxygen atoms, which play a main role and are able to lodge interchangeable cations. Layered hydroxides of transition metals are positively charged, which is designated by the formula $[M^{2+}_{1-x}M^{3+}_x(OH)_2][A^{n-}]_{x/n} \cdot mH_2O$, where x designates the molar ratio of cations of metal, whose value normally lies between 0.4 and 0.2, and n represents the charge of the organic or inorganic anion "A," such as CO_3^{2-} and SO_4^{2-}. The amount of water molecules is represented by m, which is situated in the interlayer gap; M^{2+} represents the transition metals such as Ni, Co, Mn, and Mg in +2 oxidation state; and M^{3+} represents Co, Al, Fe, and Ga in +3 oxidation state. Layered metal hydroxides have a hexagonal structure of $M(OH)_2$ layers with stacked anions between them [162,163]. The schematic diagram of positively and negatively charged hydroxides and oxides are shown in Fig. 2.8.

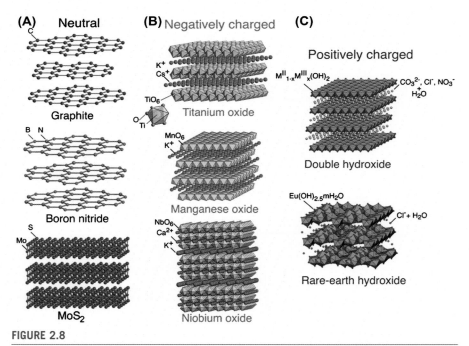

FIGURE 2.8

Different layered 2D oxides and hydroxides: (A) neutral 2D nanomaterials, (B) negatively charged oxides, and (C) positively charged hydroxides.

Reproduced with permission from R. Ma, T. Sasaki, Two-dimensional oxide and hydroxide nanosheets: controllable high-quality exfoliation, molecular assembly, and exploration of functionality, Accounts of chemical research 48(1) (2014) 136–143.

Transition metal hydroxides and oxides have certain interesting properties such as greater flexibility, tunable bandgap, diverse crystal morphologic form, high optoelectronic properties, high dielectrics, large specific surface area, and considerable transparency. These properties are mainly attributed to the theory that "s" electrons of the transition metal captured by the oxygen and the vacant orbital of the "d" electron play the rule. TMOs and transition metal—based layered double hydroxides (TMLDHs) are used for decades in various paint industries, lubricants, and cosmetics, but now, owing to their fascinating properties, they are also used as catalysts and in batteries, piezoelectric energy devices, supercapacitors, and fire retardants [164].

Layered oxides and hydroxides can be synthesized by both top-down and bottom-up approaches, with the top-down approach being the best known method to date for their synthesis. In general, the synthesis of TMOs can be done in gas phase or liquid phase. The gas-phase synthesis process mainly involves pulse layer deposition, CVD, and molecular beam epitaxy. The chemical exfoliation method is the wet synthesis process (i.e., liquid phase) using an organic intercalating agent. This is one of the most famous methods for the synthesis of 2D TMOs by exfoliation of interlayers with a gentle cutting process. In this process, organic cations such as TBA, propylammonium, or other quaternary ammonium ions are used as an intercalating agent, which forms polycationic or anionic nanosheets of 2D oxides. Different other wet chemical synthesis processes include solvothermal, hydrothermal, and microwave-assisted synthesis processes, in which oxides are grown in layers at a high temperature in a particular liquid solvent in a pressurized environment [165].

Among the different types of synthesis procedures used for TMLDHs, coprecipitation of divalent or trivalent salts using metal precursors in alkali medium is the best method that produces thin layers of TMLDH platelets of atom-level thickness. Throughout this process, several efforts have been made for using different surfactants such as dodecyl sulfate instead of an alkaline medium in order to minimize the electrostatic interaction. In order to facilitate homogeneous precipitation and high crystallinity of the structures, various nucleating agents that release ammonia, such as hexamethylenetetramine and urea, are used. This is because after the decomposition of these agents, both hydroxyl and carboxyl ions are generated as interlayer ions [166].

Exfoliation/delamination of layered double hydroxide (LDH) precursor in various polar solvents is a top-down approach of synthesis and was first discovered by Adachi-Pagano in 2000. In this method, the exfoliated $Zn-Al-NO_3$ LDH was prepared using dodecyl sulfate as the surfactant and butanol as the dispersing solvent [167]. In the delamination process, positively charged thin atomic layers of LDH are formed by ion exchange intercalation methods with a negatively charged surfactant, which enlarges the interlayer gap and results in the formation of single LDH layers. This process is thermodynamically more suitable and the application of polar solvents restricts the bulk formation to a negligible amount. Ghosh et al. have synthesized $Co(OH)_2$ platelets of hexagonal shape and incorporated them with GO. The whole process was carried out with the help of a cationic surfactant CTAB, which played a significant role in the formation of nano-architectures of the composite [168].

2.3.6 Silicene

A honeycomblike isostructural silicone allotrope with single-atom thickness is named as silicene, which has silicon atoms hexagonally arranged. Other than the carbon atoms in graphene, silicon atoms in silicene have both sp^3 and sp^2 hybridization states [169]. Because of the competition between the sp^3 and sp^2 hybridization states of silicene, the structure of silicene is very much unstable. This is the major limitation for its large-scale synthesis. Silicene has a low buckled structure with two sublattices, A and B, perpendicular to the same atomic plane. Silicene has great electronic properties because of the π-π* bands resultant of p_z orbitals, which linearly surpass the fermi level and form the "Dirac cone" [170].

Unlike graphene, silicene cannot be synthesized by the mechanical stripping method or by the chemical reduction of silicon oxide because silicon oxide is a hard insulator. For the synthesis of silicene by the chemical exfoliation method, calcium disilicide ($CaSi_2$) doped with magnesium has been used as the precursor, and this method was first put forward by Nakano et al. [171]. However, due to capping of silicon sheet surface with other oxygen atoms/impurities, the resulted silicene was of no use. After that, the large-scale synthesis of silicene sheets was discovered by Vogt et al. [172] in 2012 on Ag(111) surfaces. From that day, silicene has been synthesized by growing on a variety of substrates such as Ag(111) substrate, metal diboride substrate on silicon (111), Ag(110) silicene nanoribbon substrate, iridium (111) substrate, and Au(110) substrate [173,174].

2.3.7 Germanene

Just like silicene, germanene is member of group IV and a close relative of graphene. It is made of a single layer of germanium with sp^2/sp^3 hybridized germanium atoms arranged in a 2D honeycomblike structure. Germanene has certain electronic properties similar to those of graphene, such as quantum confinement effect, high carrier mobility, massless Dirac fermions, and tunable bandgap [170,175]. Carbon has the property to change its hybridization from sp to sp^2, but the germanium atom in germanene occupies mainly sp^3 hybridization because germanium does not have enough energy to hybridize all the p orbital with the s orbital. The Ge—Ge bond length is greatly larger than the C-C bond length, forming a p orbital overlapping with limited tendency of π bond formation. Cahangirov et al. [176], in 2009, first discovered germanene in its 2D honeycomb form. Bianco et al. [177] have synthesized germanene for the first time in its hydrogenated form by the deintercalation method, using $CaGe_2$ as the precursor, and proved that germanene has an outstanding electron mobilizing property with a bandgap of 1.53 eV. After replacing hydrogen ends with methyl functional groups the stability of germanene was found to be increased, and due to its low buckling structure, germanene can be easily functionalized and has been used in various applications. Madhushankar and his group [178] have synthesized and explored the electronic properties of germanene in field-effect transistor biosensing.

2.3.8 Phosphorene

A new member of the graphene family of group V elements other than silicene and germanene is phosphorene (Pn), also named as black phosphorus. It has a 2D single-layered or double-layered structure with weak interlayer van der Waals interaction among the stacked layers [179]. It is the most stable allotrope of phosphorus when compared with others such as violet phosphorus, red phosphorus, and white phosphorus. The different allotropes of phosphorus (Fig. 2.9) are white phosphorus made of P_4 monomeric units, red phosphorus having an amorphous texture, and violet phosphorus having a crystalline chain-like structure [180].

Phosphorene, unlike other 2D nanomaterials, has certain distinctive properties due to its unique puckered structure, arranged like an armchair, which can be converted from semiconductor to metal. Unique structural array, electronic structure, mechanical strain, superconductivity, thermoelectric properties, and the distinctive bandgap of phosphorene paved its way toward various applications. Thin-layered black phosphorus can be synthesized by CVD, liquid exfoliation, lithiation, plasma-assisted synthesis, and hydrothermal or solvothermal synthesis approach [181]. Bulk correspondent of phosphorene can be easily converted to single-layered phosphorene by the mechanical exfoliation method, which tears the stacked layers held by weak van der Waals forces. Mechanical cleavage with tape is another process of synthesis of high-purity thin atomic layers of phosphorene. Lu et al. [182] have synthesized atom-thick flakes of phosphorene by the plasma-assisted synthesis process with Ar^+ gun plasma. The liquid-phase intercalation

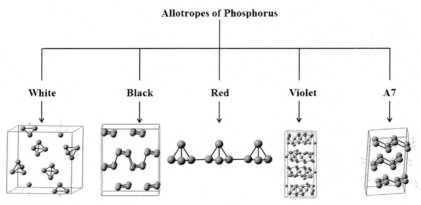

FIGURE 2.9

Different allotropic forms of phosphorene.

Reproduced with permission from C. Chowdhury, A. Datta, Exotic physics and chemistry of two-dimensional phosphorus: phosphorene, The journal of physical chemistry letters 8(13) (2017) 2909–2916.

technique is also used as the synthesis process, in which crystals of phosphorene are treated with polar or nonpolar solvents, which break the interlayer forces and result in the formation of nanoflakes. Lithiation is another top-down approach for the synthesis of phosphorene but it has certain limitations such as interaction of lithium with phosphorus [183].

2.3.9 Two-dimensional metal-organic frameworks

2D MOFs are very attractive crystalline porous materials consisting of metal bridging nodes and multipodal organic ligands bonded together through the basic coordination chemistry [184]. The metal node is connected strongly to the ligands through a coordination bond but the layers of MOF are connected by weak van der Waals forces. MOFs have attractive properties such as large surface area; great photonic, electronic, and chemical attributes; a large number of active sites; and good chemical stability. Owing to these properties, they have been used in electronic devices, gas storage, drug delivery, lithium-ion battery, supercapacitors, and different types of amperometric, capacitive, and luminescent sensors [185].

2D MOFs are mainly synthesized through the exfoliation method or the surfactant-assisted synthesis possesses. By the top-down approach, MOFs can be synthesized from their bulk material through chemical/physical exfoliation by breaking the interlayer weak forces. Herein, top-down approaches usually fail to synthesize MOFs of desired thickness and lead to aggregation of the material. Among the bottom-up approaches, surfactant-assisted synthesis and liquid-liquid or liquid-air interface technique are mainly used for the synthesis of MOFs. Amo-Ochoa and his group first synthesized ultrathin MOFs by chemical exfoliation of $[Cu_2Br(IN)_2]_n$ (IN = isonicotinato) in acetone through ultrasonication. Use of a bulky organic ligand aligns the metal planes and the ligand plane in the same direction because of the spatial effect [186]. In this strategy, using polyvinylpyrrolidone as one of the surfactants, Cao and coworkers have successfully synthesized an MOF named 2D PPF-3 (porphyrin paddlewheel framework 3). Here, $Co_2(COO)_4$ is used as the metal node, while CPP (5,10,15,20-tetrakis(4-carboxyphenyl)porphyrin) (four number) and bipyridine (one number) were used as ligands [187]. Dai et al. have synthesized MOF 2D nanocrystals modified with core-shell microsphere of SiO_2 with the help of cationic ligand-assisted synthesis. The ligand used here is 1,3-bis(4-carboxybutyl)imidazolium bromide and the metal ion used is zinc nitrate [188]. The methods used for the synthesis of MOFs have certain limitations, so more effort should be made to develop new, innovative, and cost-effective techniques.

2.4 Future aspects and conclusion

In this chapter, we have discussed about the wide field of 2D nanomaterials and their synthesis procedures including the bottom-up and top-down approaches. 2D

[53] A. Olad, H.B.K. Hagh, Graphene oxide and amine-modified graphene oxide incorporated chitosan-gelatin scaffolds as promising materials for tissue engineering, Composites Part B: Engineering 162 (2019) 692−702.

[54] M.J. Deka, D. Chowdhury, Surface charge induced tuning of electrical properties of CVD assisted graphene and functionalized graphene sheets, Journal of Materials Science and Technology 35 (1) (2019) 151−158.

[55] H.A. Guo, S. Jou, T.Z. Mao, B.R. Huang, Y.T. Huang, H.C. Yu, C.C. Chen, Silicon-and oxygen-codoped graphene from polycarbosilane and its application in graphene/n-type silicon photodetectors, Applied Surface Science 464 (2019) 125−130.

[56] R. Vinoth, I.M. Patil, A. Pandikumar, B.A. Kakade, N.M. Huang, D.D. Dionysios, B. Neppolian, Synergistically enhanced electrocatalytic performance of an N-doped graphene quantum dot-decorated 3D MoS$_2$-graphene nanohybrid for oxygen reduction reaction, ACS Omega 1 (5) (2016) 971−980.

[57] S.H. Choi, K.Y. Jung, Y.C. Kang, Amorphous GeOx-coated reduced graphene oxide balls with sandwich structure for long-life lithium-ion batteries, ACS Applied Materials and Interfaces 7 (25) (2015) 13952−13959.

[58] M.R. Kaiser, Z. Ma, X. Wang, F. Han, T. Gao, X. Fan, C. Wang, Reverse microemulsion synthesis of sulfur/graphene composite for lithium/sulfur batteries, ACS Nano 11 (9) (2017) 9048−9056.

[59] L. Kong, C. Zhang, J. Wang, W. Qiao, L. Ling, D. Long, Free-standing T-Nb$_2$O$_5$/graphene composite papers with ultrahigh gravimetric/volumetric capacitance for Li-ion intercalation pseudocapacitor, ACS Nano 9 (11) (2015) 11200−11208.

[60] X. Zhang, P. Suresh Kumar, V. Aravindan, H.H. Liu, J. Sundaramurthy, S.G. Mhaisalkar, S. Madhavi, Electrospun TiO$_2$−graphene composite nanofibers as a highly durable insertion anode for lithium ion batteries, The Journal of Physical Chemistry C 116 (28) (2012) 14780−14788.

[61] H. Yang, G.H. Guai, C. Guo, Q. Song, S.P. Jiang, Y. Wang, C.M. Li, NiO/graphene composite for enhanced charge separation and collection in p-type dye sensitized solar cell, The Journal of Physical Chemistry C 115 (24) (2011) 12209−12215.

[62] M. Hou, M. Xu, B. Li, Enhanced electrical conductivity of cellulose nanofiber/graphene composite paper with a sandwich structure, ACS Sustainable Chemistry and Engineering 6 (3) (2018) 2983−2990.

[63] L. David, R. Bhandavat, G. Singh, MoS$_2$/graphene composite paper for sodium-ion battery electrodes, ACS Nano 8 (2) (2014) 1759−1770.

[64] C. Zhang, R.Z. Zhang, Y.Q. Ma, W.B. Guan, X.L. Wu, X. Liu, C.P. Pan, Preparation of cellulose/graphene composite and its applications for triazine pesticides adsorption from water, ACS Sustainable Chemistry and Engineering 3 (3) (2015) 396−405.

[65] P. Zhu, A.S. Nair, P. Shengjie, Y. Shengyuan, S. Ramakrishna, Facile fabrication of TiO$_2$−graphene composite with enhanced photovoltaic and photocatalytic properties by electrospinning, ACS Applied Materials and Interfaces 4 (2) (2012) 581−585.

[66] L. Jin, Z. Zeng, S. Kuddannaya, D. Wu, Y. Zhang, Z. Wang, Biocompatible, freestanding film composed of bacterial cellulose nanofibers−graphene composite, ACS Applied Materials and Interfaces 8 (1) (2015) 1011−1018.

[67] T. Varga, L. Vásárhelyi, G. Ballai, H. Haspel, A. Oszko, A. Kukovecz, Z. Konya, Noble-metal-free iron nitride/nitrogen-doped graphene composite for the oxygen reduction reaction, ACS Omega 4 (1) (2019) 130−139.

[68] D. Jariwala, V.K. Sangwan, L.J. Lauhon, T.J. Marks, M.C. Hersam, Emerging device applications for semiconducting two-dimensional transition metal dichalcogenides, ACS Nano 8 (2) (2014) 1102−1120.

[69] R. Lv, J.A. Robinson, R.E. Schaak, D. Sun, Y. Sun, T.E. Mallouk, M. Terrones, Transition metal dichalcogenides and beyond: synthesis, properties, and applications of single-and few-layer nanosheets, Accounts of Chemical Research 48 (1) (2014) 56−64.

[70] W. Choi, N. Choudhary, G.H. Han, J. Park, D. Akinwande, Y.H. Lee, Recent development of two-dimensional transition metal dichalcogenides and their applications, Materials Today 20 (3) (2017) 116−130.

[71] U. Gupta, C.N.R. Rao, Hydrogen generation by water splitting using MoS_2 and other transition metal dichalcogenides, Nano Energy 41 (2017) 49−65.

[72] H. Wang, H. Yuan, S.S. Hong, Y. Li, Y. Cui, Physical and chemical tuning of two-dimensional transition metal dichalcogenides, Chemical Society Reviews 44 (9) (2015) 2664−2680.

[73] D. Voiry, A. Mohite, M. Chhowalla, Phase engineering of transition metal dichalcogenides, Chemical Society Reviews 44 (9) (2015) 2702−2712.

[74] M. Pumera, Z. Sofer, A. Ambrosi, Layered transition metal dichalcogenides for electrochemical energy generation and storage, Journal of Materials Chemistry A 2 (24) (2014) 8981−8987.

[75] W. Jie, J. Hao, Two-dimensional layered gallium selenide: preparation, properties, and applications, Advanced 2D Materials (2016) 1−36.

[76] X. Li, H. Zhu, Two-dimensional MoS_2: properties, preparation, and applications, Journal of Materiomics 1 (1) (2015) 33−44.

[77] J.A. Wilson, A.D. Yoffe, The transition metal dichalcogenides discussion and interpretation of the observed optical, electrical and structural properties, Advances in Physics 18 (73) (1969) 193−335.

[78] H. Li, Y. Shi, M.H. Chiu, L.J. Li, Emerging energy applications of two-dimensional layered transition metal dichalcogenides, Nano Energy 18 (2015) 293−305.

[79] J. Suh, T.E. Park, D.Y. Lin, D. Fu, J. Park, H.J. Jung, R. Sinclair, Doping against the native propensity of MoS_2: degenerate hole doping by cation substitution, Nano Letters 14 (12) (2014) 6976−6982.

[80] K.K. Tiong, T.S. Shou, Anisotropic electrolyte electroreflectance study of rhenium-doped MoS_2, Journal of Physics: Condensed Matter 12 (23) (2000) 5043.

[81] M. Chhowalla, H.S. Shin, G. Eda, L.J. Li, K.P. Loh, H. Zhang, The chemistry of two-dimensional layered transition metal dichalcogenide nanosheets, Nature Chemistry 5 (4) (2013) 263.

[82] H.S. Song, A.P. Tang, G.R. Xu, L.H. Liu, M.J. Yin, Y.J. Pan, One-step convenient hydrothermal synthesis of MoS_2/RGO as a high-performance anode for sodium-ion batteries, International Journal of Electrochemical Science 13 (2018) 4720−4730.

[83] D. Chakravarty, D.J. Late, Microwave and hydrothermal syntheses of WSe 2 micro/nanorods and their application in supercapacitors, RSC Advances 5 (28) (2015) 21700−21709.

[84] C.V.M. Gopi, A.E. Reddy, J.S. Bak, I.H. Cho, H.J. Kim, One-pot hydrothermal synthesis of tungsten diselenide/reduced graphene oxide composite as advanced electrode materials for supercapacitors, Materials Letters 223 (2018) 57−60.

[85] K.J. Huang, J.Z. Zhang, Y. Fan, Preparation of layered MoSe$_2$ nanosheets on Ni-foam substrate with enhanced supercapacitor performance, Materials Letters 152 (2015) 244–247.

[86] X. Zheng, Y. Zhu, Y. Sun, Q. Jiao, Hydrothermal synthesis of MoS$_2$ with different morphology and its performance in thermal battery, Journal of Power Sources 395 (2018) 318–327.

[87] M.B. Lejbini, P. Sangpour, Hydrothermal synthesis of α-Fe$_2$O$_3$-decorated MoS$_2$ nanosheets with enhanced photocatalytic activity, Optik 177 (2019) 112–117.

[88] K.K. Upadhyay, T. Nguyen, T.M. Silva, M.J. Carmezim, M.F. Montemor, Pseudocapacitive response of hydrothermally grown MoS2 crumpled nanosheet on carbon fiber, Materials Chemistry and Physics 216 (2018) 413–420.

[89] Y.H. Chang, W. Zhang, Y. Zhu, Y. Han, J. Pu, J.K. Chang, T. Takenobu, Monolayer MoSe$_2$ grown by chemical vapor deposition for fast photodetection, ACS Nano 8 (8) (2014) 8582–8590.

[90] Y. Zhao, H. Lee, W. Choi, W. Fei, C.J. Lee, Large-area synthesis of monolayer MoSe$_2$ films on SiO$_2$/Si substrates by atmospheric pressure chemical vapor deposition, RSC Advances 7 (45) (2017) 27969–27973.

[91] C.M. Hyun, J.H. Choi, S.W. Lee, J.H. Park, K.T. Lee, J.H. Ahn, Synthesis mechanism of MoS$_2$ layered crystals by chemical vapor deposition using MoO$_3$ and sulfur powders, Journal of Alloys and Compounds 765 (2018) 380–384.

[92] R.D. Mahyavanshi, G. Kalita, K.P. Sharma, M. Kondo, T. Dewa, T. Kawahara, M. Tanemura, Synthesis of MoS$_2$ ribbons and their branched structures by chemical vapor deposition in sulfur-enriched environment, Applied Surface Science 409 (2017) 396–402.

[93] S. Han, X. Luo, Y. Cao, C. Yuan, Y. Yang, Q. Li, S. Ye, Morphology evolution of MoS$_2$ nanorods grown by chemical vapor deposition, Journal of Crystal Growth 430 (2015) 1–6.

[94] X. Lu, M.I.B. Utama, J. Lin, X. Gong, J. Zhang, Y. Zhao, W. Zhou, Large-area synthesis of monolayer and few-layer MoSe$_2$ films on SiO$_2$ substrates, Nano Letters 14 (5) (2014) 2419–2425.

[95] B.B. Wang, M.K. Zhu, I. Levchenko, K. Zheng, B. Gao, S. Xu, K. Ostrikov, Effects of hydrogen on the structural and optical properties of MoSe$_2$ grown by hot filament chemical vapor deposition, Journal of Crystal Growth 475 (2017) 1–9.

[96] A. Gupta, V. Arunachalam, S. Vasudevan, Liquid-phase exfoliation of MoS$_2$ nanosheets: the critical role of trace water, The Journal of Physical Chemistry Letters 7 (23) (2016) 4884–4890.

[97] Y. Li, X. Yin, W. Wu, Preparation of few-layer MoS$_2$Nanosheets via an efficient shearing exfoliation method, Industrial and Engineering Chemistry Research 57 (8) (2018) 2838–2846.

[98] S. Marqus, H. Ahmed, M. Ahmed, C. Xu, A.R. Rezk, L.Y. Yeo, Increasing exfoliation yield in the synthesis of MoS$_2$ quantum dots for optoelectronic and other applications through a continuous multicycle acoustomicrofluidic approach, ACS Applied Nano Materials 1 (6) (2018) 2503–2508.

[99] S.A. Ansari, H. Fouad, S.G. Ansari, M.P. Sk, M.H. Cho, Mechanically exfoliated MoS$_2$ sheet coupled with conductive polyaniline as a superior supercapacitor electrode material, Journal of Colloid and Interface Science 504 (2017) 276–282.

[100] H. Yang, A. Giri, S. Moon, S. Shin, J.M. Myoung, U. Jeong, Highly scalable synthesis of MoS_2 thin films with precise thickness control via polymer-assisted deposition, Chemistry of Materials 29 (14) (2017) 5772−5776.

[101] Q.D. Truong, M. KempaiahDevaraju, Y. Nakayasu, N. Tamura, Y. Sasaki, T. Tomai, I. Honma, Exfoliated MoS_2 and $MoSe_2$ nanosheets by a supercritical fluid process for a hybrid Mg−Li-ion battery, ACS Omega 2 (5) (2017) 2360−2367.

[102] H. Lin, J. Wang, Q. Luo, H. Peng, C. Luo, R. Qi, C.G. Duan, Rapid and highly efficient chemical exfoliation of layered MoS_2 and WS2, Journal of Alloys and Compounds 699 (2017) 222−229.

[103] A. Jawaid, D. Nepal, K. Park, M. Jespersen, A. Qualley, P. Mirau, R.A. Vaia, Mechanism for liquid phase exfoliation of MoS_2, Chemistry of Materials 28 (1) (2015) 337−348.

[104] M.A. Lukowski, A.S. Daniel, F. Meng, A. Forticaux, L. Li, S. Jin, Enhanced hydrogen evolution catalysis from chemically exfoliated metallic MoS_2 nanosheets, Journal of the American Chemical Society 135 (28) (2013) 10274−10277.

[105] X.W. Huang, J.J. Wei, T. Liu, X.L. Zhang, S.M. Bai, H.H. Yang, Silk fibroin-assisted exfoliation and functionalization of transition metal dichalcogenide nanosheets for antibacterial wound dressings, Nanoscale 9 (44) (2017) 17193−17198.

[106] T. Oztas, H.S. Sen, E. Durgun, B. Ortac, Synthesis of colloidal 2D/3D MoS_2 nanostructures by pulsed laser ablation in an organic liquid environment, The Journal of Physical Chemistry C 118 (51) (2014) 30120−30126.

[107] V.Y. Fominski, R.I. Romanov, D.V. Fominski, A.V. Shelyakov, Preparation of $MoSe_{x>3}$/Mo-NPs catalytic films for enhanced hydrogen evolution by pulsed laser ablation of $MoSe_2$ target, Nuclear Instruments and Methods in Physics Research Section B: Beam Interactions with Materials and Atoms 416 (2018) 30−40.

[108] V.Y. Fominski, R.I. Romanov, D.V. Fominski, A.V. Shelyakov, Regulated growth of quasi-amorphous MoS_x thin-film hydrogen evolution catalysts by pulsed laser deposition of Mo in reactive H_2S gas, Thin Solid Films 642 (2017) 58−68.

[109] W. Chen, W. Wu, Z. Pan, X. Wu, H. Zhang, PEG400-assisted synthesis of oxygen-incorporated MoS_2 ultrathin nanosheets supported on reduced graphene oxide for sodium ion batteries, Journal of Alloys and Compounds 763 (2018) 257−266.

[110] J. Li, D. Wang, H. Ma, Z. Pan, Y. Jiang, M. Li, Z. Tian, Ionic liquid assisted hydrothermal synthesis of hollow core/shell MoS_2 microspheres, Materials Letters 160 (2015) 550−554.

[111] Y. Zhang, W. Zeng, Y. Li, Hydrothermal synthesis and controlled growth of hierarchical 3D flower-like MoS_2 nanospheres assisted with CTAB and their NO_2 gas sensing properties, Applied Surface Science 455 (2018) 276−282.

[112] W. Guo, Y. Chen, L. Wang, J. Xu, D. Zeng, D.L. Peng, Colloidal synthesis of $MoSe_2$ nanonetworks and nanoflowers with efficient electrocatalytic hydrogen-evolution activity, Electrochimica Acta 231 (2017) 69−76.

[113] L. Xu, X. Zhou, X. Xu, L. Ma, J. Luo, L. Zhang, Triethylenetetramine-assisted hydrothermal synthesis of sulfur-doped few-layer $MoSe_2$/nitrogenated graphene hybrids and their catalytic activity for hydrogen evolution reaction, Advanced Powder Technology 27 (4) (2016) 1560−1567.

[114] J. Chen, Y. Xia, J. Yang, Graphene/surfactant-assisted synthesis of edge-terminated molybdenum disulfide with enlarged interlayer spacing, Materials Letters 210 (2018) 248−251.

[115] B. Kirubasankar, P. Palanisamy, S. Arunachalam, V. Murugadoss, S. Angaiah, 2D $MoSe_2$-$Ni(OH)_2$ nanohybrid as an efficient electrode material with high rate capability for asymmetric supercapacitor applications, Chemical Engineering Journal 355 (2019) 881–890.

[116] L. Song, L. Ci, H. Lu, P.B. Sorokin, C. Jin, J. Ni, P.M. Ajayan, Large scale growth and characterization of atomic hexagonal boron nitride layers, Nano Letters 10 (8) (2010) 3209–3215.

[117] A. Pakdel, Y. Bando, D. Golberg, Nano boron nitride flatland, Chemical Society Reviews 43 (3) (2014) 934–959.

[118] T. Soma, A. Sawaoka, S. Saito, Characterization of wurtzite type boron nitride synthesized by shock compression, Materials Research Bulletin 9 (6) (1974) 755–762.

[119] Q. Weng, X. Wang, X. Wang, Y. Bando, D. Golberg, Functionalized hexagonal boron nitride nanomaterials: emerging properties and applications, Chemical Society Reviews 45 (14) (2016) 3989–4012.

[120] C. Zhou, C. Lai, C. Zhang, G. Zeng, D. Huang, M. Cheng, Y. Yang, Semiconductor/boron nitride composites: synthesis, properties, and photocatalysis applications, Applied Catalysis B: Environmental 238 (2018) 6–18.

[121] D. Golberg, Y. Bando, Y. Huang, T. Terao, M. Mitome, C. Tang, C. Zhi, Boron nitride nanotubes and nanosheets, ACS Nano 4 (6) (2010) 2979–2993.

[122] H. Wang, W. Wang, H. Wang, F. Zhang, Y. Li, Z. Fu, Synthesis of boron nitride nanotubes by combining citrate-nitrate combustion reaction and catalytic chemical vapor deposition, Ceramics International 44 (12) (2018) 13959–13966.

[123] W.M. da Silva, H. Ribeiro, T.H. Ferreira, L.O. Ladeira, E.M. Sousa, Synthesis of boron nitride nanostructures from catalyst of iron compounds via thermal chemical vapor deposition technique, Physica E: Low-Dimensional Systems and Nanostructures 89 (2017) 177–182.

[124] X. Hao, S. Dong, W. Fang, J. Zhan, L. Li, X. Xu, M. Jiang, A novel hydrothermal route to synthesize boron nitride nanocrystals, Inorganic Chemistry Communications 7 (4) (2004) 592–594.

[125] T. Chen, Q. Zhang, Z. Xie, C. Tan, P. Chen, Y. Zeng, W. Lv, Carbon nitride modified hexagonal boron nitride interface as highly efficient blue LED light-driven photocatalyst, Applied Catalysis B: Environmental 238 (2018) 410–421.

[126] C. Peng, P. Wei, X. Chen, Y. Zhang, F. Zhu, Y. Cao, F. Peng, A hydrothermal etching route to synthesis of 2D MXene (Ti_3C_2, Nb_2C): enhanced exfoliation and improved adsorption performance, Ceramics International 44 (15) (2018) 18886–18893.

[127] L.C. Nistor, G. Epurescu, M. Dinescu, G. Dinescu, Boron nitride nano-structures produced by pulsed laser ablation in acetone, in: IOP Conference Series: Materials Science and Engineering, vol. 15 (1), IOP Publishing, 2010, p. 012067.

[128] M. Maleki, A. Beitollahi, J. Javadpour, N. Yahya, Dual template route for the synthesis of hierarchical porous boron nitride, Ceramics International 41 (3) (2015) 3806–3813.

[129] Q. Liu, C. Hu, X. Wang, One-pot solvothermal synthesis of water-soluble boron nitride nanosheets and fluorescent boron nitride quantum dots, Materials Letters 234 (2019) 306–310.

[130] C.G. Yin, Y. Ma, Z.J. Liu, J.C. Fan, P.H. Shi, Q.J. Xu, Y.L. Min, Multifunctional boron nitride nanosheet/polymer composite nanofiber membranes, Polymer 162 (2019) 100–107.

[131] J. Wu, H. Chen, L. Zhao, X. He, W. Fang, W. Li, X. Du, Synthesis of boron nitride nanotubes by catalytic pyrolysis of organic-inorganic hybrid precursor, Ceramics International 43 (6) (2017) 5145–5149.

[132] H. Wang, W. Wang, H. Wang, F. Zhang, Y. Li, Z. Fu, Urchin-like boron nitride hierarchical structure assembled by nanotubes-nanosheets for effective removal of heavy metal ions, Ceramics International 44 (11) (2018) 12216–12224.

[133] D. Seo, J. Kim, S.H. Park, Y.U. Jeong, Y.S. Seo, S.H. Lee, J. Kim, Synthesis of boron nitride nanotubes using thermal chemical vapor deposition of ball milled boron powder, Journal of Industrial and Engineering Chemistry 19 (4) (2013) 1117–1122.

[134] M.L. Yola, N. Atar, Simultaneous determination of β-agonists on hexagonal boron nitride nanosheets/multi-walled carbon nanotubes nanocomposite modified glassy carbon electrode, Materials Science and Engineering: C 96 (2019) 669–676.

[135] C. Zhi, Y. Bando, C. Tang, H. Kuwahara, D. Golberg, Large-scale fabrication of boron nitride nanosheets and their utilization in polymeric composites with improved thermal and mechanical properties, Advanced Materials 21 (28) (2009) 2889–2893.

[136] Y. Maruyama, T. Kurozumi, K. Omori, H. Otsubo, T. Sato, T. Watanabe, Ammonothermal synthesis of rhombohedral boron nitride, Materials Letters 232 (2018) 110–112.

[137] M.L. Yola, Electrochemical activity enhancement of monodisperse boron nitride quantum dots on graphene oxide: its application for simultaneous detection of organophosphate pesticides in real samples, Journal of Molecular Liquids 277 (2019) 50–57.

[138] J. Di, J. Xia, M. Ji, B. Wang, S. Yin, Q. Zhang, H. Li, Advanced photocatalytic performance of graphene-like BN modified BiOBr flower-like materials for the removal of pollutants and mechanism insight, Applied Catalysis B: Environmental 183 (2016) 254–262.

[139] J. Pang, Y. Chao, H. Chang, H. Li, J. Xiong, Q. Zhang, H. Li, Silver nanoparticle-decorated boron nitride with tunable electronic properties for enhancement of adsorption performance, ACS Sustainable Chemistry & Engineering 6 (4) (2018) 4948–4957.

[140] S. Marchesini, C.M. McGilvery, J. Bailey, C. Petit, Template-free synthesis of highly porous boron nitride: insights into pore network design and impact on gas sorption, ACS Nano 11 (10) (2017) 10003–10011.

[141] D. Saha, G. Orkoulas, S. Yohannan, H.C. Ho, E. Cakmak, J. Chen, S. Ozcan, Nanoporous boron nitride as exceptionally thermally stable adsorbent: role in efficient separation of light hydrocarbons, ACS Applied Materials and Interfaces 9 (16) (2017) 14506–14517.

[142] L. David, S. Bernard, C. Gervais, P. Miele, G. Singh, Facile synthesis and high rate capability of silicon carbonitride/boron nitride composite with a sheet-like morphology, The Journal of Physical Chemistry C 119 (5) (2015) 2783–2791.

[143] X. He, J. Lin, W. Zhai, Y. Huang, Q. Li, C. Yu, C. Tang, Efficient energy transfer in terbium complexes/porous boron nitride hybrid luminescent materials, The Journal of Physical Chemistry C 121 (36) (2017) 19915–19921.

[144] A.Y. Liu, M.L. Cohen, Prediction of new low compressibility solids, Science 245 (4920) (1989) 841–842.

[145] A. Thomas, A. Fischer, F. Goettmann, M. Antonietti, J.O. Müller, R. Schlögl, J.M. Carlsson, Graphitic carbon nitride materials: variation of structure and morphology and their use as metal-free catalysts, Journal of Materials Chemistry 18 (41) (2008) 4893–4908.

[146] Y. Zheng, J. Liu, J. Liang, M. Jaroniec, S.Z. Qiao, Graphitic carbon nitride materials: controllable synthesis and applications in fuel cells and photocatalysis, Energy & Environmental Science 5 (5) (2012) 6717−6731.

[147] X. Li, J. Zhang, L. Shen, Y. Ma, W. Lei, Q. Cui, G. Zou, Preparation and characterization of graphitic carbon nitride through pyrolysis of melamine, Applied Physics A 94 (2) (2009) 387−392.

[148] B. Jürgens, E. Irran, J. Senker, P. Kroll, H. Müller, W. Schnick, Melem (2, 5, 8-triamino-tri-s-triazine), an important intermediate during condensation of melamine rings to graphitic carbon nitride: synthesis, structure determination by X-ray powder diffractometry, solid-state NMR, and theoretical studies, Journal of the American Chemical Society 125 (34) (2003) 10288−10300.

[149] G. Goglio, D. Foy, G. Demazeau, State of Art and recent trends in bulk carbon nitrides synthesis, Materials Science and Engineering: R: Reports 58 (6) (2008) 195−227.

[150] I. Tanaka, Y. Sakamoto, Low-temperature synthesis of carbon nitride by microwave plasma CVD, Japanese Journal of Applied Physics 55 (1S) (2015) 01AA15.

[151] J. Feng, T. Chen, S. Liu, Q. Zhou, Y. Ren, Y. Lv, Z. Fan, Improvement of g-C_3N_4 photocatalytic properties using the Hummers method, Journal of Colloid and Interface Science 479 (2016) 1−6.

[152] J. Lin, Z. Pan, X. Wang, Photochemical reduction of CO_2 by graphitic carbon nitride polymers, ACS Sustainable Chemistry and Engineering 2 (3) (2013) 353−358.

[153] J. Sun, B.V. Schmidt, X. Wang, M. Shalom, Self-Standing carbon nitride-based hydrogels with high photocatalytic activity, ACS Applied Materials and Interfaces 9 (3) (2017) 2029−2034.

[154] X. An, C. Hu, H. Lan, H. Liu, J. Qu, Strongly coupled metal oxide/reassembled carbon nitride/Co−Pi heterostructures for efficient Photoelectrochemical water splitting, ACS Applied Materials and Interfaces 10 (7) (2018) 6424−6432.

[155] Y. Oh, V.D. Le, U.N. Maiti, J.O. Hwang, W.J. Park, J. Lim, S.O. Kim, Selective and regenerative carbon dioxide capture by highly polarizing porous carbon nitride, ACS Nano 9 (9) (2015) 9148−9157.

[156] J. Barrio, M. Shalom, Ultralong nanostructured carbon nitride wires and self-standing C-rich filters from supramolecular microspheres, ACS Applied Materials and Interfaces 10 (46) (2018) 39688−39694.

[157] J.E. Ellis, D.C. Sorescu, S.C. Burkert, D.L. White, A. Star, Uncondensed graphitic carbon nitride on reduced graphene oxide for oxygen sensing via a photoredox mechanism, ACS Applied Materials and Interfaces 9 (32) (2017) 27142−27151.

[158] L.F. Gao, T. Wen, J.Y. Xu, X.P. Zhai, M. Zhao, G.W. Hu, H.L. Zhang, Iron-doped carbon nitride-type polymers as homogeneous organocatalysts for visible light-driven hydrogen evolution, ACS Applied Materials and Interfaces 8 (1) (2015) 617−624.

[159] J.A. Singh, S.H. Overbury, N.J. Dudney, M. Li, G.M. Veith, Gold nanoparticles supported on carbon nitride: influence of surface hydroxyls on low temperature carbon monoxide oxidation, ACS Catalysis 2 (6) (2012) 1138−1146.

[160] A. Kumar, R.K. Yadav, N.J. Park, J.O. Baeg, Facile one-pot two-step synthesis of novel in situ selenium-doped carbon nitride nanosheet photocatalysts for highly enhanced solar fuel production from CO_2, ACS Applied Nano Materials 1 (1) (2017) 47−54.

[161] B. Liu, L. Ye, R. Wang, J. Yang, Y. Zhang, R. Guan, X. Chen, Phosphorus-doped graphitic carbon nitride nanotubes with amino-rich surface for efficient CO_2 capture, enhanced photocatalytic activity, and product selectivity, ACS Applied Materials and Interfaces 10 (4) (2018) 4001−4009.

[162] R. Ma, T. Sasaki, Two-dimensional oxide and hydroxide nanosheets: controllable high-quality exfoliation, molecular assembly, and exploration of functionality, Accounts of Chemical Research 48 (1) (2014) 136−143.

[163] R. Ma, T. Sasaki, Nanosheets of oxides and hydroxides: ultimate 2D charge-bearing functional crystallites, Advanced Materials 22 (45) (2010) 5082−5104.

[164] M.S. Burke, L.J. Enman, A.S. Batchellor, S. Zou, S.W. Boettcher, Oxygen evolution reaction electrocatalysis on transition metal oxides and (oxy) hydroxides: activity trends and design principles, Chemistry of Materials 27 (22) (2015) 7549−7558.

[165] X. Long, Z. Wang, S. Xiao, Y. An, S. Yang, Transition metal based layered double hydroxides tailored for energy conversion and storage, Materials Today 19 (4) (2016) 213−226.

[166] C. Yuan, H.B. Wu, Y. Xie, X.W. Lou, Mixed transition-metal oxides: design, synthesis, and energy-related applications, Angewandte Chemie International Edition 53 (6) (2014) 1488−1504.

[167] M. Adachi-Pagano, C. Forano, J.P. Besse, Delamination of layered double hydroxides by use of surfactants, Chemical Communications (1) (2000) 91−92.

[168] D. Ghosh, S. Giri, C.K. Das, Preparation of CTAB-assisted hexagonal platelet Co(OH)$_2$/graphene hybrid composite as efficient supercapacitor electrode material, ACS Sustainable Chemistry & Engineering 1 (9) (2013) 1135−1142.

[169] J. Zhao, H. Liu, Z. Yu, R. Quhe, S. Zhou, Y. Wang, Y. Yao, Rise of silicene: a competitive 2D material, Progress in Materials Science 83 (2016) 24−151.

[170] A. Dimoulas, Silicene and germanene: silicon and germanium in the "flatland", Microelectronic Engineering 131 (2015) 68−78.

[171] H. Nakano, T. Mitsuoka, M. Harada, K. Horibuchi, H. Nozaki, N. Takahashi, T. Nonaka, Y. Seno, H. Nakamura, Soft synthesis of single-crystal silicon monolayer sheets, Angewandte Chemie International Edition 45 (38) (2006) 6303−6306.

[172] P. Vogt, P. De Padova, C. Quaresima, J. Avila, E. Frantzeskakis, M.C. Asensio, G. Le Lay, Silicene: compelling experimental evidence for graphenelike two-dimensional silicon, Physical Review Letters 108 (15) (2012) 155501.

[173] A. Fleurence, R. Friedlein, T. Ozaki, H. Kawai, Y. Wang, Y. Yamada-Takamura, Experimental evidence for epitaxial silicene on diboride thin films, Physical Review Letters 108 (24) (2012) 245501.

[174] J. Gao, J. Zhao, Initial geometries, interaction mechanism and high stability of silicene on Ag (111) surface, Scientific Reports 2 (2012) 861.

[175] A. Molle, D. Tsoutsou, A. Dimoulas, Group IV semiconductor 2D materials, 2D Materials for Nanoelectronics 17 (2016) 349.

[176] S. Cahangirov, M. Topsakal, E. Aktürk, H. Şahin, S. Ciraci, Two-and one-dimensional honeycomb structures of silicon and germanium, Physical Review Letters 102 (23) (2009) 236804.

[177] E. Bianco, S. Butler, S. Jiang, O.D. Restrepo, W. Windl, J.E. Goldberger, Stability and exfoliation of germanane: a germanium graphane analogue, ACS Nano 7 (5) (2013) 4414−4421.

[178] B.N. Madhushankar, A. Kaverzin, T. Giousis, G. Potsi, D. Gournis, P. Rudolf, B.J. van Wees, Electronic properties of germanane field-effect transistors, 2D Materials 4 (2) (2017) 021009.

[179] L. Kou, C. Chen, S.C. Smith, Phosphorene: fabrication, properties, and applications, The Journal of Physical Chemistry Letters 6 (14) (2015) 2794−2805.

[180] C. Chowdhury, A. &Datta, Exotic physics and chemistry of two-dimensional phosphorus: phosphorene, The Journal of Physical Chemistry Letters 8 (13) (2017) 2909−2916.

[181] A. Khandelwal, K. Mani, M.H. Karigerasi, I. Lahiri, Phosphorene-the two-dimensional black phosphorous: properties, synthesis and applications, Materials Science and Engineering: B 221 (2017) 17−34.

[182] W. Lu, H. Nan, J. Hong, Y. Chen, C. Zhu, Z. Liang, Z. Zhang, Plasma-assisted fabrication of monolayer phosphorene and its Raman characterization, Nano Research 7 (6) (2014) 853−859.

[183] V. Sresht, A.A. Padua, D. &Blankschtein, Liquid-phase exfoliation of phosphorene: design rules from molecular dynamics simulations, ACS Nano 9 (8) (2015) 8255−8268.

[184] M. Zhao, Y. Wang, Q. Ma, Y. Huang, X. Zhang, J. Ping, Y. Zhao, Ultrathin 2D metal−organic framework nanosheets, Advanced Materials 27 (45) (2015) 7372−7378.

[185] J.L. Rowsell, O.M. Yaghi, Metal−organic frameworks: a new class of porous materials, Microporous and Mesoporous Materials 73 (1−2) (2004) 3−14.

[186] P. Amo-Ochoa, L. Welte, R. González-Prieto, P.J.S. Miguel, C.J. Gómez-García, E. Mateo-Martí, F. Zamora, Single layers of a multifunctional laminar Cu (I, II) coordination polymer, Chemical Communications 46 (19) (2010) 3262−3264.

[187] F. Cao, M. Zhao, Y. Yu, B. Chen, Y. Huang, J. Yang, C. Tan, Synthesis of two-dimensional CoS1. 097/nitrogen-doped carbon nanocomposites using metal−organic framework nanosheets as precursors for supercapacitor application, Journal of the American Chemical Society 138 (22) (2016) 6924−6927.

[188] Q. Dai, J. Ma, S. Ma, S. Wang, L. Li, X. Zhu, X. &Qiao, Cationic ionic liquids organic ligands based metal−organic frameworks for fabrication of core−shell microspheres for hydrophilic interaction liquid chromatography, ACS Applied Materials and Interfaces 8 (33) (2016) 21632−21639.

Properties of two-dimensional nanomaterials

3

N.B. Singh[1], Saroj K. Shukla[2]

[1]*Professor, Department of Chemistry and Biochemistry, Research and Technology Development Centre, SBSR and RTDC, Sharda University, Greater Noida, Uttar Pradesh, India;* [2]*Department of Polymer Science, Bhaskaryacharya College of Applied Sciences, University of Delhi, Delhi, India*

3.1 Introduction

Nanostructured materials have been classified into four categories: (1) zero-dimensional, (2) one-dimensional, (3) two-dimensional (2D), and (4) three-dimensional nanomaterials. Layered materials exhibit properties, entirely different from their bulk counterparts, when made ultrathin. Graphene, a most important 2D material, was isolated from graphite in 2004 and it opened a new field of 2D materials [1]. The field especially beyond graphene is growing rapidly. Atomically thin 2D materials are graphene, chlorographene, fluorographene, silicene, silicane, fluorosilicene, silicon carbide, germane, germanene, fluorogermanene, chlorogermanene, boron nitride, etc. [2,3]. 2D materials have properties from insulators to semiconductors (SCs) to metals and even to superconductors [3,4]. Restricting the size of materials in one or more dimensions changes the properties of 2D materials [5]. Because of their unique properties and versatile functionalities, these 2D nanomaterials have applications in a number of areas such as energy storage and conversion, electronics, electrochemical and biomedical applications, and optoelectronics. This chapter discusses various properties of 2D nanomaterials in detail.

3.2 Types of two-dimensional nanomaterials

There are different types of 2D nanomaterials such as carbon-based nanomaterials (graphene, graphene oxide (GO), reduced graphene oxide (rGO)), clay materials, transition metal oxides (TMOs), and transition metal dichalcogenides (TMDs). Other 2D materials are graphitic C_3N_4, hexagonal boron nitride (hBN), black phosphorus, and elemental monolayers such as arsenene, germanene, silicene, and antimonene. 2D nanomaterials are normally single layer or a few layers of atomic thickness (Fig. 3.1) [6].

Gupta et al. categorized 2D materials into three categories (Table 3.1) [7].

2D materials are divided in the following three categories [7].

Two-Dimensional Nanostructures for Biomedical Technology. https://doi.org/10.1016/B978-0-12-817650-4.00003-6

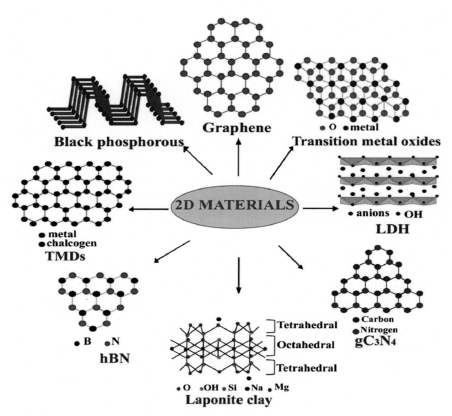

FIGURE 3.1

Structures of some two-dimensional (2D) materials. *LDH*, layered double hydroxide; *TMDs*, transition metal dichalcogenides.

Reproduced with permission from J. Ashtami, S. Anju, P.V. Mohanan, 2D materials for next generation healthcare applications, International Journal of Pharmaceutics, 551 (2018), 309–321.

(i) Layered van der Waals solids

These materials are common to be exfoliated into mono- and few-layered nanosheets. If M = Bi, Mo, V, Ta, Ti, W, Nb, Hf, Zr and X = Se, S, Te, then TMDs (MX_2 [X−M−X layer]) are well known for layering van der Waals solids, particularly MoS_2, $MoSe_2$, and WS_2.

(ii) Layered ionic solids

These are another category of layered 2D solid materials. The best example of layered ionic solids is exfoliated europium hydroxide. Examples include $K_{1.5}Eu_{0.5}Ta_3O_{10}$; $La_{0.90}Eu_{0.05}Nb_2O_7$; $KLnNb_2O_7$, $RbLnTa_2O_7$, and

Table 3.1 Types of 2D materials.

2D chalcogenides	MoS_2, WS_2, $MoSe_2$	Semiconducting dichalcogenides: $MoTe_2$, WTe_2, ZrS_2, $ZrSe_2$, and so on	Metallic dichalcogenides: $NbSe_2$, NbS_2, TaS_2, TiS_2, $NiSe_2$, and so on Layered semiconductors: $GaSe$, $GaTe$, $InSe$,Bi_2Se_3, and so on	
Graphene family	Graphene	hBN White graphene	BCN	Fluorographene | Graphene oxide
2D oxides	Micas, BSCCO Layered Cu oxides	MoO_3, WO_3 TiO_2, MnO_2,V_2O_5, TaO_3, RuO_2, and so on	Pervoskite type: $LaNb_2O_7$, (Ca,Sr) Nb_3O_{10}, $Bi_4Ti_3O_{12}$,$CaTa_2TiO_{10}$, and so on	Hydroxides: $Ni(OH)_2$, $Eu(OH)_2$, and so on Others

2D, *two-dimensional*; BCN, *boron carbon nitride*; BSCCO, *bismuth strontium calcium copper oxide.*
Reproduced with permission from A. Gupta, S. Tamilselvan, S. Seal, Recent development in 2D materials beyond graphene, Progress in Materials Science 73 (2015), 44–126.

$K_2Ln_2Ti_3O_{10}$ (where Ln is the lanthanide ion); $KCa_2Nb_3O_{10}$; and hydroxide of metals, e.g., $Co_{2/3}Fe_{1/3}(OH)_2^{1/3+}$ and $Eu(OH)_{2.5}(DS)_{0.5}$.

(iii) Surface-assisted nonlayered solids

Chemical vapor deposition (CVD) and epitaxial growth on substrate methods are used for the synthesis of thin layered materials of this type. Analogues of graphene and silicene are the most common examples of surface-assisted nonlayered solids.

3.3 Preparation of two-dimensional materials

Top-down and bottom-up approaches are used to synthesize 2D materials (Fig. 3.2) [3,6,8].

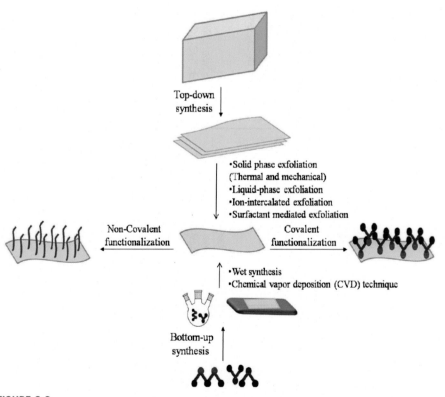

FIGURE 3.2

Preparative methods for two-dimensional materials and functionalization [6].

Reproduced with permission from J. Ashtami, S. Anju, PV Mohanan, 2D materials for next generation healthcare applications, International Journal of Pharmaceutics 551 (2018), 309–321.

3.3.1 Top-down method

Basically the top-down method involves the exfoliation method. Thin layer sheets are obtained by splitting layers of bulk materials by the direct exfoliation method. Liquid-phase, surfactant-assisted, mechanical treatments, and ion intercalation methods are used to synthesize 2D materials by direct exfoliation. The method is used in either solid or liquid state [9]. Mechanical exfoliation is considered to be exfoliation in solid phase. Making thin layers by peeling off from bulk structures is done in this technique. Synthesis of graphene is normally done by this method. In the thermal exfoliation technique, fast gas-generating decomposition and high internal pressure method is involved. However, the method has a number of disadvantages such as structural defects and low yield [6]. The micromechanical exfoliation technique has also been used to prepare few-atom-thick sheets of 2D layered inorganic materials [7]. The method does not use any chemical and depends on applied sheer force during peeling. Structural integrity and high crystallinity are preserved during the process. Liquid-phase exfoliation methods are considered to be better than solid-phase exfoliation. Layer to layer interactions normally dominate the layer—solvent interactions. Some liquid exfoliation processes are given in Fig. 3.3 [7].

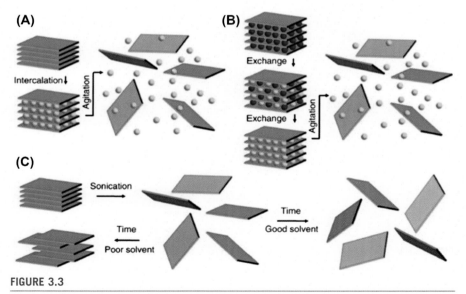

FIGURE 3.3

Various exfoliation processes for the synthesis of two-dimensional materials.

Reproduced with permission from A. Gupta, S. Tamilselvan, S. Seal, Recent development in 2D materials beyond graphene, Progress in Materials Science 73 (2015), 44–126.

3.3.2 Bottom-up approach

The bottom-up method includes solvothermal/hydrothermal/self-assembly/template synthesis of nanocrystals. Uses of various precursors are reported in this regard (Fig. 3.4) [10].

In the bottom-up approach, 2D nanomaterials are prepared by allowing the reactions between atoms or molecules. Two main bottom-up approaches are wet synthesis and CVD method. For the synthesis of good-quality sheets of graphene, the CVD technique is normally used. Atomic layer deposition and metal-organic CVD methods are also used for such synthesis.

In the synthesis of layered double hydroxides, coprecipitation, ion-exchange, and reconstruction methods are normally used. These are eco-friendly and inexpensive methods. In the coprecipitation method, the size and crystallinity of the particles

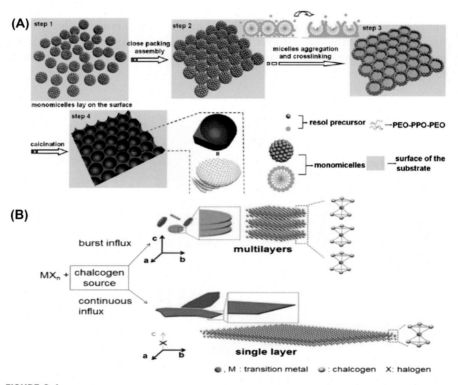

FIGURE 3.4

Preparation of 2D nanosheets by the wet chemical method: (A) ordered mesoporous graphene nanosheets and (B) single- and multilayer transition metal sulfide nanosheets. *PEO*, poly(ethylene oxide); *PPO*, poly(propylene oxide).

Reproduced with permission from Y. Han, Y. Ge, Y. Chao, C. Wang, G. Wallace Gordon,.Recent progress in 2D materials for flexible supercapacitors, Journal of Energy Chemistry 27 (2018), 57–72.

are influenced by pH, temperature, and time. The hydrothermal method is another wet technique for the preparation of 2D materials such as TMOs, hBN, and TMDs [6]. The hydrothermal method is normally used for the preparation of layered 2D materials. One such example is the chemical reaction between sulfur and/or selenium in hydrazine monohydrate solution and ammonium molybdate ($(NH_4)_6Mo_7O_{24} \cdot 4H_2O$) at 150–180°C for 48 h, forming a monolayer of MoS_2 and $MoSe_2$. No restacking tendencies were observed by using this method [7].

Using the wet chemical reaction, 2D materials are made. Boric acid (H_3BO_3) and urea ($CO(NH_2)_2$) react at 900°C under nitrogen atmosphere and form boron nitride as follows:[11]

$$2H_3BO_3 \rightarrow B_2O_3 + 3H_2O$$

$$CO(NH_2)_2 \rightarrow NH_3 + HNCO$$

$$B_2O_3 + 2NH_3 \rightarrow 2BN + 3H_2O$$

Concentration adjustments control the layer thickness.

3.4 Characterization methods

A number of characterization techniques to know about the composition, thickness, size, defects, crystallinity, oxidation states, properties, etc. of 2D materials have been developed. A large number of techniques such as microscopic techniques including optical microscopy, scanning electron microscopy, transmission electron microscopy (TEM), and atomic force microscopy; Raman spectroscopy; X- ray diffraction (XRD); and X-ray photoelectron spectroscopy have frequently been used for characterizing 2D materials [7]. However, each technique has its merits and demerits. Therefore, very often, a combination of techniques is used. Here some of the techniques have been discussed in brief.

Graphene has been studied extensively by using Raman spectroscopic technique (Fig. 3.5), particularly to know the quality and number of graphene layers [12].

XRD study is an effective method for characterizing crystalline materials and determining the crystal structure. Different XRD patterns of graphene, GO, and graphite are shown in Fig. 3.6 [13]. A sharp peak of pristine graphite at $2\theta = 26.6°$ is shifted to 13.9° in graphite oxide. Graphene nanosheets show no peak.

TEM is one of the most powerful methods used for the study of structural quality and the number of layers in graphene. TEM picture of graphene in dimethyl sulfoxide is shown in Fig. 3.7 [13]. The number of layers of graphene can be seen in Fig. 3.7.

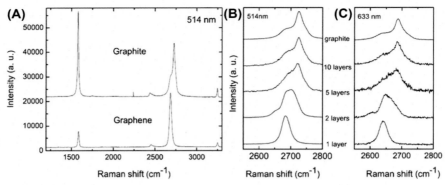

FIGURE 3.5

Raman spectra: (A) graphene and bulk graphite and (B,C) peaks for varying number of graphene layers at 514 nm [12].

Reproduced with permission from M.H. Kang, D. Lee, J. Sung, J. Kim, B.H. Kim, J. Park, Structure and chemistry of 2DMaterials, Comprehensive Nanoscience and Nanotechnology, second ed., vol. 2, 55—90.

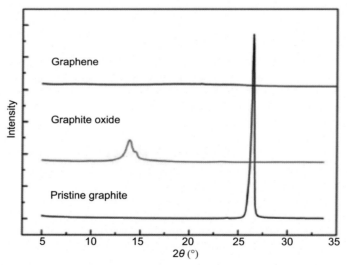

FIGURE 3.6

X-ray diffraction pattern.

Reproduced with permission from S. Yotsarayuth, A. Onsuda, T. Kriengkri, W. Chatchawal, Synthesis, characterization, and applications of graphene and derivatives, chapter nine — synthesis, characterization, and applications of graphene and derivatives, Carbon-Based Nanofillers and Their Rubber Nanocomposites (2019), 259—283.

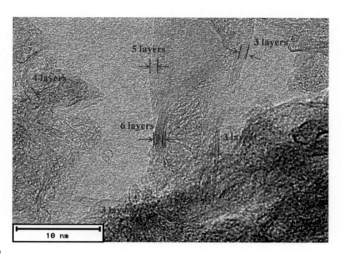

FIGURE 3.7

Transmission electron microscopic picture of graphene in dimethyl sulfoxide [13].

Reproduced with permission from S. Yotsarayuth, A. Onsuda, T. Kriengkri, W. Chatchawal, Synthesis, charac-
terization, and applications of graphene and derivatives, chapter nine — synthesis, characterization, and ap-
plications of graphene and derivatives, Carbon-Based Nanofillers and Their Rubber Nanocomposites (2019),
259–283.

3.5 Properties of two-dimensional nanomaterials

2D nanomaterials have attractive properties that are different from those of other
types of materials. Optical and electronic properties are different due to the confinement of electrons. This plays an important role in determining the band structure.
Some other property changes are due to geometry effects and the high surface-
bulk ratio [5]. Graphene, a zero-gap SC, has typical electronic properties. It also
has good optical, mechanical, thermal, and electrochemical properties. These properties are superior to the properties of other carbon allotropes such as diamond,
graphite, carbon nanotubes and fullerene (Table 3.2) [14].

General properties of 2D materials are discussed in the following.

3.5.1 Mechanical properties

Mechanical properties of 2D materials play important roles in various applications.
Atomic configurations control the mechanical properties of 2D nanomaterials and
these can be divided into three classes based on the number of atomic planes
(Fig. 3.8) [15]. Table 3.3 lists the mechanical properties of 2D graphene-
analogous nanomaterials.

Some properties, particularly mechanical properties, of certain 2D materials
have been studied in detail. Fracture strengths and Young's moduli of graphenelike

Table 3.2 The properties of graphene and other carbon allotropes.

Carbon allotropes	Diamond	Fullerene (C_{60})	Graphite	Graphene	Carbon nanotube
Hybridized form	sp^3	Mainly sp^2	sp^2	sp^2	Mainly sp^2
Crystal system	Octahedral	Tetragonal	Hexagonal	Hexagonal	Icosahedral
Dimension	Three	Zero	Three	Two	One
Experimental specific surface area ($m^2 g^{-1}$)	20–160	80–90	~10–20	~1500	~1300
Density ($g\ cm^{-3}$)	3.5–3.53	1.72	2.09–2.33	>1	>1
Optical properties	Isotropic	Nonlinear optical response	Uniaxial	97.7% Of optical transmittance	Structure-dependent properties
Thermal conductivity ($Wm^{-1}K^{-1}$)	900–2320	0.4	1500–2000, 5–10	4840–5300	3500
Hardness	Ultrahigh	High	High	Highest (single layer)	High
Tenacity	—	Elastic	Flexible nonelastic	Flexible elastic	Flexible elastic
Electronic properties	Insulator, semiconductor	Insulator	Electric conductor	Semimetal, zero-gap semiconductor	Metallic and semiconducting
Electric conductivity ($5\ cm^{-1}$)	—	10^{-10}	Anisotropic, $2-3*10^4,6$	2000	Structure dependent

Reproduced with permission from W. Zhong-Shuai, G. Zhou, L.C. Yin, W. Ren, F. Li, C. Hui-Ming, Graphene/metal oxide composite electrode materials for energy storage, Nano Energy 1 (2012) 107–131.

3.5.2 Thermal properties

Different thermal properties, such as conductivity, heat capacity, and inertia, of different 2D materials have been found to be of great importance for thermal management and thermoelectric energy generation. Some important factors that affect thermal properties are size, isotopic doping, surface functionalization, strain, and edge configurations [9]. The effect of size on thermal conductivity of MoS_2 has been described and the results indicated that initially, an increase in size increases conductivity, but after a certain period, it becomes constant [18]. Thermal conductivities of some 2D materials are given in Table 3.4 [15].

The rapid growth in integrated circuits (ICs) requires a faster switching speed, several transistors, and higher integration density. The power consumption in ICs is now rapidly increasing. Local temperature increase causes degradation of device performance. The thermal conductivity of graphene is because of the combination of various effects such as boundary, mass disorder, structural defect, and interface. Thermal conductivity of both graphene nanoribbons and carbon nanotubes can be reduced by isotopic doping [19]. Thermal management is an attempt to control the working temperature of devices such as ICs. Different thermal managements in 2D materials are shown in Fig. 3.9 [20].

Unusual thermal properties may be a basis to understand new phonon transport physics and may lead to novel applications in various emerging fields [21].

3.5.3 Electric properties

2D materials can exist in a number of crystalline forms and the electric conductivity changes. They may be metallic/semiconducting/semimetallic/insulator [15]. Because of this the bandgap is tunable. Some examples of different 2D materials with energy gap and electric characteristics are given in Table 3.5 [15].

The electric properties of 2D materials differ in the presence and absence of different gases. The bandgaps and the technical variables that modify the values are directly associated with electric properties. GO and rGO having zero bandgap are highly selective and sensitive sensors, without modifying their electric properties. These materials have high electric conductivity and low Johnson electric noise. However, due to the change in concentration of carriers induced by gases, electric conductivities change significantly. Dopants in 2D materials change the bandgap. The electric conductivity of 2D materials due to semiconducting properties/dopants/defects/functionalization is directly related to the bandgap space. If the bandgap is small, the electric conductivity of the sensing material will be high [22].

Electric properties of 2D materials can be changed by self-assembled monolayers (SAMs). Electric properties are changed when 2D materials are placed on top of SAMs [23]. During the preparation of 2D materials of atomic thickness, wrinkles or folds are formed. Wrinkled structures improve the properties of 2D materials. Single-layer graphene has hexagonal structures, and some counterparts possess AB

Table 3.4 Thermal conductivities of some two-dimensional materials [15].

Material	Thermal conductivity ($W \cdot m^{-1} K^{-1}$)
Graphene	4800–5300
	2500
	1800
	2200
	1600
	3300
Borophene	75(Z), 150(A)
Germanene	10.52
	2.4
Silicene	28
	9.4
	19.34
	28.3
	8.64
	40.1
	27.7(Z), 25.4(A)
g-C_3N_4 (s-triazine)	7.7
g-C_3N_4 (tri-triazine)	3.5
$MoTe_2$	19
Phosphorene	110(Z), 36(A)
	30.15(Z), 13.65(A)
	112.5(Z), 24.1(A)
	110.7(Z), 63.6(A)
	42.55(Z), 9.89(A)
Antimonene	7.9
	15.1
g-C_3N_4 (s-triazine)	7.7
g-C_3N_4 (tri-triazine)	3.5
hBN	484 (bilayer)
	600
	400
	300
C_3N	128
	820
C_2N	82.2
	40
	64.8
MoS_2	85–112
	82
	84
	83
	83
	90
	101–110

A, armchair; b, buckled phase; Z, zigzag.

Reproduced with permission from B. Liu, K. Zhou, Recent progress on graphene-analogous 2D nanomaterials: properties, modeling and applications, Progress in Materials Science 100 (2019) 99–169.

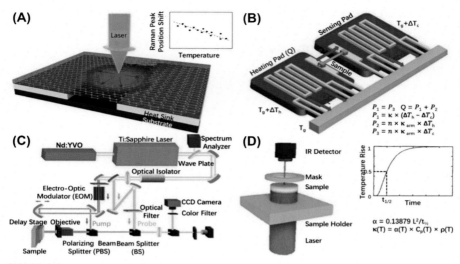

FIGURE 3.9

Different thermal measurements. *CCD*, charge-coupled device; *IR*, infrared; *Nd:YVO*, neodymium-doped yttrium orthovanadate.

Reproduced with permission from H. Song, J. Liu, B. Liu, J. Wu, C. Hui-MingF. King, Two-dimensional materials for thermal management applications, Joule 2 (2018) 442–463.

or ABC stacking lattices. It is found that different levels of roughness of wrinkles produce contrasting electric properties [24].

Low electron scattering in 2D materials is responsible for improved electric properties for several advanced applications. Electric properties of materials such as graphene and transition metal sulfide can also be changed by introducing heterogeneous atoms into the lattice. The electric properties of MoS_2 are changed due to doping [25]. Charge transfer occurs if 2D materials are present in contact with other materials with different fermi levels. This is also referred to as "contact" or "charge transfer doping." Generally, the material required for contact doping should be highly thin with high transparency. In this regard, monolayer is a requisite parameter for producing uniform doping of a 2D material [26]. The addition of SAM tunes the electric properties of TMDs. The other approaches for doping are covalent functionality, fixed charge layer deposition, and substitutional deposition [27]. 2D materials have very good electric properties but the theoretic and experimental values differ. These differences may be due to extrinsic scattering of carriers, traps at the interface, defects, and remote surface optical phonons [28].

3.5.4 Thermoelectric properties

2D materials directly convert heat into electricity due to bandgap engineering for harnessing waste heat. The conversion of heat into electricity is based on Seebeck

Table 3.5 Bandgap and conducting behavior of some two-dimensional nanomaterials.

Material	Energy gap (eV)	Characteristics
Graphene	0	Semimetal
Arsenene	1.66(w)	Direct semiconductor
Arsenene	2.49(b)	Indirect semiconductor
Borophene	0	Metal
Germanene	0.033	Semimetal
Silicene	0.002	Semimetal
Phosphorene	1.83	Direct semiconductor
Stanene	0.101	Semimetal
hBN	5.56	Insulator
$MoTe_2$	1.28	Direct semiconductor
Antimonene	1.18(w)	Direct semiconductor
	2.28(b)	Indirect semiconductor
Bismuthene	0.36(w), 0.99(b)	Direct semiconductor
WTe_2	1.03	Direct semiconductor
MoS_2	2.02	Direct semiconductor
$MoSe_2$	1.72	Direct semiconductor
WS_2	1.98	Direct semiconductor
WSe_2	1.63	Direct semiconductor
$g\text{-}C_3N_4$ (s-triazine)	3.3	Indirect semiconductor
$g\text{-}C_3N_4$ (tri-triazine)	2.71	Direct semiconductor
C_2N	2.47	Direct semiconductor
C_3N	1.04	Indirect semiconductor

b, buckled phase; w, symmetric washboard puckered phase.
Reproduced with permission from B. Liu B., K. Zhou, Recent progress on graphene-analogous 2D nanomaterials: properties, modeling and applications, Progress in Materials Science 100 (2019) 99–169.

effect (S), which is an indicator for conversion of change of heat into voltage [28,29]. Thus modulation of S is an interesting strategy to improve thermoelectric devices. If a material has high electric conductivity (σ) to minimize the internal energy loss and a high Seebeck coefficient (S) to produce high voltage and low thermal conductivity (κ) to maintain the temperature gradient, it is considered to possess good thermoelectric property. In this context the innovations of 2D materials have drastically improved the performance of thermoelectric devices. The Seebeck coefficients of some 2D materials are given in Table 3.6.

Furthermore, several modifications in 2D materials are reported to improve the thermoelectric performance and durability. Generally, thermoelectric properties of materials are estimated on the basis of both the electric and thermal properties. Effects of defects on the thermoelectric performance of graphene were reported by Anno.et al. [31]. Oxygen plasma treatment introduced defects in graphene, which

Table 3.6 Seebeck coefficient of some two-dimensional materials.

S.N.	Materials	Seebeck coefficient	Reference
1	MoS_2	$8.5\,mW\,m^{-1}\,K^{-2}$	Hippalgaonkar et al. [30]
2	Graphene	$\sim 100\,\mu V\,K^{-1}$	Anno et al. [31]
3	Mo-based MXenes	$3.09 \times 10^{-4}\,W\,m^{-1}\,K^{-2}$	Kim et al. [32]
4	Polyaniline/graphene	$0.05{-}1.47\,\mu W\,m^{-1}\,K^{-2}$	Du et al. [33]
5	PEDOT:PSS/graphene	$14.6\,\mu V\,K^{-1}$	Du et al. [34]
6	Graphene/polyaniline (tubular)	—	Wang et al. [21]
7	Sb_2Te_3	$371\,\mu W\,m^{-1}\,K^{-2}$	Hewitt et al. [35]

reduce the Seebeck coefficient as well as its electric conductivity. Increasing defect density modifies the thermoelectric power. However, higher defect densities reduce thermal conductivity more than electric conductivity of pristine graphene [31]. Similarly, several composites are prepared to improve the thermoelectric behavior of 2D materials, along with improvements in devices.

3.5.5 Electronic properties

Electronic property is influenced by localized/delocalized presence of electron in a solid state. It has intensively boosted the electronic industries in terms of economy and efficiency by improving different technologies. 2D materials indicate that electrons or holes have quantized energy levels for one spatial dimension. Some of the important electronic properties of 2D materials are energy level, transport, phonon scattering and excitation, etc. These properties are also modified by doping, making composite as well as ion irradiation. It is reported that the resistance of graphene layer is decreased up to 3×10^{11} ions cm^{-2} due to the fivefold increase in electron and hole mobilities of graphene layer. The irradiation of ion also induced an increase in electron and hole mobilities up to 1×10^{11} ions cm^2 [36]. A limitation of graphene is bandgapless electron conduction. It also restricts its application in the fabrication of graphene-based electronic devices, such as field-effect transistors. Thus different heterostructures have been designed for bandgap engineering. Some of important graphene-based heterostructures are G/hBN, G/As, G/Sb, G/ZnO, and G/MoS_2. These graphene-based heterostructures show many novel properties far beyond their single components [37]. Another important 2D material that has also been successfully designed by mechanical exfoliation is "phosphorene." It shows some extraordinary and superior electronic properties than graphene for the preparation of nanoelectronic devices. It is a direct bandgap material with bandgap around 1.45 eV, an electron carrier mobility of 1000 $cm^2\,Vs^{-1}$, and a large current ratio of

104 [38]. Furthermore, thickness is important to control the properties of 2D materials.

Cooperative effects of different properties yield better properties than individual ones. Optically induced electronics are referred to as optoelectronics, and mechanically induced electricity is called as piezoelectricity. These properties have a significant impact on the application of 2D materials. These properties are used for the development of 2D-material-based devices such as light-emitting diodes. Optoelectronic properties are light controlled and are used for the development of several devices such as photodiodes, phototransistors, photomultipliers, optoisolators, light-emitting diodes, optical switches, and integrated optical circuit [39]. The bandgap-designed 2D materials show several improvements in different optoelectronic devices.

In graphene heterojunctions, the essential component for nanoelectronics is the connectivity of metallic and semiconducting regions. The metallic regions are introduced either by making graphene nanoribbons with a specified width in such a way that there is no energy bandgap or alternatively by using graphene sheets. By introducing nanoholes, semiconducting regions can be obtained.

3.5.6 Magnetic properties

These are one of the most important properties and have applications in electric motors, telephones, computer peripherals, and medical diagnosis. Furthermore, these properties have significantly improved different technologies. Initially, it was proved that ferromagnetism in atomically thin layered 2D materials at room temperature has broadened their prospects for device applications. In single-layer 2D materials, intrinsic magnetic order has been studied for applications in low-power ultracompact spintronics. It also explored the formation devices made purely from 2D materials after exfoliation of graphene. In this regard, several unusual magnetism are reported for the development of different devices. It has been observed that pristine monolayer $Ni(OH)_2$ has no macromagnetism, with antiferromagnetic coupling between two nearest Ni atoms. On the other hand, in the $Ni(OH)_2$-BN 2D heterostructure, a larger magnetic moment with ferromagnetic coupling is found. The magnetic properties of $Ni(OH)_2$-XN heterostructures, which are tunable, open a new door for designing the spintronic devices in 2D stacked nanostructures [40,41]. Furthermore, ferromagnetic 2D materials with rich electronic and optical properties have been used for the fabrication of compact magnetic, magnetoelectronic, and magneto-optical devices [42]. Cr-based 2D materials (Cr_2Ge_2Te) called CGT have shown ferromagnetism. The observed magnetic behavior for CGT 2D multilayers can be controlled by using an external magnetic field and it opens new dimensions for their application in ultracompact spintronics. Similarly, postgraphene 2D transition materials show superior magnetic and optical properties for use in the field of electronics, catalysis, photonics, and sensing. Unusual magnetic properties, such as ferromagnetism in graphene, are reported due to edge defects [43]. The adsorption of molecules also has a significant impact on the virgin magnetism of graphite.

3.5.7 Piezoelectric properties

Materials producing electric current when placed under mechanical stress are called piezoelectric materials. Also if an electric current is applied, they slightly change shape (a maximum of 4%). Some of the materials showing piezoelectric properties are bone, proteins, crystals (e.g., quartz), and ceramics (e.g., lead zirconate titanate). There are many materials that are centrosymmetric, but not piezoelectric, in their bulk form. When thinned down to monolayer, they show piezoelectricity. Some of those materials are hBN, many TMDs, TMOs, II-VI SCs, III-V SCs, III-monochalcogenides (III-MCs), and IV-MCs [44]. Many 2D materials that are non-piezoelectric can be converted to piezoelectric by modifying their surface or structure.

The crystal structures of 2D materials showing piezoelectric properties are shown in Fig. 3.10 [44]. Materials having deformed hexagonal or hexagonal lattice show piezoelectricity. Many new and innovative devices have been made with 2D materials. For a wide range of applications, especially for strain sensors, gas sensors, pressure sensors, mass sensors, displacement sensors, light sensors, energy storage, energy harvesting, high-frequency switches, motors, and tweezers, 2D piezoelectric devices have been used [45].

3.5.8 Optical properties

Optical properties of materials are mostly determined by the electronic band structure and screening. The electronic and band engineering of 2D materials produce useful optical properties in 2D materials and has been used in the development of light-emitting diodes. Spin-orbital coupling plays an important role in modifying some of the attributes, such as optical properties, by changing a fundamental bandgap responsible for increasing the light absorption near the gap edges [15]. The optical properties of the IVA 2D nanomaterials are different from each other. The optical properties depending on the frequency of some 2D materials are shown in Fig. 3.11 [15]. Light-emitting properties of 2D materials are hindered by small bandgaps.

Because of large electronic bandgaps, VA 2D nanomaterials have light-absorbing and light-emitting properties. If the bandgap energy is lower than that of photonics, it can be quickly absorbed or emitted by materials having direct bandgap in their puckered phase. Additional photons are absorbed or emitted from indirect bandgaps. A bandgap ranging between ~ 1 and $\sim 2\,eV$ in 2D TMDs is responsible for emitting and absorbing light. Significant changes occur in optical properties of TMDs because of indirect-to-direct bandgap transition. Varying thicknesses of MoS_2 layers can show photodetection of light at different wavelengths. Green light was detected by MoS_2 having monolayer of bandgap energy of 1.8 eV and bilayer of bandgap energy of 1.65 eV. For red light, a triple-layer MoS_2 having a bandgap energy of 1.35 eV is well suited.

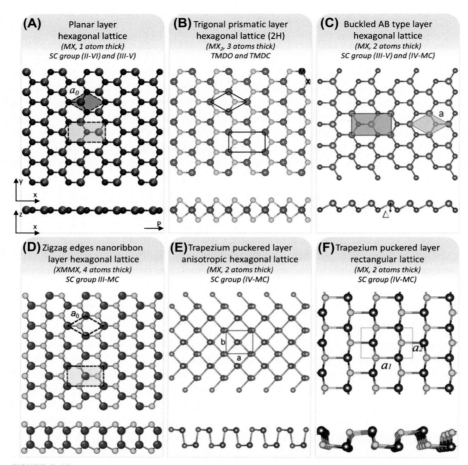

FIGURE 3.10

Monolayer structures. *MC*, monochalcogenide; *SC*, semiconductor.

Reproduced with permission from H. Ronan, Khan U., F. Christian, K. Sang-Woo, Piezoelectric properties in two-dimensional materials: simulations and experiments, Materials Today 21(6) (2018) 611–630.

3.5.9 Photonic properties

The term photonics was invented in the early 1960s and it is concerned with the generation and detection of photons through emission, modulation, signal processing, and transmission. In the recent years, graphene and graphenelike 2D materials have attracted a lot of interest in the photonics community. hBN, graphene, and TMDs with semiconducting properties and phosphorene/black phosphorus are being investigated extensively for their photonic and optoelectronic properties. 2D materials exhibit many unique properties important for device applications in nanophotonics.

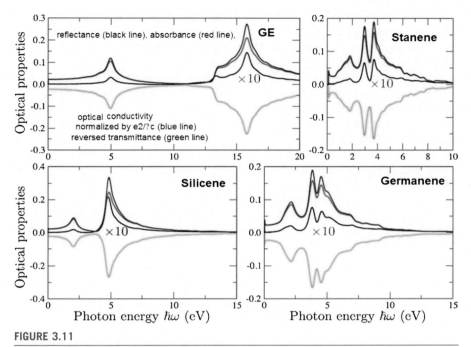

FIGURE 3.11

Frequency-dependent optical properties [15].

Reproduced with permission from B. Liu, K. Zhou, Recent progress on graphene-analogous 2D nanomaterials: properties, modeling and applications, Progress in Materials Science 100 (2019) 99–169.

3.5.10 Plasmonic properties

Plasma is the fourth state of matter and is a mixture of negative electrons and positive ions. Plasmons are a type of quasiparticle, describing collective oscillations of the electron gas in many SCs and metals. 2D nanomaterials exhibit more attractive characteristic plasmonic properties because of advanced electronic and optical properties. Graphene, a 2D nanomaterial, shows the powerful capability of tuning surface plasmons, which can be utilized in light detection applications and new light-involved electronics design [43]. Graphene can be excited to present metal-like reflection or the plasmonic effect, depending on the frequency of the incident light in this region. For plasmonic devices, encapsulation of graphene with hBN layers holds great potential. Novel optoelectronic devices can be developed by a combination of various 2D TMDs with graphene.

3.5.11 Lubricant additives

Because of the unique molecular structure of 2D nanomaterials, lubricating properties have been studied in detail [46]. Layered structure materials have widely been used as solid lubricants. van der Waals forces between the two adjacent layers allow adjacent layers to slide easily. 2D nanostructured materials have larger specific surface area than other nanomaterials. They cover a greater surface area. Recently, 2D nanomaterials have been extensively explored as novel lubricant additives (Fig. 3.12). 2D nanomaterials are found to be better than organic lubricant additives. Performances, particularly in friction reduction and antiwear behavior, were found to be better in the presence of 2D materials. These 2D materials working as lubricants are chemically more stable and emit less in the atmosphere, making the environment more sustainable.

The 2D nanomaterials used as lubricant additives have been divided into three categories: (1) graphene family, (2) metal dichalcogenides, and (3) others. Graphene has been studied extensively because of its superior properties such as chemical, mechanical, and electric properties. The lubricating performance of various types of base media can be improved by adding graphene-based nanomaterials. This is because of their ability to separate the contact and to form a protective film on the surface.

3.6 Properties of two-dimensional materials used in biomedical applications

2D nanomaterials have special properties that make them attractive as nanoagents for gene delivery, biosensors, and other biomedical applications [47]. Thus 2D

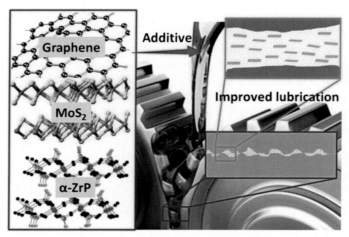

FIGURE 3.12

Lubricant additive.

Reproduced with permission from H. Xiao, S. Liu, 2D nanomaterials as lubricant additive: a review. Materials and Design 135 (2017) 319–332.

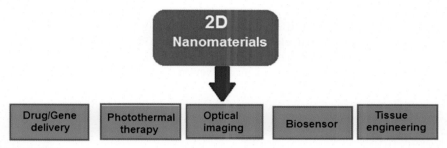

FIGURE 3.13

Biomedical applications of two-dimensional (2D) nanomaterials.

materials can be endorsed for a number of biomedical applications (Fig. 3.13). This is possible because of the unique properties of 2D nanomaterials, such as (1) very high specific surface areas, compared with other nanomaterials, enable adsorption of any amount of nucleic acid molecules; (2) ability to form biomedical nanocomposites with high mechanical and typical physicochemical properties; and (3) their thinnest structures, which are sensitive to lights, enable them to be used in optical imaging and photothermal therapies.

3.7 Future prospects

Nanotechnology, information technology, and biotechnology are considered to be the technology of the present century. 2D materials cover a number of areas in basic and applied sciences and have typical physical and chemical properties different from those of their bulk counterparts. 2D materials with typical electronic and excellent optical properties and high surface area will be very useful, in the near future, for electronic applications and use in lithium-ion batteries. They can also be used as biosensors, solar cells, supercapacitors, and catalysts. They will have a lot of potential in technology in the coming years. The price of 2D materials is quite high, but as technologies using 2D materials mature, higher demand for all types of 2D materials is likely to continue to drive the price lower. Graphene may soon be used as a technologic material. However, in the long term, other 2D materials may start to establish themselves in the marketplace.

3.8 Conclusions

2D materials are a new class of materials with special properties and open up new areas of applications. A variety of such materials are known and can be prepared by the top-down and bottom-up methods. A number of techniques have been used to characterize such materials. Carbon-based 2D materials, particularly graphene and graphene analogues, have been extensively investigated. There are a number of other 2D materials having different properties and applications. A number of properties including mechanical, electronic, optical, magnetic, and piezoelectric properties make them attractive and important. However, the study of properties and applications of 2D materials is in its infancy and requires detailed investigation before they are commercialized.

References

[1] K.S. Novoselov, A.K. Geim, S.V. Morozov, D.Y. Jiang Zhang, S.V. Dubonos, I.V. Grigorieva, A.A. Firsov, Electric field effect in atomically thin carbon films, Science 306 (2004) 666−669.

[2] P. Miró, M. Audiffred, T. Heine, An atlas of two-dimensional materials, Chemical Society Reviews 43 (2014) 6537.

[3] S.Z. Butler, et al., Progress, challenges, and opportunities in two-dimensional materials beyond graphene, ACS Nano 7 (2013) 2898−2926.

[4] G.R. Bhimanapati, et al., Recent advances in two-dimensional materials beyond graphene, ACS Nano 9 (2015) 11509−11539.

[5] R. Mas-Balleste, C. Gomez-Navarro, J. Gomez-Herrero, F. Zamora, 2D materials: to graphene and beyond, Nanoscale 3 (2011) 20−30.

[6] J. Ashtami, S. Anju, P.V. Mohanan, 2D materials for next generation healthcare applications, International Journal of Pharmaceutics 551 (2018) 309−321.

[7] A. Gupta, S. Tamilselvan, S. Seal, Recent development in 2D materials beyond graphene, Progress in Materials Science 73 (2015) 44−126.

[8] L. Wang, Q. Xiong, F. Xiao, Hongwei & Duan, 2D nanomaterials based electrochemical biosensors for cancer diagnosis, Biosensors and Bioelectronics 89 (2017) 136−151.

[9] T. Zhang, J. Liu, C. Wang, X. Leng, Y. Xiao, L. Fu, Synthesis of graphene and related two-dimensional materials for bioelectronics devices, Biosensors and Bioelectronics 89 (2017) 28−42.

[10] Y. Han, Y. Ge, Y. Chao, C. Wang, G. Wallace Gordon, Recent progress in 2D materials for flexible supercapacitors, Journal of Energy Chemistry 27 (2018) 57−72.

[11] V. Guerra, C. Wan, T. McNally, Thermal conductivity of 2D nano-structured boron nitride (BN) and its composites with polymers, Progress in Materials Science 100 (2019) 170−186.

[12] M.H. Kang, D. Lee, J. Sung, J. Kim, B.H. Kim, J. Park, Structure and chemistry of 2DMaterials, in: second ed.Comprehensive Nanoscience and Nanotechnology, vol. 2, 2019, pp. 55−90.

[13] S. Yotsarayuth, A. Onsuda, T. Kriengkri, W. Chatchawal, Synthesis, characterization, and applications of graphene and derivatives, chapter nine - synthesis, characterization, and applications of graphene and derivatives, Carbon-Based Nanofillers and Their Rubber Nanocomposites (2019) 259−283.

[14] W. Zhong-Shuai, G. Zhou, L.-C. Yin, W. Ren, F. Li, C. Hui-Ming, Graphene/metal oxide composite electrode materials for energy storage, Nano Energy 1 (2012) 107−131.

[15] B. Liu, K. Zhou, Recent progress on graphene-analogous 2D nanomaterials: properties, modeling and applications, Progress in Materials Science 100 (2019) 99−169.

[16] P. Hess, Prediction of mechanical properties of 2D solids, with related bonding configuration, RSC Advances 7 (2017) 29786–29793.

[17] A. Deji, J. Brennan Christopher, B.J. Scott, E. Philip, R. Felts Jonathan, H. Gao, R. Huang, J.-S. Kim, L. Teng, L. Yao, M. Liechti Kenneth, N. Lu, S. Park Harold, J. Reed Evan, P. Wang, I. Yakobson Boris, T. Zhang, Y.-W. Zhang, Y. Zhou, Y. Zhu, A review on mechanics and mechanical properties of 2D materials—graphene and beyond, Extreme Mechanics Letters 13 (2017) 42–77.

[18] X. Liu, G. Zhang, Q.X. Per, Y.W. Zhang, Phonon thermal conductivity of monolayer MoS_2 sheet and nanoribbons, Applied Physics Letters 102 (2013) 133113.

[19] G. Zhang, Y.-W. Zhang, Thermal properties of two-dimensional materials, Chinese Physics B 26 (3) (2017) 034401.

[20] H. Song, J. Liu, B. Liu, J. Wu, C. Hui-Ming, F. King, Two-dimensional materials for thermal management applications, Joule 2 (2018) 442–463.

[21] Y. Wang, X. Ning, D. Li, J. Zhu, Thermal properties of two dimensional layered materials, Advanced Functional Materials (2017) 1604134.

[22] R. Vargas-Bernal, Electrical properties of two-dimensional materials used in gas sensors, Sensors 19 (2019) 1295, https://doi.org/10.3390/s19061295.

[23] W.H. Lee, Y.D. Park, Tuning electrical properties of 2D materials by self-assembled monolayers, Advanced Materials Interfaces (2017) 1700316.

[24] W. Chen, X. Gui, L. Yang, H. Zhu, Z. Tang, Wrinkling of Two-Dimensional Materials: Methods, Properties and Applications, The Royal Society of Chemistry, 2018, https://doi.org/10.1039/c8nh00112j.

[25] Y. Zhao, K. Xu, F. Pan, C. Zhou, F. Zhou, Y. Chai, Doping, contact and interface engineering of two dimensional layered transition metal dichalcogenides transistors, Advanced Functional Materials 27 (19) (2017) 1603484.

[26] W.H. Lee, Y.D. Park, Tuning electrical properties of 2D materials by self assembled monolayers, Advanced Materials Interfaces 5 (1) (2018) 1700316.

[27] P. Zhao, S. Desai, M. Tosun, T. Roy, H. Fang, A. Sachid, M. Amani, C. Hu, A. Javey, 2D layered materials: from materials properties to device applications, in: Electron Devices Meeting (IEDM), December,2015, IEEE International, 2015, pp. 27–33.

[28] J. Bei, Y. Zhenyu, X. Liu, L. Yuan, L. Liao, Interface engineering for two-dimensional semiconductor transistors, Nano Today (2019).

[29] L.E. Bell, Cooling, heating, generating power, and recovering waste heat with thermoelectric systems, Science 321 (2008) 1457–1461.

[30] K. Hippalgaonkar, Y. Wang, Y. Ye, D.Y. Qiu, H. Zhu, Y. Wang, J. Moore, S.G. Louie, X. Zhang, High thermoelectric power factor in two-dimensional crystals of MoS_2, Physical Review B 95 (11) (2017) 115407.

[31] Y. Anno, Y. Imakita, K. Takei, S. Akita, T. Arie, Enhancement of graphene thermoelectric performance through defect engineering, 2D Materials 4 (2) (2017) 025019.

[32] H. Kim, B. Anasori, Y. Gogotsi, H.N. Alshareef, Thermoelectric properties of two-dimensional molybdenum-based MXenes, Chemistry of Materials 29 (15) (2017) 6472–6479.

[33] Y. Du, S.Z. Shen, W. Yang, R. Donelson, K.C. Philip, S. Casey, Simultaneous increase in conductivity and Seebeck coefficient in a polyaniline/graphene nanosheets thermoelectric nanocomposite, Synthetic Metals 161 (2012) 2688–2692.

[34] F.-P. Du, N.-N. Cao, Y.-F. Zhang, P. Fu, Y.-G. Wu, Z.-D. Lin, R. Shi, A. Amini, C. Cheng, PSS/graphene quantum dots films with enhanced thermoelectric properties via strong interfacial interaction and phase separation, Scientific Reports 8 (2018) 6441.

[35] C.A. Hewitt, Q. Li, J. Xu, D.C. Schall, H. Lee, Q. Jiang, D.L. Carroll, Ultrafast digital printing toward 4D shape changing materials, Advanced Materials 29 (7) (2017) 1605390.

[36] S. Kumar, A. Kumar, A. Tripathi, C. Tyagi, D.K. Avasthi, Engineering of electronic properties of single layer graphene by swift heavy ion irradiation, Journal of Applied Physics 123 (2018) 161533.

[37] H.V. Phuc, V.V. Ilyasov, N.N. Hieu, C.V. Nguyen, Electric-field tunable electronic properties and Schottky contact of graphene/phosphorene heterostructure, Vacuum 149 (2018) 231−237.

[38] H. Liu, A.T. Neal, Z. Zhu, Z. Luo, X. Xu, D. Tománek, P.D. Ye, Phosphorene: an unexplored 2D semiconductor with a high hole mobility, ACS Nano 8 (4) (2014) 4033−4041.

[39] B.W. Baugher, H.O. Churchill, Y. Yang, P. Jarillo-Herrero, Optoelectronic devices based on electrically tunable p−n diodes in a monolayer dichalcogenide, Nature Nanotechnology 9 (2014) 262−267.

[40] X.-L. Wei, Z.-K. Tang, G.-C. Guo, S. Ma, L.-M. Liu, Electronic and magnetism properties of two-dimensional stacked nickel hydroxides and nitrides, Scientific Reports 5 (2015) 11656.

[41] Z.K. Tang, C.J. Tong, W. Geng, D.Y. Zhang, L.M. Liu, Two-dimensional Ni (OH) 2-XS2 (X= Mo and W) heterostructures, 2D Materials 2 (3) (2015) 034014.

[42] W. Han, R.K. Kawakami, M. Gmitra, J. Fabian, Graphene spintronics, Nature Nanotechnology 9 (10) (2014) 794−807.

[43] C.N.R. Rao, H.S.S. Ramakrishna, K.S. Subrahmanyam, U. Maitra, Unusual magnetic properties of graphene and related materials, Chemical Science 3 (2012) 45.

[44] H. Ronan, U. Khan, F. Christian, K. Sang-Woo, Piezoelectric properties in two-dimensional materials: simulations and experiments, Materials Today 21 (6) (2018) 611−630.

[45] L. Yu, Z. Li, C. Cheng, H. Shan, L. Zheng, Z. Fang, Plasmonics of 2D nanomaterials: properties and applications, Advancement of Science (2017) 1600430.

[46] H. Xiao, S. Liu, 2D nanomaterials as lubricant additive: a review, Materials & Design 135 (2017) 319−332.

[47] F. Yin, B. Gu, Y. Lin, P. Nishtha, S.C. Tjin, J. Qu, S.P. Lau, Y. Ken-Tye, Functionalized 2D nanomaterials for gene delivery applications, Coordination Chemistry Reviews 347 (2017) 77−97.

Further reading

[1] M.S. Dresselhaus, G.C. Ming, Y. Tang, R. Yang, H. Lee, D. Wang, Z. Ren, J.-P. Fleurial, P. Gogna, New directions for low dimensional thermoelectric materials, Advanced Materials 19 (2007) 1043−1053.

Graphene-based nanostructures for biomedical applications

Keisham Radhapyari, PhD[1], Suparna Datta, PhD[2], Snigdha Dutta, MSc[1], Nimisha Jadon, PhD[3], Raju Khan, PhD[4]

[1]*Regional Chemical Laboratory, Central Ground Water Board, North Eastern Region, Ministry of Jal Shakti, Department of Water Resources, River Development and Ganga Rejuvenation, Guwahati, Assam, India;* [2]*Regional Chemical Laboratory, Central Ground Water Board, Eastern Region, Ministry of Jal Shakti, Department of Water Resources, River Development and Ganga Rejuvenation, Kolkota, West Bengal, India;* [3]*School of Studies in Environmental Chemistry, Jiwaji University, Gwalior, Madhya Pradesh, India;* [4]*Analytical Chemistry Group, Chemical Sciences & Technology Division, CSIR-NEIST, Jorhat, Assam, India; CSIR-Advanced Materials and Processes Research Institute (AMPRI), Bhopal, Madhya Pradesh, India*

4.1 Introduction

Nanotechnology is booming in various research fields over the past few decades, especially in the biomedical field. Nanomaterials open a new arena in biomedical diagnostics. Various nanomaterials have been developed for various applications, which show a superior quality of interest than those from the parent material [1−5]. Nanomaterials have certain unique features of nanomaterials include the ability to manipulate materials at the nanoscale, which enables to expedite, exact it use engineering to investigate physiochemical properties and their emblematic interfaces with biological systems [6−13]. Alterations in structural, morphologic, electric, optical, and chemical properties of nanostructured materials offer principally striking and innovative choices for biomedical applications [14,15].

Carbon is one of the key components that makes it essential to life on earth. Nanotechnology opened a door to novel nanosized carbon allotropes such as fullerene, carbon nanotube (CNT), and graphene with different dimensions, which are (1) zero dimension (0D), (2) one dimension (1D), and (3) two dimensions (2D), respectively. Carbon-based nanomaterials have offered immense potential in extensive applications, and graphene is the newest among the different carbon nanomaterials. In the recent years, studies involving graphene-related nanomaterials in the biological field have rapidly increased. Owing to the unique chemical characteristic of these materials, including thermal, electronic, optical, and mechanical properties, they serve as an excellent platform for evolving unique nanostructured materials to facilitate various demands in biomedical applications.

Two-Dimensional Nanostructures for Biomedical Technology. https://doi.org/10.1016/B978-0-12-817650-4.00004-8

Over the past decades, carbon nanostructures such as carbon nanofibers, carbon mesostructures, CNTs, and graphene have shown tremendous potential for application in biosensors and sensor applications [16–20]. The physicochemical properties of the 2D nanomaterial graphene with regard to structure and conformation drive its roles in biomedical applications, which in turn are superior to those of other mesoporous nanostructures with spherical morphologic form. The applications and designs of graphene-based biosensors and sensors have attracted tremendous attention [21,22] due to the shift toward graphene nanostructures, especially electrochemical biosensors because of their superior electric and electrochemical behavior and their large active surface area, which in turn make graphene-based nanostructures (GBNs) one of the popular and best materials for biomedical applications.

Death caused by cancer has caused havoc all over the world. As per the global- and country-level data, mortality rates due to cancer vary significantly with the country and earnings. In the Leading Causes of Death in Females of United States, 2015 report, cancer was ranked second [23]. Cancer has been a health challenge around the globe in the past decades because of its high occurrence and large-scale deaths. It is one of the serious questions for humanity, even though there is rapid advancement and growth in the medical sector for human health. Studies reveal that by 2020, there will be between 15 and 17 million new cases of cancer every year, of which 60% of the cases will be in developing countries [24]. In spite of tremendous efforts devoted to discover cures for cancers, the assumed 5-year survival rate of cancer patients remains low [25].

Among the various types of cancers, breast cancer is the leading cause of cancer deaths among females, with the highest diagnostic frequency. About 63,410 cases of female breast carcinoma in situ were diagnosed in 2017 in the United States [26]. In case of African American women, breast cancer is the leading cause of death among young women aged less than 45 years. The risk of breast cancer death rises by 5% for every 1-year reduction in a diagnosis, reflecting more aggressive phenol types than breast cancers occurring later in life [27].

In India, approximately 100,000 women are diagnosed with breast cancer with a case fatality ratio of 40%. India has become a country of high breast-cancer-related deaths [28–30]. It is reported that every second, Indian women are dying due to breast cancer, making the survival rates poor compared with those of developed countries. Breast cancer has been observed to be notably common in women aged 50 years or younger [31,32] and in premenopausal women [33], and it was found to be more prevalent in African American [34–37] and Asian women [38] and in those having westernized diets [39,40]. Breast cancer among younger women is related to poor survival [41]; early detection has the potential for improved survival, less-invasive treatment, and a higher quality of life, thus reducing the burden of disease and the cost of treatment. Consequently, it is urgently needed to develop new and economical ways to diagnose breast cancer and predict prognosis, so as to provide precise and personalized treatment for patients.

The recent advancement in the field of nanomaterials and nanotechnology elucidate exploring revolutionary approaches for the diagnosis of cancer [42,43]. Although there are many tools for the detection of breast cancer, such as imaging and biopsy, there is still opportunity to ameliorate. In the recent years, predicting the recurrence or progression of breast cancer and detecting it in an early stage are more and more commonly conducted using tumorous biomarking of tissues and serum.

Various diagnostic techniques were reported for common breast cancer, which can detect about 80%—90% of breast cancer in women. Immunohistochemistry (IHC), enzyme-linked immunosorbent assay (ELISA), and radioimmunoassay (RIA) are some of the leading biomarker-based breast cancer detection techniques. However, all these diagnostic techniques have various limitations such as low specificity and low sensitivity in mammography, missing of tumor cells in biopsy, insensitivity toward low-level markers and the intrinsic color of analytes leading to false detection in case of ELISA, radioactivity risk in RIA, and so on. Therefore highly sensitive noninvasive methods for breast cancer diagnosis are in high demand. Various reviews have been published highlighting the applications of graphene nanomaterials in cancer and breast cancer diagnosis. However, none has done a detailed and comprehensive study on GBN applications in the biomedical field and in cancer diagnosis.

In this chapter, we intend to address some novel endeavors dedicated to biosensors using GBNs and their importance in biomedical applications, with special reference to cancer and breast cancer. Various methods based on graphene nanocomposites offering remarkable sensitivity for early detection of breast cancer are discussed. Recent trends in techniques exclusively for breast cancer and ovarian cancer detection are also highlighted. Furthermore, the chapter will discuss the various important applications of antimicrobial properties of GBNs in the biomedical field. The importance of graphene-based nanomaterials in tissue engineering will also be illustrated briefly, discussing the great potential of graphene in the field of bone, cardiac, and cartilage tissue engineering. This chapter offers crucial insights into the new nanostructures based on graphene as a soft template and their potential applications in the biomedical field. We hope this chapter would also stimulate new ideas and approaches to the relevant ongoing biomedical research.

4.2 Overview of graphene-based nanostructures

Graphene, a carbon allotrope, is a flat single layer of graphite that contains stacked layers of carbon atoms with a honeycomb lattice. Graphene, the mother of all carbon atoms, has various unique properties, viz. mechanical, optical, structural, and thermal characteristics. Graphene is a nanosized single-atom-thick 2D structure that provides high surface area with adjustable surface chemistry that can form hybrids. Derivatives of graphene include graphene oxide (GO), reduced graphene oxide (RGO), graphene nanoribbons, graphene quantum dots (GQDs), graphene nanopores, and three-dimensional graphene foam. The structural models of several graphene derivatives with different structures are shown in Fig. 4.1 [44].

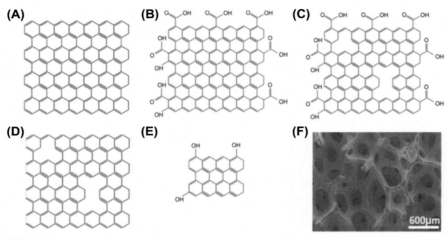

FIGURE 4.1

Schematic representation of (A) graphene, (B) graphene oxide (GO), (C) reduced graphene oxide (RGO), (D) porous graphene, (E) graphene quantum dots, and (F) three-dimensional graphene foam.

Reprinted with permission from T.A. Tabish, S. Zhang, P.G. Winyard, Developing the next generation of graphene-based platforms for cancer therapeutics: the potential role of reactive oxygen species, Redox Biology 15 (2018) 34–40.

4.2.1 Graphene

With the isolation of single-layer graphene from graphite by Andre Geim and Nosovelov in 2004 [45] through mechanically cleaving a graphite crystal using the scotch tape method, graphene has attracted attention among the scientific community to a great extent. However, graphene had come into existence since 1859 [46] and has been studied theoretically long back by Wallace [47]. It has been prepared by mechanical or chemical exfoliation of graphite via chemical vapor deposition. It has a high intrinsic mobility, large specific surface area, and high thermal conductivity. Owing to the absence of oxygen groups, graphene is considered as hydrophobic.

4.2.2 Graphene oxide

GO has been produced by exfoliation of graphite oxide and is a single layer of graphite oxide. Steps involved in the production of GO is acid-base treatment of graphite oxide followed by sonication. Several functional groups, such as oxygen and epoxide, carbonyl, hydroxyl, and phenol groups, are present on the surface of GO. The only difference between graphene and GO is the presence of oxygen atoms bound to carbon in the latter. GO is the product of a hydrophilic derivative of graphene. GO possesses both aromatic (sp^2) and aliphatic (sp^3) domains that facilitate interactions at the surface [48–50]. Graphene oxide has been synthesized by

Hummers method and has oxygenated groups on the surface of the molecule. There is no specific structure for GO, but structural and morphologic characterization gives an idea of the GO structure [51] (Fig. 4.1B).

4.2.3 Reduced graphene oxide

RGO is the product of thermal or chemical reduction of graphite oxide or GO. RGO is considered as an intermediate structure between the ideal graphene sheet and the highly oxidized GO (Fig. 4.1C).

4.2.4 Porous graphene

Porous graphene is a graphene sheet that is missing carbon atoms from its plane. Owing to its high specific surface area, hydrophobic nature, and biocompatibility, it provides fascinating materials for biological applications. Graphene nanopores usually have a pore size of 1−30 nm. Vacancies and pores can clearly be seen in the porous graphene sheet (Fig. 4.1D).

4.2.5 Graphene quantum dots

GQDs are luminescent nanocrystals with size less than 50 nm. They have potential applications in cancer diagnosis and treatment due to their attractive properties. Fig. 4.1E shows water-soluble GQDs with functional groups ($C-OH$, $C=O$, $C-O-C$, $C-H$) on their surface.

4.2.6 Three-dimensional graphene

Using chemical vapor deposition, three-dimensional graphene networks are formed in the form of foam, sponge, or aerogel that have been assembled from individual graphene sheets, which preserve the unique properties of individual graphene sheets.

4.2.7 Graphene-based nanostructures

Derivatives of graphene, such as GO and RGO, are ideal platforms for constructing GBNs for various applications. Recently, hybrid materials of noble metal nanocrystal/decorated GO or RGO with enhanced properties and functions have been explored extensively. The planar graphene has a thickness of one or several carbon atoms with limited volume that suppresses its potential in some applications. Limitations on applications of GBNs have been overcome by using various substrates viz. nanoparticles (NPs) [52], metal foams [53], and nonmetal porous structures [54] and by not using any template. Guan et al. [55] had reported the hollow nanocubes that were formed by piecing together six planar square surfaces. The foam is a cellular lightweight and porous structure that was induced by gaseous bubbles introduced during the production process. Transmission electron microscopic (TEM) and scanning electron microscopic (SEM) images of some GBNs are provided in Fig. 4.2.

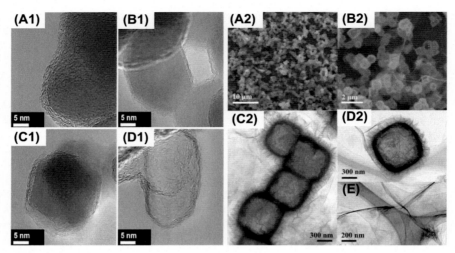

FIGURE 4.2

(A) Graphene nanostructures. Transmission electron microscopic (TEM) images of graphene nanoshells: (a) graphene-coated aluminum oxide nanoparticles, (b) graphene-deposited titanium oxide nanoparticles, (c) graphene-covered magnesium oxide nanoparticles, and (d) graphene shells after inner nanoparticle removal. (B) (a and b) Scanning electron microscopic graphs and (c—e) TEM graphs of hollow graphene nanocubes.

(A) Z. Sun, L. Zhang, F. Dang, Y. Liu, Z. Fei, Q. Shao, H. Lin, F. Guo, L. Xiang, N. Yerra, Experimental and simulation-based understanding of morphology-controlled barium titanate nanoparticles under co-adsorption of surfactants, CrystEngComm 19 (24) (2017) 3288–3298. (B) Reprinted with permission from A. Bachmatiuk, R.G. Mendes, C. Hirsch, C. Jähne, M.R. Lohe, J. Grothe, S. Kaskel, L. Fu, R. Klingeler, J. Eckert, P. Wick, M.H. Rümmeli, Few-layer graphene shells and nonmagnetic encapsulates: a versatile and nontoxic carbon nanomaterial, ACS Nano, 7 (12) (2013) 10552–10562 and X. Guan, J. Nai, Y. Zhang, P. Wang, J. Yang, L. Zheng, J. Zhang, L. Guo, CoO hollow cube/reduced graphene oxide composites with enhanced lithium storage capability. Chemistry of Materials, 26 (20) (2014) 5958–5964.

Synthesis of GBNs has been done by either a "top-down" or a "bottom-up" approach. Jana et al. [56] illustrated various approaches for the synthesis of GBNs. Both methods have their advantages and disadvantages. Reina et al. [51] emphasized on the bottom-up method for the synthesis of GBNs rather than the top-down method because of the nonuniformity of the synthesized GBNs, which may interfere with GBN-based electronic devices for biomedical applications.

4.3 Graphene-based nanostructures in biomedical applications

Graphene-based nanomaterials have shown various potential applications in the biomedical field because of their many exclusive physicochemical properties (Fig. 4.3) [57]. Advancement has been made especially in the areas of biosensors,

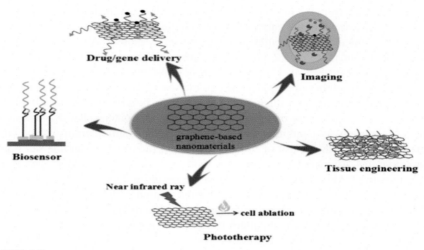

FIGURE 4.3

The potential applications of graphene-based nanomaterials in the biomedical field.

Reprinted with permission from Q. Ying, H. Feng, Y. Chenggong, L. Xuewu, L. Dong, M. Long, Z. Qiuqiong, L. Jiahui, W. Jingde, Advancements of graphene-based nanomaterials in biomedicine, Materials Science and Engineering: C, 90 (2018) 764–780.

bioimaging, cancer or other disease diagnosis and treatment, tissue engineering, drug delivery, phototherapy, and cellular growth and differentiation [58–62].

4.3.1 Sensors/biosensors for breast cancer

Biopsy, mammography, sonography, magnetic resonance imaging (MRI), molecular breast imaging, thermography, etc. are some of the common breast cancer diagnostic techniques. These methods can identify about 80%–90% of breast cancer in women. ELISA, IHC, and RIA are some of the leading biomarker-based breast cancer detection techniques. However, all these diagnostic techniques have various limitations such as low specificity and low sensitivity in mammography, missing of tumor cells in biopsy, insensitivity toward the intrinsic color of analytes and low-level markers leading to false detection in case of ELISA, radioactivity risk in RIA, and so on [63]. Therefore highly sensitive noninvasive methods for breast cancer diagnosis are in high demand.

The sensitivity of mammography ranges between 30% and 55% [64–66]. However, it is reported that whole-breast ultrasonography (US) supplemented with mammography screening has served to be better for early recognition of breast cancer in women with dense breast tissue [67]. In an average, 47% of women going through screening mammography have either extremely or heterogeneously dense breasts. The potential benefit of screening (US) with screening mammography is short exposure to ionizing radiation. Although the use of this technique increases cancer detection from 0.3 to 7.7 per 1000 women who are screened [68,69], the positive predictive value of the current US screening techniques is as low as 8.6% [70]. So contrast-enhanced ultrasound imaging, which is molecularly targeted, is now more popular in place of the prevailing ultrasound-supplemented mammography screening. Bachawal et al. [71] have used contrast microbubbles targeted at B7-H3, which is a member of the B7 family of immunoregulators, in US molecular imaging signal in transgenic mice and found this to be effective in improving the accuracy of US detection of breast cancer. When computer-aided diagnosis, i.e., the CAD system, is combined to US screening, the mammogram interpretation gets enhanced. With the use of image processing, radiologists can analyze digital mammograms. About 60%–90% of cancer biopsy results interpreted by radiologists without using CAD can become false-positive results and the cancer is found to be benign later [72]. CAD helps in distinguishing benign cells from malignant ones. The textural information related to a mammographic image is the most significant aspect to distinguish abnormal pattern from normal pattern. For the texture analysis of mammogram images, wavelet transform is predominantly a popular method in which multiresolution testing is done. Gray-level co-occurrence matrix (GLCM) is another important means of texture analysis that estimates the second-order statistical properties of images [73–75]. Beura et al. [76] have worked out a scheme that employs GLCM and 2D discrete wavelet transform subsequently to obtain feature matrix from mammograms.

Biosensors are powerful tools for diagnostic purposes in breast cancer, as they are precise, sensitive, and economical. The field of biosensors is vast and multidisciplinary. The greatest advantage of biosensors is that they have rapid response because of direct evaluation in physiologic fluids such as blood, milk, saliva, and urine. Immunosensors are part of the biosensor technology. They can apprehend the direct binding of antigen or an antibody to form an immunocomplex at the transducer surface [77] Immunosensors are the commonly used diagnostic tools for all types of cancers, including breast cancer. They may be categorized as electrochemical [78], piezoelectric [79], and optical [80]. In the diagnosis of breast cancer, electrochemical biosensors are a widely accepted technique. They alter chemical signals into an assessable amperometric signal by the use of amperometric, impedimetric, and potentiometric transducers to concurrently detect and investigate the specific compounds of interest. Glucose oxidase immobilization was used to design the first enzyme electrode. Use of electrochemical biosensors based on graphene is prevalent as cancer biomarker. They have many advantages such as large surface area, selective biomolecule detection, excellent electric conductivity, and rapid electron transfer.

4.3.1.1 Detection of carcinoembryonic antigen for breast cancer diagnosis

Carcinoembryonic antigen (CEA) is widely used to diagnose various types of cancers, such as ovarian cancer, breast cancer, and colon cancer, in healthy individuals. It is important to develop sensitive methods to detect CEA in an individual for prophylaxis and cure of the above-mentioned cancers. Graphene-based biosensors, gold nanoparticles (AuNPs), are commonly used for detection of CEA. A sensitive sandwich-type AuNP-based electrochemical immunosensor, AuNPs and graphene, was fabricated to detect CEA [81]. AuNPs speed up the electron transfer on the electrode interface because of their good biocompatibility, high specific surface area, and higher electric conductivity. Characterization details are shown in Fig. 4.4, and the developed method have a linear range from 0.1 pg mL^{-1} to 100 ng mL^{-1}, with the detection limit of CEA to be 0.0697 pg mL^{-1}.

Gold-graphene composites are also successfully used and were successfully synthesized in which HRP-anti-CEA (horseradish peroxidase graphene-labelled

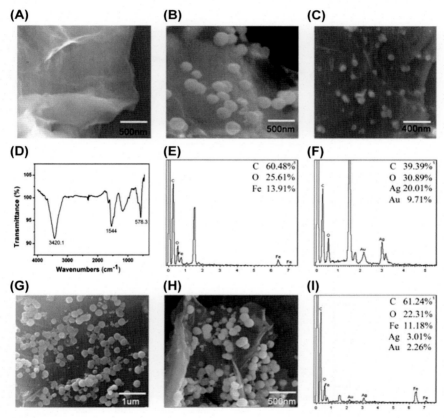

FIGURE 4.4

Scanning electron microscopic (SEM) image: (A) GO, (B) graphene sheet (GS)-Fe$_3$O$_4$, and (C) GS-Au@Ag. (D) Fourier transform infrared spectrometer analysis of GS-Fe$_3$O$_4$. Energy-dispersive X-ray (EDX) spectrum: (E) GS-Fe$_3$O$_4$ and (F) GS-Au@Ag. (G,H) SEM images of GS-Fe$_3$O$_4$/Au@Ag with different scale. (I) EDX spectrum of GS-Fe$_3$O$_4$/Au@Ag.

Reprinted with permission from Y. Li, Y. Zhang, F. Li, M. Li, L. Chen, Y. Dong, Q. Wei, Sandwich-type amperometric immunosensor using functionalized magnetic graphene loaded gold and silver core-shell nanocomposites for the detection of carcinoembryonic antigen, Journal of Electroanalytical Chemistry 795 (2017) 1–9.

anti-CEA antibody) and HRP have been successively adsorbed on the Au-graphene modified glassy carbon electrode (GCE). Differential pulse voltammetry (DPV) technique was used for the quantification of CEA concentration ranging from 0.10 to 80 ng mL^{-1} with a detection limit of 0.04 ng mL^{-1} (S/N = 3). Such a new

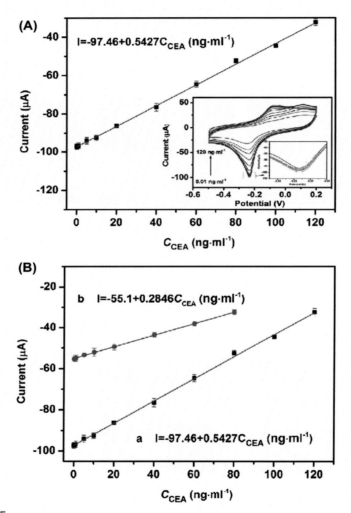

FIGURE 4.5

(A) Calibration plot of the current response via concentration of carcinoembryonic antigen (CEA). (B) Electrochemical responses of the immunoassay with (a) rCu$_2$OeGO and (b) rCu$_2$O toward various concentrations of CEA.

Reprinted with permission from T. Feng, X. Chen, X. Qiao, Z. Sun, H. Wang, Y. Qi, C. Hong, Graphene oxide supported rhombic dodecahedral Cu2O nanocrystals for the detection of carcinoembryonic antigen, Analytical Biochemistry 494 (2016) 101–107.

graphene-nanostructure-based technique was in excellent accordance with the results determined by ELISA, which shows superior constancy, selectivity, and reproducibility [82]. Feng et al. [83] designed a simple rhombic dodecahedral electrochemical immunosensor, that is, Cu_2O nanocrystals GO AuNPs (Rcu_2O-GO-AuNPs). Cyclic voltammetry (CV) (Fig. 4.5) was utilized for quantifying this method and it shows a large linear range from 0.01 to 120 ng mL^{-1} with a small limit of detection of 0.004 ng mL^{-1}. The authors claimed that the developed method may be a prospective alternative for the diagnostic utilization of GO-supported unique-morphology materials in biosensors and biomedicine applications.

Nile blue A (NB) hybridized electrochemically reduced graphene oxide (NB-ERGO) [84] was designed for the detection of CEA. Because of the arrangement of antigen-antibody immune complex, the low response currents of NB were directly proportional to the concentrations of CEA and the immunosensor determination was based on this fact. The projected immunosensor has been used for the determination of CEA in clinical serum samples with agreeable results ranging from 0.001 to 40 ng mL^{-1}. An interesting nanogold-based mesoporous carbon foam (Au/MCF) coupled with a signal amplification by C-Au synergistic silver enhancement has been developed to detect CEA [85]. The newly designed graphene-based biosensor holds immense prospective for ultrasensitive electrochemical biosensing, with detection limit as low as 0.024 pg mL^{-1}. Wang et al. [86] synthesized three types of aptamer-functionalized silver and silver/gold nanoclusters (DNA-AgNCs and DNA-Ag/AuNCs) for the detection of mucin 1 in a linear range from 1.33 to 200 ng mL^{-1} having a detection limit of 0.18 ng mL^{-1}, CEA had a linear range from 6.7 ng mL^{-1} to 13.3 μg mL^{-1} resulting in a detection limit of 3.18 ng mL^{-1}, and cancer antigen 125 (CA 125) acquired a linear range of 2 ng mL^{-1} to 6.7 μg mL^{-1} showing a detection limit of 1.26 ng mL^{-1}.

4.3.1.2 Detection of carbohydrate antigen 15-3 for breast cancer diagnosis

A new and responsive label-free immunoassay that is based on gold nanospears (Au NSs) electrochemically assembled onto thiolated GQDs (CysA/GQDs) was successfully designed for sensitive voltammetric detection of carbohydrate antigen 15-3 (CA15-3), a breast cancer tumor marker [87] in serum that is very crucial for cancer prognosis. A linear dynamic range with a detection limit of 0.16−125 U mL^{--1} and 0.11 U mL^{-1}, respectively, was shown by this immunosensor. The mentioned immunosensor exhibited tremendous analytical performance to detect CA15-3.

CA15-3 is employed as a tumor biomarker that plays a vital role for the detection of breast cancer. For healthy individuals the serum CA15-3 concentration is less than 30 U mL^{-1}. But in case of breast cancer patients the level of CA15-3 is above 100 U mL^{-1} [88]. Chemiluminescence immunoassay, ELISA, sandwich-type immunosensor, and optical immunoassay are some of the detection techniques of CA15-3. For immobilization of CA15-3, Au NPs or GQDs can form the ideal platform and this can be a responsive assay of CA15-3 antibodies because antigen-antibody interaction is not sufficient to produce a sensitive electrochemical test signal directly. Such works have been carried out by Hasanzadeh et al. [87]. They have studied Au NS-based label-free immunoassay in which the Au NSs were electrochemically adsorbed onto CysA/GQDs, thus making the CA15-3 antibody detection simple and rapid.

4.3.1.3 Detection of MCF-7 for breast cancer diagnosis

Wang et al. [89] designed an electrochemical biosensor which has polyadenine (polydA)-aptamer modified gold electrode and polydA-aptamer functionalized AuNPs/GO hybrid for the particular identification of breast cancer cells MCF-7. They have employed the DPV technique. Aptamer functionalized silver NCs (Ag NCs) have also been widely used in aptamer-based techniques in breast cancer detection.

4.3.1.4 Detection of circulating microRNA and tumor cells for breast cancer diagnosis

Circulating microRNAs (miRNAs) are potential biomarkers for the detection of breast cancer [90−92]. So this is attributed to the enhanced tissue specificity, immense stability, and peculiar expression in diverse tumor types of miRNA. The expression profiles of miRNA are capable of outlining the molecular breast cancer subtypes [93]. A sensitive HRP/AuNPs-barcode/miRNA-NAH/graphene-based biosensor [94] was also developed for detecting up to 6 fM and a linear range of 0.01−700 pM. To detect serum miR-199a-5p, an innovative electrochemical nanobiosensor was developed by utilizing modified GCE with gold nanorod (GNR) and GO on which a thiolated probe was mutilated. To study the electrochemical characteristics and the behavior of nanobiosensors, the electrochemical impedance spectroscopy (EIS) was used, which exhibits the linear range of the calibration curve to vary from 15 fM to 148 pM, the detection limit is 4.5 fM and the standard deviation is of 2.9% [95]. Azimzadeh et al. [58] have developed a new electrochemical nanobiosensor to detect miRNA in the early stages of breast cancer. The method is designed on a GO functionalized GNR (GO/GNR) adapted GCE employing Oracet Blue as an electroactive label. The method has great sensitivity as well as being selective, simple, and cost effective when compared with previously reported electrochemical nanobiosensors. The proposed nanobiosensor could be employed to directly detect the miR-155 in plasma without the need of sample extraction and/or amplification, which can be used in clinical purposes such as early detection, prognostic trends in the breast cancer patients, and/or as an indicator required for treatment response to drugs.

Circulating tumor cells (CTCs) are effective biomarkers for the detection of specific tumor metastasis because they are less invasive and more reliable when compared with biopsy, radiographic photography, serum tumor marker−based detection, etc. Tian et al. [96] developed a sensitive electrochemical method for the detection of CTCs in which RGO/AuNP composites are used as a support material, along with a catalyst, which is CuO nanozyme. The detection of MCF-7 CTCs was done by an electrochemical cytosensor that has effective surface identification between MUC-1 aptamer and specific mucin 1 protein (MUC-1) that is overexpressed on the MCF-7 cell membranes. The CuO nanozyme acts as a signal-amplifying nanoprobe in a sensitive electrochemical cytosensor for the detection of CTCs. This method has a large detection range from 50 to 7×10^3 cells mL^{-1} and a detection limit of mere 27 cells mL^{-1}. The method using combined MUC-1 and MUC-1 aptamer was applied to real serum samples.

4.3.1.5 Detection of SKBR-3 and triple negative breast cancer cells for breast cancer diagnosis

SKBR-3 breast cancer cell has Her2 antigen on its plasma membrane and is one of the most significant breast cancer cells. A highly sensitive, selective, and stable electrochemical immunosensor designed as a sandwich type was developed for the detection of SKBR-3 breast cancer cells [97]. For the fabrication of the sensor, green synthesized RGO is used as a platform for the immobilization of primary Herceptin antibody (anti-HCT). A variety of RGO-tetrasodium 1,3,6,8-pyrenetetrasulfonic acid/metal hexacyanoferrate (RGO-TPA/MHCFnano) nanocomposites includes RGO-TPA/FeHCF, RGO-TPA/CoHCF, RGO-TPA/NiHCF, and RGO-TPA/CuHCF. They were utilized as electrochemical labels of secondary Herceptin antibody (Fig. 4.6). The DPV

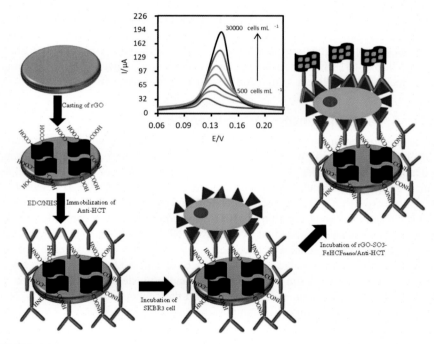

FIGURE 4.6

Schematic diagram for SKBR-3 detection. *EDC/NHS, N*-ethyl-*N'*-(3-(dimethylamino) propyl)carbodiimide/*N*-hydroxysuccinimide; *HCF*, hexacyanoferrate; *HCT*, Herceptin; *rGO*, reduced graphene oxide.

Reprinted with permission from M.A. Tabrizi, M. Shamsipur, R. Saber, S. Sarkar, N. Zolfaghari, An ultrasensitive sandwich-type electrochemical immunosensor for the determination of SKBR-3 breast cancer cell using RGO-TPA/FeHCF nanolabeled Anti-HCT as a signal tag, Sensors and Actuators B: Chemical 243 (2017) 823–830.

technique was employed to detect SKBR-3 breast cancer cells, with a linear range of 500–30,000 cells mL^{-1} and a limit of detection of 21 cells mL^{-1}.

Tao and Auguste [98] have successfully developed nanoprobe-GO-based sensor elements that can be obstructed in the existence of breast cells to yield fluorescent readouts. This sensor can detect and differentiate normal, metastatic, and cancerous breast cells and also triple negative breast cancer cells. They have successfully detected breast cancer cells and discriminated between estrogen receptor positive, human epidermal growth factor receptor-2 positive, and triple negative phenotypes.

4.3.1.6 Detection of H_2O_2 from living cancer cells for breast cancer diagnosis

A trimetallic AuPtPd/RGO nanocomposite [99] has been developed, which can perform as a high electrochemical catalyst for the reduction of H_2O_2 and sensitively monitor the release of H_2O_2 from living cancer cells. Trimetallic nanomaterials that are constituted of Pt, Au, and Pd NPs and RGO serve as an improved sensing platform by the use of this basic and original construction. The preparation of PtAuPd/RGO nanocomposites was customized on a GCE by physical adsorption.

Various techniques developed to detect breast cancer on the basis of graphene nanostructure [100–104] and non-graphene-based materials [105–133] are illustrated in Table 4.1, which includes method/specification, target molecule, nanomaterial/electrode used, linear range, and detection with respective references.

4.3.2 Sensors/biosensors for oral cancer

The sixth most common cancer in the world is oral cancer. It is found to be more prevailing in men than in women. It may metastasize to other parts of body leading to death if early stage detection fails. An oral cancer biosensor was developed by Kumar et al. [134], which is a nanostructured metal oxide (NMO). It is based on 2D electroactive RGO. The nanostructured hafnium oxide (nHfO2) serves as a model NMO. The reduced agglomeration of $nHfO_2$ was acquired by using controlled hydrothermal synthesis and was studied by means of NP tracking analysis. SEM, X-ray diffraction, and TEM techniques have been applied for the characterization of the sensor. The biosensor revealed a high sensitivity of 18.24 μA mL ng^{-1} with linear detection limit ranging from 0 to 30 ng mL^{-1} and with detection limit as low as 0.16 ng mL^{-1}. Good results were achieved with the concentration of CYFRA-21-1 obtained through ELISA in saliva samples of oral cancer patients.

4.3.3 Application of graphene nanostructures in ovarian cancer

Ovarian cancer is the fifth most common malignancy in women worldwide and responsible for the highest number of deaths of women in respect of all cancers. It is the sixth most common malignancy and fifth most common disease in women worldwide. Over the past decades, overall 4% of all cancers (approx. 200,000) new cases of ovarian cancer are diagnosed each year worldwide [135,136]. Detection of

Table 4.1 Breast cancer detection by graphene-based nanostructures and other non-graphene-based techniques.

Method/Specification	Target molecule	Nanomaterial/Electrode used	Linear range	Detection limit	References
Graphene-based sensors/biosensors					
HRP-anti-CEA	CEA	AuNP-graphene and GCE	0.05–350 ng mL^{-1}	0.01 ng mL^{-1}	[100]
Anti-CEA	CEA	–Thionine-RGO AuNP-Thi-graphene/GCE	10–500 pg mL^{-1}	4 pg mL^{-1}	[101]
	MUC1	Aptamer/graphene/Au/GCE with aptamer/thionine (TH)/Platimun Iron alloy (PtFe) conjugate	100–5×10^7 Cells mL^{-1}	38 Cells mL^{-1}	[102]
Anti-CA15-3/	CA15-3	NGS-GCE	0.1–20 U mL^{-1}	0.012 U mL^{-1}	[103]
S6 aptamer/Au/ZnO/	HER-2	GITO electrode	1×10^2–1×10^6 Cells mL^{-1}	58 Cells mL^{-1}	[104]
HRP/AuNP-barcode/ miRNA-NAH/	miR-21	AuNP/graphene/Au electrode	0.01–700 pM	6 fM	[94]
Electrochemical biosensor	CEA	DNA-Ag/AuNCs			[86]
Electrochemical immunoassay	CEA	Au–graphene/HRP-anti-CEA	0.10–80 ng mL^{-1}	0.04 ng mL^{-1}	[82]
Electrochemical immunosensor	CEA	rCu$_2$OeGOeAuNPs	0.01–120 ng mL^{-1}	0.004 ng mL^{-1}	[83]
Electrochemical immunosensor	CEA	AuNP/NB-ERGO	0.001–40 ng mL^{-1}	0.00045 ng mL^{-1}	[84]
Electrochemical immunosensor	CEA	Au/MCF-mediated silver	0.05 pg mL^{-1} to 1 ng mL^{-1}	0.024 pg mL^{-1}	[85]
Electrochemical immunosensor	CEA	Au@Ag/Fe$_3$O$_4$-GS/Ni^{2+}	0.1 pg mL^{-1} to 100 ng mL^{-1}	0.0697 pg mL^{-1}	[81]
Electrochemical nanobiosensor	miRNA	GCE/GO/GNR	15 fM to 148 pM	4.5 fM	[95]

Continued

Table 4.1 Breast cancer detection by graphene-based nanostructures and other non-graphene-based techniques.—cont'd

Method/Specification	Target molecule	Nanomaterial/Electrode used	Linear range	Detection limit	References
Electrochemical immunosensor	CA15-3	CysA/Au NSs/GQDs	0.16–125 U mL^{-1}	0.11 U mL^{-1}	[87]
Electrochemical nanobiosensor	miRNA	GNRs/GO/GCE	2.0 fM to 8.0 pM	0.6 fM	[58]
Electrochemical immunosensor	SKBR-3	RGO-TPA/FeHCF	500–30,000 Cells mL^{-1}	21 Cells mL^{-1}	[97]
Fluorescent nanodots	Triple negative breast cancer cells	—		200 Cells	[98]
Electrochemical cytosensor	CTCs	RGO/AuNP composites	50–7 × 10^3 Cells mL^{-1}	27 Cells mL^{-1}	[96]
Non-graphene-based techniques					
cDNA/CHIT-co-PANI (chitosan-co-polyaniline)	BRCA1	Indium tin oxide	0.05–25 fM	0.05 fM	[105]
BRCA1/BSA/anti-BRCA1/ BMIM.BF4	BRCA1	Mesoporous carbon nanosphere-toluidine blue nanocomposite and GCE	0.01–15 ng mL^{-1}	3.97 ng mL^{-1}	[106]
ssDNA probe/	BRCA1	Au electrode	1 × 10^{-19}–1×10^7 M	4.6×10^{-20} M	[107]
ssDNA probe/polyethylene glycol	BRCA1	AuNP and GCE	50.0 fM–1.0 fM	1.7 fM	[108]
ssDNA probe/1-pyrenebutyric acid-N-hydroxysuccinimideester (PANHS)/	BRCA1	—	10^{-16}×10^{-10} M	3.7×10^{-17} M	[109]

HRP/anti-CEA/	CEA	AuNPs/ZnONPs/Au electrode	$0.1–70$ ng mL^{-1}	0.01 ng mL^{-1}	[110]
Anti-CEA	CEA	/AuNPs/Azure I/Nf-MWCNT	$0.1–40$ ng mL^{-1}	0.03 ng mL^{-1}	[111]
DNA probe	MUC1	Gold SPEs	$0–10$ ng mL^{-1}	0.95 ng mL^{-1}	[112]
Aptamer-HRP/MUC1/aptamer1/	MUC1	Au electrode	$100^{-1}×10^{7}$ Cells	100 Cells	[113]
MUC1aptamer/MCF-7/anti-CEA	MUC1	CdSNPs	$10^{4}–10^{7}$ Cells mL^{-1}	$3.3×10^{2}$ Cells mL^{-1}	[114]
Anti-CA15-3	CA15-3	PtNCs/OrgSi@CS-CNTs/GCE	$0.1–160$ U mL^{-1}	0.04 U mL^{-1}	[115]
	CA15-3	ZnO nanorods on quartz crystal microbalance	$0.5–26$ U mL^{-1}	0.5 U mL^{-1}	[116]
Vascular endothelial growth factor receptor-1 (R1-VEGF/EDC-NHS3-MPA/)	VEGF	Au	$10–70$ pg mL^{-1}	10 pg mL^{-1}	[117]
Aptamer-methylene blue/	VEGF	Au electrode	50 pM–0.15 nM	5 pM	[118]
epidermal growth factor receptor antibody (EA/anti-EGFRab)/DTSP/	EGFR	Au electrode	1 pg mL^{-1}–100 ng mL^{-1}	1 pg mL^{-1}	[119]
PG/PDITC/Cys/	EGFR	AuNPs/Au electrode	1 pg mL^{-1}–1 mg mL^{-1}	0.34 pg mL^{-1}	[120]
Anti-c-ErbB2/polypyrrole-NHS/	HER-2	GCE	100–10,000 Cells mL^{-1}	100 Cells mL^{-1}	[121]
Anti-HER-2/HER-2/anti-HER-2-S-AP/SPCE	HER-2	Screen-printed carbon electrodes	15–100 ng mL^{-1}	4.4 ng mL^{-1}	[122]
AP-Ab2/HER-2ECD/Ab1/proteinA/MB/SPE	HER-2 ECD	–	0–15 ng mL^{-1}	6 ng mL^{-1}	[123]
Anti-HER-3/HER-3/BSA/HER-3/4-ATP/	HER-3	Au electrode	$0.4–2.4$ pg mL^{-1}	0.4 pg mL^{-1}	[124]
CV		DNA/PICA/GCE	$3.34×10^{-9}$ to $10.6×10^{-9}$ M	$1.0×10^{-9}$ M	[125]

Continued

Table 4.1 Breast cancer detection by graphene-based nanostructures and other non-graphene-based techniques.—cont'd

Method/Specification	Target molecule	Nanomaterial/Electrode used	Linear range	Detection limit	References
DPV	MCF-7	DNA/polyaniline/GCE	22.5×10^{-9} to 2.25×10^{-12} M	1.0×10^{-12} M	[126]
Electrochemiluminescent detection		—	$500-2 \times 10^{7}$ Cells mL^{-1}	230 Cells mL^{-1}	[127]
CV		—	$1 \times 10^{5}-1 \times 10^{8}$ Cells mL^{-1}	1×10^{5} Cells mL^{-1}	[128]
Photoluminescence		—	$250-10^{4}$ Cells mL^{-1}	201 and 85 Cells mL^{-1}	[129]
Electrochemiluminescent detection		—	100–2500	30 Cells mL^{-1}	[130]
Fluorescence resonance energy transfer		—	50–3000 Cells mL^{-1}	36	[131]
Colorimetric detection via dual-aptamer target binding strategy		—	$10-10^{5}$ Cells mL^{-1}	10 Cells mL^{-1}	[132]
Colorimetric aptasensor based on aggregation of Au NPs		—	$10-10^{5}$ Cells mL^{-1}	10 Cells mL^{-1}	[133]

AuNP, gold nanoparticle; BSA, bovine serum albumin; CA15-3, carbohydrate antigen 15-3; CEA, carcinoembryonic antigen; CTCs, circulating tumor cells; CV, cyclic voltammetry; DPV, differential pulse voltammetry; DTSP, dithiobissuccinimidyl propionate; EGFR, epidermal growth factor receptor; GCE, glassy carbon electrode; GITO, graphene indium tin oxide; GNR, gold nanorod; GO, graphene oxide; GQD, graphene quantum dot; GS, graphene sheet; HRP, horse-radish peroxidase; MCF, mesoporous carbon foam; miRNA, microRNA; MWCNT, multiwalled carbon nanotube; NB-ERGO, Nile blue A hybridized electro-chemically reduced graphene oxide; NCs, nanoclusters; NGS, N-doped graphene sheet; NHS, N-hydroxysuccinimide; NSs, nanosheets; PDITC, 1,4- phenylene diisothiocyanate; PICA, poly(indole-6-carboxylic acid); RGO, reduced graphene oxide; SPE, screen printed electrode; ssDNA, single-stranded DNA; TPA, tetrasodium 1,3,6,8-pyrenetetrasulfonic acid; VEGF, vascular endothelial growth factor.

ovarian cancer is somewhat complex and hypercritical due to its unique pattern of metastatic spread. It spreads locally within the peritoneal cavity and is superficially invasive, which is unlike other cancer types in which metastasis occurs via the blood stream to distant sites. Techniques for the initial stage identification of uterine cancer are in high demand because between 70% and 75% of ovarian carcinomas are not discovered till they have reached an advanced stage III. Imaging is the prevalent diagnostic tool for detecting ovarian cancer, including MRI, CT scan, and US, which are important clinical diagnostic tools available for ovarian cancer. However, in their conventional forms, these radiologic approaches are not tumor specific and not useful for intraoperative applications. Fluorescence imaging techniques are more specific and sensitive in the detection of ovarian cancer. The use of optical imaging systems and tumor-specific fluorophores have become more common for ovarian cancer diagnosis.

An alternative approach to conventional chemotherapy is the application of graphene nanocomposites to reduce and eradicate only targeted cancer stem cells (CSCs). Small populations of cells are known as CSCs that possess the ability of self-renewal and form tumors and are responsible for recurrence, drug resistivity, and chemoresistivity that remarkably affect cancer therapy. Nowadays, an effective anticancer therapeutic strategy is the CSC-focused therapy but a major challenge in cancer therapy approach is eradication of CSCs, which could be done by applying nanocomposites for targeting CSCs. Choi et al. synthesized biomolecule-mediated RGO-silver NP nanocomposites (RGO-Ag) by using R-phycoerythrin and examined in ovarian cancer cells and ovarian cancer stem cells (OvCSCs). RGO-Ag exhibit notable toxicity for both ovarian cancer cells and OvCSCs and manifest remarkable cytotoxicity toward highly tumorigenic $ALDH^+CD133^+$ cells. The number of A2780 and $ALDH^+CD133^+$ colonies was significantly reduced after 3 weeks incubation of OvCSCs with RGO-Ag and the results revealed that the applied nanocomposite was highly toxic to OvCSCs and generated free oxygen radicals that reduced cell viability, caused lactate dehydrogenase leakage, reduced the potential of mitochondrial membrane, and increased apoptotic gene expression that resulted in dysfunction of mitochondria and apoptosis initiation. When RGO-Ag is combined with salinomycin, it increased apoptosis to fivefold higher levels than with individual treatment. A very low concentration of RGO-Ag/salinomycin is sufficient for selective destruction of OvCSCs and sensitizing tumor cells. RGO-Ag and its combinations are found to be the novel nanotherapeutic molecules for selective killing tumorigenic $ALDH^+CD133^+$ cells and destructing CSCs [137].

Graphene and its nanocomposites possess unique physicochemical properties that led to their utilization for cancer therapy. Cisplatin (Cis) has the ability to significantly induce selective death of cancer cells and hence is widely used as a chemotherapeutic drug for a variety of cancers. Yuan et al. investigate the combined impact of Cis and an RGO-silver NP nanocomposite (RGO-AgNP) on cervical cancer (HeLa) cells. They synthesized RGO, AgNPs, and RGO-AgNP nanocomposites by using C-phycocyanin. The anticancer properties of Cis and the synthesized nanocomposites were examined by using cellular assay series such as LDH leakage, cell proliferation, cell viability, and reactive oxygen species generation. At cellular levels, the concentrations of oxidative and antioxidative stress markers such as superoxide dismutase, malondialdehyde, catalase, and glutathione were also evaluated. Real-time reverse-transcription polymerase chain reaction was used for measuring the expression of autophagy, proapoptotic, and antiapoptotic genes. Inhibition of tumor cell viability by cisplatin and other synthesized GO nanocomposites is a dose-dependent process and showed significant impact on apoptosis, cytotoxicity, and cell proliferation. The expression of apoptotic and autophagy genes has been significantly improved in the presence of Cis-RGO-AgNP composite and it also remarkably increased the generation of reactive oxygen species that are responsible for the accumulation of autophagosomes and autophagolysosomes in HeLa cells; this proves that RGO-AgNPs could be a powerful synergistic agent with Cis or any other chemotherapeutic agent and successfully used for cervical cancer treatment [138].

Chemotherapeutic treatment of cancer has now become limited because of its inability to target specific cancer tissues and its negative impact on other healthy tissues and organs. Saifullah et al. have prepared a nanocomposite GO-PEG-PCA-FA (GO-polyethylene glycol-protocatechuic acid-folic acid) delivery system with PCA and GO-PEG nanocarrier coated with FA for specifically targeting cancer cells. The designed GO-PEG-PCA-FA delivery system was shown to be less toxic toward normal fibroblast 3T3 cells but showed significant efficiency for anticancer activity as compared with free drug PCA against ovarian cancer. Cytotoxic and apoptotic studies were carried out by treating free drug PCA, GO-PEG, GO-PEG-PCA, and GO-PEG-PCA-FA with normal fibroblast (3T3) cells. Different concentrations of the prepared delivery system were incubated with 3T3 cells for a maximum of 72 h. Standard MTT (3-[4,5-dimethylthiazol-2-yl]-2,5-diphenyl tetrazolium bromide) assay protocol was used for determining cell viability. All the prepared delivery systems, including free drug PCA, were observed to be biocompatible and nontoxic and the cell viability achieved was more than 80% (after 72 h incubation period). The study confirmed that the designed nanocomposite anticancer delivery system is nontoxic for normal cells but suitable for specifically targeting the cancer cells. The GO-PEG-PCA-FA anticancer delivery system was found to deliver high activity toward cancer cells rather than free drug and other anticancer delivery systems [139]. Small interfering RNA (siRNA) has been explored as a promising therapeutic agent for gene therapy. Du et al. modified GO with PEG, polyethylenimine (PEI), and (FA) for targeted delivery of siRNA that restricted ovarian cancer cell growth, and the efficacy of such a complex was evaluated by a series of in vitro experiments. Agarose-gel electrophoresis demonstrated that siRNA can be adsorbed onto the surface of PEG-GO-PEI-FA by electrostatic interaction. The surface morphology of the material was determined using an atomic force microscope, and the particle size and potential of the material were tested by dynamic light scattering. By atomic force microscopy, it was confirmed that GO bears a flaky shape and after attachment of PEG, PEI, and FA to GO, the particle size of PEG-GO-PEI-FA increased and it took a microsphere shape. Laser confocal microscopy demonstrated that siRNA-adsorbed PEG-GO-PEI-FA could target folate receptor—overexpressing ovarian cancer cells. Compared with PEG-GO-PEI/siRNA without folate modification, PEG-GO-PEI-FA/siRNA showed a more pronounced

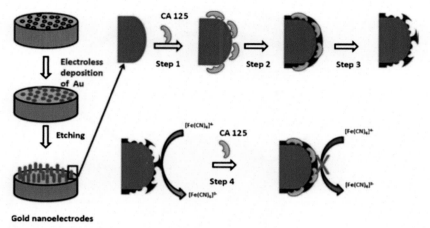

Gold nanoelectrodes

FIGURE 4.7

Molecular imprinted protein nanosensor fabrication and template protein detection. Step1, adsorption of cancer antigen 125 (CA 125) onto the nanoelectrode surface; step 2, electrochemical polymerization of phenol; step 3, template protein removal; and step 4, CA 125 binding and signal generation.

Reprinted with permission from S. Viswanathan, C. Rani, S. Ribeiro, C. Delerue Matos, Molecular imprinted nanoelectrodes for ultra-sensitive detection of ovarian cancer marker, Biosensors and Bioelectronics 33 (2012) 179.

inhibitory effect on the growth of ovarian cancer cells [140]. Vishwanathan et al. developed a sensor to detect epithelial ovarian cancer by protein-imprinted polymer on three-dimensional gold nanoelectrode ensemble (GNEE) that detects CA 125, a protein biomarker associated with ovarian cancer (Fig. 4.7). During therapy, CA 125 was used to follow up women patients before or after treatment for epithelial ovarian cancer. CV, DPV, and EIS techniques were used to fabricate and analyze CA 125−imprinted GNEE. Immunospecific capture of CA 125 molecules was done on the surfaces of very thin, protein-imprinted sites on GNEE and the captured molecules were detected as a reduction in the faradic current from the redox marker. The developed sensor showed good sensitivity in the concentration range of 0.5−400 U mL^{-1} and the detection limit was found to be 0.5 U mL^{-1}. Spiked and unknown real human blood serum was analyzed and the presence of undesirable proteins in the serum did not significantly affect the sensitivity. Hence, molecular imprinting with nanomaterials provides an alternative approach to trace the detection of ovarian cancer biomarker proteins [141].

4.3.4 Graphene nanostructure in microbial applications

Graphene-based nanocomposites are the promising material for the development of antimicrobial surfaces because of their superior biocompatibility, antibacterial properties, and tolerable cytotoxic effects on mammalian cells. Antibacterial

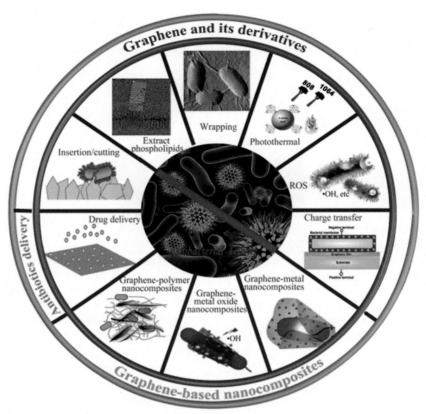

FIGURE 4.8

The antibacterial applications of graphene-based nanomaterials. *ROS*, reactive oxygen species.

Reprinted with permission from H. Ji, H. Sun, X. Qu, Antibacterial applications of graphene-based nano-materials: recent 2 achievements and challenges, Advanced Drug Delivery Reviews 105 (Part B) (2016) 176–189.

nanomaterials attached to the graphene substrate are more stable and well dispersed. Ji et al. [142] had discussed and highlighted the various antibacterial applications of graphene-based nanocomposite in their review. They have illustrated the antibacterial mechanism of graphene and the various recent advances of antibacterial activity of graphene till 2016 (Fig. 4.8]).

Graphene exerts its antibacterial actions via physical damage and chemical damage. Physical damage includes direct contact of the sharp edges of graphene with the bacterial membrane. Chen et al. [143] reported that the growth of *Escherichia coli* can be effectively inhibited by direct contact with graphene nanomaterials. Other important phenomena of physical damage are the destructive extraction of lipid molecules [143] and trapping of microorganisms by graphene nanosheets [144], in which the bacteria are trapped in graphene sheets. Akhavan et al. [145] reported that the trapped bacteria are inactive because they are isolated from their environment and cannot proliferate. The damage of bacteria by chemical ways is caused by oxidative stress with the generation of reactive oxygen species and charge transfer. Graphene nanocomposites were used as a support to disperse and stabilize various nanomaterials, metal oxides, polymer, and metals with high antibacterial efficiency due to the synergistic effect. The limitations of graphene nanocomposites, due to their poor solubility and processability, could be overcome by incorporating graphene into the polymer matrix. Lu et al. [146] have used chitosan to fabricate antibacterial hybrids to decrease bacterial cell viability.

Furthermore, limitations of curly GO sheets were resolved by using expanded single-sheet GO and keeping the surface functional. In one study, Kim et al. [147] fabricated continuous uniform poly(vinyl alcohol) (PVA)-chitosan-GO nanofibers by using an electrospinning technique and combining electron-beam irradiation. The nanofibers prepared by this method displayed excellent antibacterial activity against *E. coli*, indicating their potential in biomedical applications.

4.3.5 Graphene nanostructure applications in tissue engineering

With more fascinating and exceptional properties of graphene materials, such as exceptional mechanical and electric properties, revealed and incorporated into bioengineering applications, more works had been done on the applications of graphene in tissue engineering [148,149]. Tissue engineering is an interdisciplinary science that strives to manifest biological substitutes to maintain, restore, or improve function of a tissue or whole organ.

A number of studies have been reported that discussed the applications of graphene for bone [150], cardiac [151], neural [152], cartilage [153,154], musculoskeletal tissue engineering, etc. Ku and Park [155] in their study had explored the application of graphene in musculoskeletal tissue engineering using mouse myoblast C2C12 cell lines. They have shown that GO, compared with RGO, had a higher myotube fusion/maturation index and better upregulated the expression of myogenic genes viz. MyoD, myogenin, troponin T, and myosin heavy chain.

Shin et al. [156] have elaborately discussed various applications of graphene-based materials for tissue engineering in their review. Furthermore, Goenka et al. [157] also mentioned the applications of graphene-based nanomaterials in drug delivery and tissue engineering.

Graphene-based materials showed great potential to be applied in the field of bone tissue engineering. Díez-Pascual and Díez-Vicente [158] had developed a poly(propylene fumarate)/PEG-modified GO (PPF/PEG-GO) nanocomposite having enough strength and stiffness to provide effective support for bone tissue formation. Furthermore, the antibacterial activity of the nanocomposite was also investigated against gram-positive *Staphylococcus aureus* and *Staphylococcus epidermidis* and gram-negative *Pseudomonas aeruginosa* and *E. coli* microorganisms.

The antibacterial activity rose sharply upon increasing the GO concentration; the biocide effect was shown to be stronger in gram-positive bacteria. It is also interesting to know from cell viability data that PPF/PEG-GO composites do not induce toxicity over human dermal fibroblasts.

In another study [159], GO was incorporated into PVA, which shows improved mechanical properties of the nanocomposite. Here selective laser sintering was used to fabricate nanocomposite scaffolds with an interconnected porous structure. The highest improvements in the mechanical properties, i.e., 60%, 152%, and 69% in improvement of compressive strength, are obtained. The study shows that GO/PVA nanocomposite scaffolds are good candidates for bone tissue engineering.

4.4 Conclusion

In this chapter, we have discussed various studies describing GBN for biomedical applications. Special attention has been given to applications on cancer detection because of cancer's importance in the society. The chapter also emphasized that incorporation of graphene-based nanomaterials in biosensor technologies has shown great promise due to their high surface area, electric conductivity, electron transfer rate, and capacity to immobilize a variety of different biomolecules. The development of biosensors that are sensitive, stable, and specific to their target molecule and that can be processed rapidly is promising for their use in the biomedical field. We have also discussed many biosensors that have changed the world of breast cancer diagnosis offering rapid, simple, and cost-effective routes. The chapter also discussed about the fascinating and exceptional properties of graphene materials and their bioengineering applications, especially microbial and tissue engineering. Even though graphene is an excellent electrode material for sensing applications in the biomedical field, novel methods for well-controlled synthesis and processing of graphene need more attention and should be investigated in the future. Many more studies need to be conducted to examine the safety and reliability of the sensors/biosensors and to achieve uniform and reliable results.

References

[1] T. Liu, K. Yu, L. Gao, H. Chen, N. Wang, L. Hao, T. Li, H. He, Z. Guo, A graphene quantum dot decorated $SrRuO_3$ mesoporous film as an efficient counter electrode for high-performance dye-sensitized solar cells, Journal of Materials Chemistry A 5 (34) (2017) 17848—17855.

[2] Z. Sun, L. Zhang, F. Dang, Y. Liu, Z. Fei, Q. Shao, H. Lin, J. Guo, L. Xiang, N. Yerra, Experimental and simulation-based understanding of morphology-controlled barium titanate nanoparticles under co-adsorption of surfactants, CrystEngComm 19 (24) (2017) 3288—3298.

[3] J. Zhao, S. Ge, D. Pan, Y. Pan, V. Murugadoss, R. Li, W. Xi, Y. Lu, T. Wu, E.K. Wujcik, Q. Shao, X. Mai, Z. Guo, Microwave hydrothermal synthesis of In_2O_3-ZnO

nanocomposites and their enhanced photoelectrochemical properties, Journal of the Electrochemical Society 166 (5) (2019) H3074–H3083.

[4] Y. Li, B. Zhou, G. Zheng, X. Liu, T. Li, C. Yan, Cheng, K. Dai, C. Liu, C. Shen, Z. Guo, Continuously prepared highly conductive and stretchable SWNT/MWNT synergistically composited electrospun thermoplastic polyurethane yarns for wearable sensing, Journal of Materials Chemistry C 6 (9) (2018) 2258–2269.

[5] Y. Feng, S. Ge, J. Li, S. Li, H. Zhang, Y. Chen, Z. Guo, Synthesis of 3, 4, 5-trihydroxy-2-[(hydroxyimino) methyl] benzoic acid as a novel rust converter, Green Chemistry Letters and Reviews 10 (4) (2017) 455–461.

[6] G.U. Flechsig, J. Peter, G. Hartwich, J. Wang, P. Gründler, DNA hybridization detection at heated electrodes, Langmuir 21 (17) (2005) 7848–7853.

[7] J.J. Gooding, R. Wibowo, J. Liu, W. Yang, D. Losic, S. Orbons, F.J. Mearns, J.G. Shapter, D.B. Hibbert, Protein electrochemistry using aligned carbon nanotube arrays, Journal of the American Chemical Society 125 (30) (2003) 9006–9007.

[8] S. Grinberg, C. Linder, V. Kolot, T. Waner, Z. Wiesman, E. Shaubi, E. Heldman, Novel cationic amphiphilic derivatives from vernonia oil: synthesis and self-aggregation into bilayer vesicles, nanoparticles, and DNA complexants, Langmuir 21 (17) (2005) 7638–7645.

[9] W. Huang, S. Taylor, K. Fu, Y. Lin, D. Zhang, T.W. Hanks, A.M. Rao, Y.P. Sun, Attaching proteins to carbon nanotubes via diimide-activated amidation, Nano Letters 2 (4) (2002) 311–314.

[10] C.R. Kagan, C.B. Murray, M. Nirmal, M.G. Bawendi, Electronic energy transfer in CdSe quantum dot solids, Physical Review Letters 76 (9) (1996) 1517.

[11] S. Krishnamoorthy, T. Bei, E. Zoumakis, G.P. Chrousos, A.A. Iliadis, Morphological and binding properties of interleukin-6 on thin ZnO films grown on (1 0 0) silicon substrates for biosensor applications, Biosensors and Bioelectronics 22 (5) (2006) 707–714.

[12] H. Nakao, H. Hayashi, F. Iwata, H. Karasawa, K. Hirano, S. Sugiyama, T. Ohtani, Fabricating and aligning π-conjugated polymer-functionalized DNA nanowires: atomic force microscopic and scanning near-field optical microscopic studies, Langmuir 21 (17) (2005) 7945–7950.

[13] S.G. Wang, R. Wang, P.J. Sellin, Q. Zhang, DNA biosensors based on self-assembled carbon nanotubes, Biochemical and Biophysical Research Communications 325 (4) (2004) 1433–1437.

[14] T.K. Jain, M.A. Morales, S.K. Sahoo, D.L. Leslie-Pelecky, V. Labhasetwar, Iron oxide nanoparticles for sustained delivery of anticancer agents, Molecular Pharmaceutics 2 (3) (2005) 194–205.

[15] W.T. Liu, Nanoparticles and their biological and environmental applications, Journal of Bioscience and Bioengineering 102 (1) (2006) 1–7.

[16] J.H. Jung, D.S. Cheon, F. Liu, K.B. Lee, T.S. Seo, A graphene oxide based immunobiosensor for pathogen detection, Angewandte Chemie International Edition 49 (33) (2010) 5708–5711.

[17] E. Rand, A. Periyakaruppan, Z. Tanaka, D.A. Zhang, M.P. Marsh, R.J. Andrews, K.H. Lee, B. Chen, M. Meyyappan, J.E. Koehne, A carbon nanofiber based biosensor for simultaneous detection of dopamine and serotonin in the presence of ascorbic acid, Biosensors and Bioelectronics 42 (2013) 434–438.

[18] A. Walcarius, Electrocatalysis, sensors and biosensors in analytical chemistry based on ordered mesoporous and macroporous carbon-modified electrodes, TrAC Trends in Analytical Chemistry 38 (2012) 79−97.

[19] N. Yang, X. Chen, T. Ren, P. Zhang, D. Yang, Carbon nanotube based biosensors, Sensors and Actuators B: Chemical 207 (2015) 690−715.

[20] K. Radhapyari, P. Kotoky, M.R. Das, R. Khan, Graphene polyaniline nanocomposite-based biosensor for detection of antimalarial drug artesunate in pharmaceutical formulation and biological fluids, Talanta 111 (2013) 47−53.

[21] M. Pumera, Graphene-based nanomaterials and their electrochemistry, Chemical Society Reviews 39 (11) (2010) 4146−4157.

[22] S. Sabury, S.H. Kazemi, F. Sharif, Graphene−gold nanoparticle composite: application as a good scaffold for construction of glucose oxidase biosensor, *Materials Science and Engineering*: C 49 (2015) 297−304.

[23] https://www.cdc.gov/women/lcod/2015/index.htm dated 1.02.2019.

[24] F. Bray, A. Jemal, N. Grey, J. Ferlay, D. Forman, Global cancer transitions according to the human development index (2008−2030): a population-based study, The Lancet Oncology 13 (8) (2012) 790−801.

[25] R.L. Siegel, K.D. Miller, A. Jemal, Cancer statistics, 2016, CA: A Cancer Journal for Clinicians 66 (1) (2016) 7−30.

[26] Division of Cancer Prevention and Control, Centers for Disease Control and Prevention. Breast Cancer Statistics. https://www.cdc.gov/cancer/breast/statistics/index.htm.

[27] [American Cancer Society. Breast Cancer Facts and Figures 2015e2016. https://www.cancer.org/content/dam/cancer-org/research/cancer-facts-andstatistics/breastcancer-facts-and-figures/breast-cancer-facts-and-figures2015-2016.pdf.

[28] A. Dogra, D.C. Doval, M. Sardana, S.K. Chedi, A. Mehta, Clinicopathological characteristics of triple negative breast cancer at a tertiary care hospital in India, Asian Pacific Journal of Cancer Prevention 15 (24) (2014) 10577−10583.

[29] P. Shetty, India faces growing breast cancer epidemic, The Lancet 379 (9820) (2012) 992−993.

[30] V.L. Gaopande, S.S. Joshi, M.M. Kulkarni, S.S. Dwivedi, A clinicopathologic study of triple negative breast cancer, Journal of the Scientific Society 42 (1) (2015) 12.

[31] P. Suresh, U. Batra, D.C. Doval, Epidemiological and clinical profile of triple negative breast cancer at a cancer hospital in North India, Indian Journal of Medical and Paediatric Oncology: Official Journal of Indian Society of Medical and Paediatric Oncology 34 (2) (2013) 89.

[32] A.A. Thike, J. Iqbal, P.Y. Cheok, A.P.Y. Chong, G.M.K. Tse, B. Tan, N.S. Wong, P.H. Tan, Triple negative breast cancer: outcome correlation with immunohistochemical detection of basal markers, The American Journal of Surgical Pathology 34 (7) (2010) 956−964.

[33] S. Sen, R. Gayen, S. Das, S. Maitra, A. Jha, M. Mahanta, A clinical and pathological study of triple negative breast carcinoma: experience of a tertiary care centre in eastern India, Journal of the Indian Medical Association 110 (10) (2012) 686−689.

[34] S. Badve, D.J. Dabbs, S.J. Schnitt, F.L. Baehner, T. Decker, V. Eusebi, S.B. Fox, S. Ichihara, J. Jacquemier, S.R. Lakhani, J. Palacios, E.A. Rakha, A.L. Richardson, F.C. Schmitt, P.H. Tan, G.M. Tse, B. Weigelt, I.O. Ellis, J.S. Reis-Filho, Basal-like and triple-negative breast cancers: a critical review with an emphasis on the implications for pathologists and oncologists, Modern Pathology 24 (2) (2011) 157.

[35] V.S. Jamdade, N. Sethi, N.A. Mundhe, P. Kumar, M. Lahkar, N. Sinha, Therapeutic targets of triple-negative breast cancer: a review, British Journal of Pharmacology 172 (17) (2015) 4228–4237.

[36] A.M. Brewster, M. Chavez-MacGregor, P. Brown, Epidemiology, biology, and treatment of triple-negative breast cancer in women of African ancestry, The Lancet Oncology 15 (13) (2014) e625–e634.

[37] G.J. Morris, S. Naidu, A.K. Topham, F. Guiles, Y. Xu, P. McCue, G.F. Schwartz, P.K. Park, A.L. Rosenberg, K. Brill, E.P. Mitchell, Differences in breast carcinoma characteristics in newly diagnosed African–American and Caucasian patients: a single-institution compilation compared with the National Cancer Institute's Surveillance, Epidemiology, and end results database, Cancer: Interdisciplinary International Journal of the American Cancer Society 110 (4) (2007) 876–884.

[38] M. Singh, Y. Ding, L.Y. Zhang, D. Song, Y. Gong, S. Adams, D.S. Ross, J.H. Wang, S. Grover, D.C. Doval, C. Shao, Z.L. He, V. Chang, W.W. Chin, F.M. Deng, B. Singh, D. Zhang, R.L. Xu, P. Lee, Distinct breast cancer subtypes in women with early-onset disease across races, American Journal of Cancer Research 4 (4) (2014) 337.

[39] J.G. Elmore, D.L. Miglioretti, L.M. Reisch, M.B. Barton, W. Kreuter, C.L. Christiansen, S.W. Fletcher, Screening mammograms by community radiologists: variability in false-positive rates, Journal of the National Cancer Institute 94 (18) (2002) 1373–1380.

[40] S. Taplin, L. Abraham, W.E. Barlow, J.J. Fenton, E.A. Berns, P.A. Carney, G.R. Cutter, E.A. Sickles, D. Carl, J.G. Elmore, Mammography facility characteristics associated with interpretive accuracy of screening mammography, Journal of the National Cancer Institute 100 (12) (2008) 876–887.

[41] M. Pollán, Epidemiology of breast cancer in young women, Breast Cancer Research and Treatment 123 (1) (2010) 3–6.

[42] F. Yin, B. Gu, Y. Lin, N. Panwar, S.C. Tjin, J. Qu, S.P. Lau, K.,T. Yong, Functionalized 2D nanomaterials for gene delivery applications, Coordination Chemistry Reviews 347 (2017) 77–97.

[43] G. Song, L. Cheng, Y. Chao, K. Yang, Z. Liu, Emerging nanotechnology and advanced materials for cancer radiation therapy, Advanced Materials 29 (32) (2017) 1700996.

[44] T.A. Tabish, S. Zhang, P.G. Winyard, Developing the next generation of graphene-based platforms for cancer therapeutics: the potential role of reactive oxygen species, Redox Biology 15 (2018) 34–40.

[45] K.S. Novoselov, A.K. Geim, S.V. Morozov, D. Jiang, Y. Zhang, S.V. Dubonos, I.V. Grigorieva, A.A. Firsov, Electric field effect in atomically thin carbon films, Science 306 (5696) (2004) 666–669.

[46] D.R. Dreyer, S. Park, C.W. Bielawski, R.S. Ruoff, The chemistry of graphene oxide, Chemical Society Reviews 39 (1) (2010) 228–240.

[47] P.R. Wallace, The band theory of graphite, Physical Review 71 (9) (1947) 622.

[48] H.C. Schniepp, J.L. Li, M.J. McAllister, H. Sai, M. Herrera-Alonso, D.H. Adamson, R.K. Prud'homme, R. Car, D.A. Saville, I.A. Aksay, Functionalized single graphene sheets derived from splitting graphite oxide, The Journal of Physical Chemistry B 110 (17) (2006) 8535–8539.

[49] T. Kuila, S. Bose, A.K. Mishra, P. Khanra, N.H. Kim, J.H. Lee, Chemical functionalization of graphene and its applications, Progress in Materials Science 57 (7) (2012) 1061–1105.

[50] V. Georgakilas, J.N. Tiwari, K.C. Kemp, J.A. Perman, A.B. Bourlinos, K.S. Kim, R. Zboril, Noncovalent functionalization of graphene and graphene oxide for energy materials, biosensing, catalytic, and biomedical applications, Chemical Reviews 116 (9) (2016) 5464–5519.

[51] G. Reina, J.M. González-Domínguez, A. Criado, E. Vázquez, A. Bianco, M. Prato, Promises, facts and challenges for graphene in biomedical applications, Chemical Society Reviews 46 (15) (2017) 4400–4416.

[52] A. Bachmatiuk, R.G. Mendes, C. Hirsch, C. Jähne, M.R. Lohe, J. Grothe, S. Kaskel, L. Fu, R. Klingeler, J. Eckert, P. Wick, M.H. Rümmeli, Few-layer graphene shells and nonmagnetic encapsulates: a versatile and nontoxic carbon nanomaterial, ACS Nano 7 (12) (2013) 10552–10562.

[53] Z. Chen, W. Ren, L. Gao, B. Liu, S. Pei, H.M. Cheng, Three-dimensional flexible and conductive interconnected graphene networks grown by chemical vapour deposition, Nature Materials 10 (6) (2011) 424–428.

[54] K. Kim, T. Lee, Y. Kwon, Y. Seo, J. Song, J.K. Park, H. Lee, J.,Y. Park, H. Ihee, S.J. Cho, R. Ryoo, Lanthanum-catalysed synthesis of microporous 3D graphene-like carbons in a zeolite template, Nature 535 (7610) (2016) 131.

[55] X. Guan, J. Nai, Y. Zhang, P. Wang, J. Yang, L. Zheng, J. Zhang, L. Guo, CoO hollow cube/reduced graphene oxide composites with enhanced lithium storage capability, Chemistry of Materials 26 (20) (2014) 5958–5964.

[56] A. Jana, E. Scheer, S. Polarz, Synthesis of graphene–transition metal oxide hybrid nanoparticles and their application in various fields, Beilstein Journal of Nanotechnology 8 (1) (2017) 688–714.

[57] Q. Ying, H. Feng, Y. Chenggong, L. Xuewu, L. Dong, M. Long, Z. Qiuqiong, L. Jiahui, W. Jingde, Advancements of graphene-based nanomaterials in biomedicine, Materials Science and Engineering: C 90 (2018) 764–780.

[58] M. Azimzadeh, M. Rahaie, N. Nasirizadeh, K. Ashtari, H. Naderi-Manesh, An electrochemicalnanobiosensor for plasma miRNA-155, based on graphene oxide and gold nanorod, for early detection of breast cancer, Biosensors and Bioelectronics 77 (2016) 99–106.

[59] T. Hu, L. Zhang, W. Wen, X. Zhang, S. Wang, Enzyme catalytic amplification of miRNA-155 detection with graphene quantum dot-based electrochemical biosensor, Biosensors and Bioelectronics 77 (2016) 451–456.

[60] J. Lin, Y. Huang, P. Huang, Graphene-based nanomaterials in bioimaging, in: Biomedical Applications of Functionalized Nanomaterials, Elsevier, 2018, pp. 247–287.

[61] M.B. Lundeberg, Y. Gao, A. Woessner, C. Tan, P. Alonso-González, K. Watanabe, T. Taniguchi, J. Hone, R. Hillenbrand, F.H. Koppens, Thermoelectric detection and imaging of propagating graphene plasmons, Nature Materials 16 (2) (2017) 204.

[62] W. Guo, X. Zhang, X. Yu, S. Wang, J. Qiu, W. Tang, L. Li, H. Liu, Z.L. Wang, Self-powered electrical stimulation for enhancing neural differentiation of mesenchymal stem cells on graphene–poly (3, 4-ethylenedioxythiophene) hybrid microfibers, ACS Nano 10 (5) (2016) 5086–5095.

[63] S. Mittal, H. Kaur, N. Gautam, A.K. Mantha, Biosensors for breast cancer diagnosis: a review of bioreceptors, biotransducers and signal amplification strategies, Biosensors and Bioelectronics 88 (2017) 217–231.

[64] W.A. Berg, Z. Zhang, D. Lehrer, R.A. Jong, E.D. Pisano, R.G. Barr, M.J. Morton, Detection of breast cancer with addition of annual screening ultrasound or a single

screening MRI to mammography in women with elevated breast cancer risk, JAMA 307 (13) (2012) 1394—1404.

[65] N.F. Boyd, H. Guo, L.J. Martin, L. Sun, J. Stone, E. Fishell, R.A. Jong, G.H. Chiarelli, S. Minkin, M.J. Yaffe, Mammographic density and the risk and detection of breast cancer, New England Journal of Medicine 356 (3) (2007) 227—236.

[66] T.M. Kolb, J. Lichy, J.H. Newhouse, Comparison of the performance of screening mammography, physical examination, and breast US and evaluation of factors that influence them: an analysis of 27,825 patient evaluations, Radiology 225 (1) (2002) 165—175.

[67] J.R. Scheel, J.M. Lee, B.L. Sprague, C.I. Lee, C.D. Lehman, Screening ultrasound as an adjunct to mammography in women with mammographically dense breasts, American Journal of Obstetrics and Gynecology 212 (1) (2015) 9—17.

[68] W. Buchberger, A. Niehoff, P. Obrist, P. DeKoekkoek-Doll, M. Dünser, Clinically and mammographically occult breast lesions: detection and classification with high-resolution sonography, in: Seminars in Ultrasound, CT and MRI, vol. 21 (4), WB Saunders, 2000, pp. 325—336.

[69] I. Leconte, C. Feger, C. Galant, M. Berlière, B.V. Berg, W. D'Hoore, B. Maldague, Mammography and subsequent whole-breast sonography of nonpalpable breast cancers: the importance of radiologic breast density, American Journal of Roentgenology 180 (6) (2003) 1675—1679.

[70] W.A. Berg, J.D. Blume, J.B. Cormack, E.B. Mendelson, D. Lehrer, M. Böhm-Vélez, E.D. Pisano, R.A. Jong, W.P. Evans, M.J. Morton, M.C. Mahoney, Combined screening with ultrasound and mammography vs mammography alone in women at elevated risk of breast cancer, JAMA 299 (18) (2008) 2151—2163.

[71] S.V. Bachawal, K.C. Jensen, K.E. Wilson, L. Tian, A.M. Lutz, J.K. Willmann, Breast cancer detection by B7-H3—targeted ultrasound molecular imaging, Cancer Research 75 (12) (2015) 2501—2509.

[72] H.D. Cheng, X.J. Shi, R. Min, L.M. Hu, X.P. Cai, H.N. Du, Approaches for automated detection and classification of masses in mammograms, Pattern Recognition 39 (4) (2006) 646—668.

[73] A.S. Kurani, D.H. Xu, J. Furst, D.S. Raicu, Raicu. Co-occurrence matrices for volumetric data, in: 7th IASTED International Conference on Computer Graphics and Imaging, Kauai, 2004.

[74] R.M. Haralick, K. Shanmugam, Textural features for image classification, IEEE Transactions on systems, man, and cybernetics (6) (1973) 610—621.

[75] F. Albregtsen, Statistical Texture Measures Computed from Gray Level Coocurrence Matrices, in: Image *Processing Laboratory*, vol. 5, Department of Informatics, University of Oslo, 2008.

[76] S. Beura, B. Majhi, R. Dash, Mammogram classification using two-dimensional discrete wavelet transform and gray-level co-occurrence matrix for detection of breast cancer, Neurocomputing 154 (2015) 1—14.

[77] B.L. Liu, M.A. Saltman, Immunosensor technology: historical perspective and future outlook, Laboratory Medicine 27 (2) (1996) 109—115.

[78] M. Santandreu, S. Alegret, E. Fabregas, Determination of β-HCG using amperometric immunosensors based on a conducting immunocomposite, Analytica Chimica Acta 396 (2—3) (1999) 181—188.

[79] B. Zhang, Q. Mao, X. Zhang, T. Jiang, M. Chen, F. Yu, W. Fu, A novel piezoelectric quartz micro-array immunosensor based on self-assembled monolayer for

determination of human chorionic gonadotropin, Biosensors and Bioelectronics 19 (7) (2004) 711−720.

[80] N. Nakamura, T.K. Lim, J.M. Jeong, T. Matsunaga, Flow immunoassay for detection of human chorionic gonadotrophin using a cation exchange resin packed capillary column, Analytica Chimica Acta 439 (1) (2001) 125−130.

[81] Y. Li, Y. Zhang, F. Li, M. Li, L. Chen, Y. Dong, Q. Wei, Sandwich-type amperometric immunosensor using functionalized magnetic graphene loaded gold and silver core-shell nanocomposites for the detection of carcinoembryonic antigen, Journal of Electroanalytical Chemistry 795 (2017) 1−9.

[82] L. Zhu, L. Xu, N. Jia, B. Huang, L. Tan, S. Yang, S. Yao, Electrochemical immunoassay for carcinoembryonic antigen using gold nanoparticle−graphene composite modified glassy carbon electrode, Talanta 116 (2013) 809−815.

[83] T. Feng, X. Chen, X. Qiao, Z. Sun, H. Wang, Y. Qi, C. Hong, Graphene oxide supported rhombic dodecahedral Cu_2O nanocrystals for the detection of carcinoembryonic antigen, Analytical Biochemistry 494 (2016) 101−107.

[84] Y.S. Yu, Gao, X.F. Zhu, J.K. Xu, L.M. Lu, W.M. Wang, T.T. Yang, H.K. Xing, Y.F. Yu, Label-free electrochemical immunosensor based on Nile blue A-reduced graphene oxide nanocomposites for carcinoembryonic antigen detection, Analytical Biochemistry 500 (2016) 80−87.

[85] D. Lin, J. Wu, H. Ju, F. Yan, Nanogold/mesoporous carbon foam-mediated silver enhancement for graphene-enhanced electrochemical immunosensing of carcinoembryonic antigen, Biosensors and Bioelectronics 52 (2014) 153−158.

[86] Y. Wang, S. Wang, C. Lu, X. Yang, Three kinds of DNA-directed nanoclusters cooperating with graphene oxide for assaying mucin 1, carcinoembryonic antigen and cancer antigen 125, Sensors and Actuators B: Chemical 262 (2018) 9−16.

[87] M. Hasanzadeh, S. Tagi, E. Solhi, A. Mokhtarzadeh, N. Shadjou, A. Eftekhari, S. Mahboob, An innovative immunosensor for ultrasensitive detection of breast cancer specific carbohydrate (CA 15-3) in unprocessed human plasma and MCF-7 breast cancer cell lysates using gold nanospear electrochemically assembled onto thiolated graphene quantum dots, International Journal of Biological Macromolecules 114 (2018) 1008−1017.

[88] J. Wu, Z. Fu, F. Yan, H. Ju, Biomedical and clinical applications of immunoassays and immunosensors for tumour markers, TrAC Trends in Analytical Chemistry 26 (7) (2007) 679−688.

[89] K. Wang, M.Q. He, F.H. Zhai, R.H. He, Y.L. Yu, A novel electrochemical biosensor based on polyadenine modified aptamer for label-free and ultrasensitive detection of human breast cancer cells, Talanta 166 (2017) 87−92.

[90] S. Asaga, C. Kuo, T. Nguyen, M. Terpenning, A.E. Giuliano, D.S. Hoon, Direct serum assay for microRNA-21 concentrations in early and advanced breast cancer, Clinical Chemistry 57 (1) (2011) 84−91.

[91] E.J. Jung, L. Santarpia, J. Kim, F.J. Esteva, E. Moretti, A. Buzdar, D. Leo, X.F. Le, R.C. Bast Jr., S.T. Parl, L. Pusztai, Plasma microRNA 210 levels correlate with sensitivity to trastuzumab and tumor presence in breast cancer patients, Cancer 118 (10) (2012) 2603−2614.

[92] E. van Schooneveld, M.C. Wouters, I. Van der Auwera, D.J. Peeters, H. Wildiers, P.A. Van Dam, I. Vergote, P.B. Vermeulen, L.Y. Dirix, S.J. Van Laere, Expression profiling of cancerous and normal breast tissues identifies microRNAs that are

differentially expressed in serum from patients with (metastatic) breast cancer and healthy volunteers, Breast Cancer Research 14 (1) (2012) R34.

[93] I. Van der Auwera, W. Yu, L. Suo, L. Van Neste, P. Van Dam, E.A. Van Marck, P. Pauwels, P.B. Vermeulen, L.Y. Dirix, S.J. Van Laere, Array-based DNA methylation profiling for breast cancer subtype discrimination, PLoS One 5 (9) (2010) e12616.

[94] H. Yin, Y. Zhou, C. Chen, L. Zhu, S. Ai, An electrochemical signal 'off—on'sensing platform for microRNA detection, The Analyst 137 (6) (2012) 1389—1395.

[95] A. Ebrahimi, I. Nikokar, M. Zokaei, E. Bozorgzadeh, Design, development and evaluation of microRNA-199a-5p detecting electrochemical nanobiosensor with diagnostic application in Triple Negative Breast Cancer, Talanta 189 (2018) 592—598.

[96] L. Tian, J. Qi, K. Qian, O. Oderinde, Q. Liu, C. Yao, W. Song, Y. Wang, Copper (II) oxide nanozyme based electrochemical cytosensor for high sensitive detection of circulating tumor cells in breast cancer, Journal of Electroanalytical Chemistry 812 (2018) 1—9.

[97] M.A. Tabrizi, M. Shamsipur, R. Saber, S. Sarkar, N. Zolfaghari, An ultrasensitive sandwich-type electrochemical immunosensor for the determination of SKBR-3 breast cancer cell using RGO-TPA/FeHCF nanolabeled Anti-HCT as a signal tag, Sensors and Actuators B: Chemical 243 (2017) 823—830.

[98] Y. Tao, D.T. Auguste, Array-based identification of triple-negative breast cancer cells using fluorescent nanodot-graphene oxide complexes, Biosensors and Bioelectronics 81 (2016) 431—437.

[99] W. Dong, Y. Ren, Z. Bai, Y. Yang, Z. Wang, C. Zhang, Q. Chen, Trimetallic AuPtPd nanocomposites platform on graphene: applied to electrochemical detection and breast cancer diagnosis, Talanta 189 (2018) 79—85.

[100] Z. Zhong, W. Wu, D. Wang, D. Wang, J. Shan, Y. Qing, Z. Zhang, Nanogold-enwrapped graphene nanocomposites as trace labels for sensitivity enhancement of electrochemical immunosensors in clinical immunoassays: carcinoembryonic antigen as a model, Biosensors and Bioelectronics 25 (10) (2010) 2379—2383.

[101] F.Y. Kong, M.T. Xu, J.J. Xu, H.Y. Chen, A novel lable-free electrochemical immunosensor for carcinoembryonic antigen based on gold nanoparticles—thionine—reduced graphene oxide nanocomposite film modified glassy carbon electrode, Talanta 85 (5) (2011) 2620—2625.

[102] M. Yan, G. Sun, F. Liu, J. Lu, J. Yu, X. Song, An aptasensor for sensitive detection of human breast cancer cells by using porous GO/Au composites and porous PtFe alloy as effective sensing platform and signal amplification labels, Analytica Chimica Acta 798 (2013) 33—39.

[103] H. Li, J. He, S. Li, A.P. Turner, Electrochemical immunosensor with N-doped graphene-modified electrode for label-free detection of the breast cancer biomarker CA 15-3, Biosensors and Bioelectronics 43 (2013) 25—29.

[104] F. Liu, Y. Zhang, J. Yu, S. Wang, S. Ge, X. Song, Application of ZnO/graphene and S6 aptamers for sensitive photoelectrochemical detection of SK-BR-3 breast cancer cells based on a disposable indium tin oxide device, Biosensors and Bioelectronics 51 (2014) 413—420.

[105] A. Tiwari, S. Gong, Electrochemical detection of a breast cancer susceptible gene using cDNA immobilized chitosan-co-polyaniline electrode, Talanta 77 (3) (2009) 1217—1222.

[106] H. Fan, Y. Zhang, D. Wu, H. Ma, X. Li, Y. Li, H. Wang, H. Li, B. Du, Q. Wei, Construction of label-free electrochemical immunosensor on mesoporous carbon

nanospheres for breast cancer susceptibility gene, Analytica Chimica Acta 770 (2013) 62−67.

[107] A. Benvidi, A.D. Firouzabadi, M.D. Tezerjani, S.M. Moshtaghiun, M. Mazloum-Ardakani, A. Ansarin, A highly sensitive and selective electrochemical DNA biosensor to diagnose breast cancer, Journal of Electroanalytical Chemistry 750 (2015) 57−64.

[108] W. Wang, X. Fan, S. Xu, J.J. Davis, X. Luo, Low fouling label-free DNA sensor based on polyethylene glycols decorated with gold nanoparticles for the detection of breast cancer biomarkers, Biosensors and Bioelectronics 71 (2015) 51−56.

[109] A. Benvidi, M.D. Tezerjani, S. Jahanbani, M.M. Ardakani, S.M. Moshtaghioun, Comparison of impedimetric detection of DNA hybridization on the various biosensors based on modified glassy carbon electrodes with PANHS and nanomaterials of RGO and MWCNTs, Talanta 147 (2016) 621−627.

[110] P. Norouzi, V.K. Gupta, F. Faridbod, M. Pirali-Hamedani, B. Larijani, M.R. Ganjali, Carcinoembryonic antigen admittance biosensor based on Au and ZnO nanoparticles using FFT admittance voltammetry, Analytical Chemistry 83 (5) (2011) 1564−1570.

[111] A.L. Sun, G.R. Chen, Q.L. Sheng, J.B. Zheng, Sensitive label-free electrochemical immunoassay based on a redox matrix of gold nanoparticles/Azure I/multi-wall carbon nanotubes composite, Biochemical Engineering Journal 57 (2011) 1−6.

[112] A. Florea, Z. Taleat, C. Cristea, M. Mazloum-Ardakani, R. Săndulescu, Label free MUC1 aptasensors based on electrodeposition of gold nanoparticles on screen printed electrodes, Electrochemistry Communications 33 (2013) 127−130.

[113] X. Zhu, J. Yang, M. Liu, Y. Wu, Z. Shen, G. Li, Sensitive detection of human breast cancer cells based on aptamer−cell−aptamer sandwich architecture, Analytica Chimica Acta 764 (2013) 59−63.

[114] T. Li, Q. Fan, T. Liu, X. Zhu, J. Zhao, G. Li, Detection of breast cancer cells specially and accurately by an electrochemical method, Biosensors and Bioelectronics 25 (12) (2010) 2686−2689.

[115] W. Li, R. Yuan, Y. Chai, S. Chen, Reagent less amperometric cancer antigen 15-3 immunosensor based on enzyme-mediated direct electrochemistry, Biosensors and Bioelectronics 25 (11) (2010) 2548−2552.

[116] X. Wang, H. Yu, D. Lu, J. Zhang, W. Deng, Label free detection of the breast cancer biomarker CA15.3 using ZnO nanorods coated quartz crystal microbalance, Sensors and Actuators B: Chemical 195 (2014) 630−634.

[117] M.K. Sezgintürk, A new impedimetric biosensor utilizing vegf receptor-1 (flt-1): early diagnosis of vascular endothelial growth factor in breast cancer, Biosensors and Bioelectronics 26 (10) (2011) 4032−4039.

[118] S. Zhao, W. Yang, R.Y. Lai, A folding-based electrochemical aptasensor for detection of vascular endothelial growth factor in human whole blood, Biosensors and Bioelectronics 26 (5) (2011) 2442−2447.

[119] A. Vasudev, A. Kaushik, S. Bhansali, Electrochemical immunosensor for label free epidermal growth factor receptor (EGFR) detection, Biosensors and Bioelectronics 39 (1) (2013) 300−305.

[120] R. Elshafey, A.C. Tavares, M. Siaj, M. Zourob, Electrochemical impedance immunosensor based on gold nanoparticles−protein G for the detection of cancer marker epidermal growth factor receptor in human plasma and brain tissue, Biosensors and Bioelectronics 50 (2013) 143−149.

[121] B. Seven, M. Bourourou, K. Elouarzaki, J.K. Constant, C. Gondran, M. Holzinger, S. Cosnier, S. Timur, Impedimetric biosensor for cancer cell detection, Electrochemistry Communications 37 (2013) 36−39.

[122] R.C. Marques, S. Viswanathan, H.P. Nouws, C. Delerue-Matos, M.B. González-García, Electrochemical immunosensor for the analysis of the breast cancer biomarker HER2 ECD, Talanta 129 (2014) 594−599.

[123] Q.A.M. Al-Khafaji, M. Harris, S. Tombelli, S. Laschi, A.P.F. Turner, M. Mascini, G. Marrazza, An electrochemical immunoassay for HER2 detection, Electroanalysis 24 (4) (2012) 735−742.

[124] M.N. Sonuç, M.K. Sezgintürk, Ultrasensitive electrochemical detection of cancer associated biomarker HER3 based on anti-HER3 biosensor, Talanta 120 (2014) 355−361.

[125] X. Li, J. Xia, S. Zhang, Label-free detection of DNA hybridization based on poly (indole-5-carboxylic acid) conducting polymer, Analytica Chimica Acta 622 (1−2) (2008) 104−110.

[126] N. Zhu, Z. Chang, P. He, Y. Fang, Electrochemically fabricated polyaniline nanowire-modified electrode for voltammetric detection of DNA hybridization, Electrochimica Acta 51 (18) (2006) 3758−3762.

[127] M. Su, H. Liu, L. Ge, Y. Wang, S. Ge, J. Yu, M. Yan, Aptamer-Based electrochemiluminescent detection of MCF-7 cancer cells based on carbon quantum dots coated mesoporous silica nanoparticles, Electrochimica Acta 146 (2014) 262−269.

[128] S.K. Arya, K.Y. Wang, C.C. Wong, A.R.A. Rahman, Anti-EpCAM modified LC-SPDP monolayer on gold microelectrode based electrochemical biosensor for MCF-7 cells detection, Biosensors and Bioelectronics 41 (2013) 446−451.

[129] X. Hua, Z. Zhou, L. Yuan, S. Liu, Selective collection and detection of MCF-7 breast cancer cells using aptamer-functionalized magnetic beads and quantum dots based nano-bio-probes, Analytica Chimica Acta 788 (2013) 135−140.

[130] W. Wei, D.F. Li, X.H. Pan, S.Q. Liu, Electro chemiluminescent detection of Mucin 1 protein and MCF-7 cancer cells based on the resonance energy transfer, The Analyst 137 (9) (2012) 2101−2106.

[131] H.S. Choo, K.C. Lee, Preparation of poly [Styrene (ST)-co-allyloxy-2-hydroxypropane sulfonic acid sodium salt (COPS-I)] colloidal crystalline photonic crystals, Journal of Nanoscience and Nanotechnology 15 (10) (2015) 7685−7692.

[132] K. Wang, D. Fan, Y. Liu, E. Wang, Highly sensitive and specific colorimetric detection of cancer cells via dual-aptamer target binding strategy, Biosensors and Bioelectronics 73 (2015) 1−6.

[133] Y.S. Borghei, M. Hosseini, M. Dadmehr, S. Hosseinkhani, M.R. Ganjali, R. Sheikhnejad, Visual detection of cancer cells by colorimetric aptasensor based on aggregation of gold nanoparticles induced by DNA hybridization, Analytica Chimica Acta 904 (2016) 92−97.

[134] S. Kumar, S. Ashish, Kumar, S. Augustine, S. Yadav, B.K. Yadav, R.P. Chauhan, A.K. Dewan, B.D. Malhotra, Effect of Brownian motion on reduced agglomeration of nanostructured metal oxide towards development of efficient cancer biosensor, Biosensors and Bioelectronics 102 (2018) 247−255.

[135] J. Ferlay, G. Randi, C. Bosetti, F. Levi, E. Negri, P. Boyle, C. La Vecchia, Declining mortality from bladder cancer in Europe, BJU International 101 (1) (2008) 11−19.

[136] R. Siegel, D. Naishadham, A. Jemal, Cancer statistics, 2012, Ca: A Cancer Journal for Clinicians 62 (1) (2012) 10−29.

[137] Y.J. Choi, S. Gurunathan, J.H. Kim, Graphene oxide—silver nanocomposite enhances cytotoxic and apoptotic potential of salinomycin in human ovarian cancer stem cells (OvCSCs): a novel approach for cancer therapy, International Journal of Molecular Sciences 19 (3) (2018) 710.

[138] Y.G. Yuan, S. Gurunathan, Combination of graphene oxide—silver nanoparticle nanocomposites and cisplatin enhances apoptosis and autophagy in human cervical cancer cells, International Journal of Nanomedicine 12 (2017) 6537.

[139] B. Saifullah, K. Buskaran, R. Shaikh, F. Barahuie, S. Fakurazi, M. MohdMoklas, M. Hussein, Graphene oxide—PEG—protocatechuic acid nanocomposite formulation with improved anticancer properties, Nanomaterials 8 (10) (2018) 820.

[140] S. Du, Y. Wang, J. Ao, K. Wang, Z. Zhang, L. Yang, X. Liang, Targeted delivery of siRNA to ovarian cancer cells using functionalized graphene oxide, Nano Life 8 (1) (2018) 1850001.

[141] S. Viswanathan, C. Rani, S. Ribeiro, C. Delerue Matos, Molecular imprinted nanoelectrodes for ultra-sensitive detection of ovarian cancer marker, Biosensors and Bioelectronics 33 (2012) 179.

[142] H. Ji, H. Sun, X. Qu, Antibacterial applications of graphene-based nanomaterials: recent 2 achievements and challenges, Advanced Drug Delivery Reviews 105 (Part B) (2016) 176—189.

[143] J. Chen, H. Peng, X. Wang, F. Shao, Z. Yuan, H. Han, Graphene oxide exhibits broad spectrum antimicrobial activity against bacterial phytopathogens and fungal conidia by intertwining and membrane perturbation, Nanoscale 6 (810) (2014) 1879—1889.

[144] O. Akhavan, E. Ghaderi, A. Esfandiar, Wrapping bacteria by graphene nanosheets for isolation from environment, reactivation by sonication, and inactivation by near-infrared irradiation, The Journal of Physical Chemistry B 115 (2011) 6279—6288.

[145] B. Lu, T. Li, H. Zhao, X. Li, C. Gao, S. Zhang, E. Xie, Graphene-based composite materials beneficial to wound healing, Nanoscale 4 (2012) 2978—2982.

[146] Y. Liu, M. Park, H.K. Shin, B. Pant, J. Choi, Y.W. Park, J.Y. Lee, S.J. Park, H.Y. Kim, Facile preparation and characterization of poly(vinyl alcohol)/chitosan/graphene oxide biocomposite nanofibers, Journal of Industrial and Engineering Chemistry 20 (2014) 4415—4420.

[147] E. Murray, B.C. Thompson, S. Sayyar, G. Wallace, Enzymatic degradation of graphene/polycaprolactone materials for tissue engineering, Polymer Degradation and Stability 111 (2014) 71—77.

[148] E. Nishida, H. Miyaji, H. Takita, I. Kanayama, M. Tsuji, T. Akasaka, T. Sugaya, R. Sakagami, M. Kawanami, Graphene oxide coating facilitates the bioactivity of scaffold material for tissue engineering, Japanese Journal of Applied Physics 53 (2014) 06JD04.

[149] T.R. Nayak, H. Andersen, V.S. Makam, C. Khaw, S. Bae, X. Xu, P.-L.R. Ee, J.-H. Ahn, B.H. Hong, G. Pastorin, Graphene for controlled and accelerated osteogenic differentiation of human mesenchymal stem cells, ACS Nano 5 (2011) 4670—4678.

[150] J. Park, S. Park, S. Ryu, S.H. Bhang, J. Kim, J.K. Yoon, Y.H. Park, S.P. Cho, S. Lee, B.H. Hong, B.S. Kim, Graphene-regulated cardiomyogenic differentiation process of mesenchymal stem cells by enhancing the expression of extracellular matrix proteins and cell signalling molecules, Advanced Healthcare Materials 3 (2014) 176—181.

[151] J. Park, B. Kim, J. Han, J. Oh, S. Park, S. Ryu, S. Jung, J.-Y. Shin, B.S. Lee, B.H. Hong, D. Choi, B.-S. Kim, Graphene oxide flakes as a cellular adhesive: prevention of

reactive oxygen species mediated death of implanted cells for cardiac repair, ACS Nano 9 (2015) 4987—4999.

[152] G. Keller, H.R. Snodgrass, Human embryonic stem cells: the future is now, Nature Medicine 5 (1999) 151—152.

[153] H.H. Yoon, S.H. Bhang, T. Kim, T. Yu, T. Hyeon, B.S. Kim, Dual Roles of graphene oxide in chondrogenic differentiation of Adult stem cells: cell-Adhesion substrate and growth factor-delivery carrier, Advanced Functional Materials 24 (2014) 6455—6464.

[154] J. Liao, Y. Qu, B. Chu, X. Zhang, Z. Qian, Biodegradable CSMA/PECA/graphene porous hybrid scaffold for cartilage tissue engineering, Scientific Reports 5 (2015) 9879.

[155] S.H. Ku, C.B. Park, Myoblast differentiation on graphene oxide, Biomaterials 34 (2013) 2017—2023.

[156] S.R. Shin, Y.-C. Li, H.L. Jang, P. Khoshakhlagh, M. Akbari, A. Nasajpour, S.Z. Yu, T. Ali, K. Ali, Graphene-based materials for tissue engineering, Advanced Drug Delivery Reviews 105 (2016) 255—274.

[157] S. Goenka, V. Sant, S. Sant, Graphene-based nanomaterials for drug delivery and tissue engineering, Journal of Controlled Release 173 (2014) 75—88.

[158] A.M. Díez-Pascual, A.L. Díez-Vicente, Poly (propylene fumarate)/polyethylene glycol-modified graphene oxide nanocomposites for tissue engineering, ACS Applied Materials and Interfaces 8 (28) (2016) 17902—17914.

[159] C. Shuai, P. Feng, C. Gao, X. Shuai, T. Xiao, S. Peng, Graphene oxide reinforced poly (vinyl alcohol): nanocomposite scaffolds for tissue engineering applications, *RSC Advances* 5 (2015) 25416—25423.

Clay nanostructures for biomedical applications

5

Pallabi Saikia, PhD

Assistant Professor, Department of Chemistry, School of Basic Sciences, Assam Kaziranga University, Jorhat, Assam, India

ABBREVIATIONS

AFM	Atomic force microscopy
BTMA	Benzyltrimethylammonium
ChiPgA	Chitosan/polygalacturonic acid
CTAB	Cetyltrimethylammonium bromide
DHTA	12-(Methacryloyloxy)dodecyl 6-(hexanoyloxy)trimethylammonium bromide
DMTA	Dynamic mechanical thermal analysis
DSC	Differential scanning calorimetry
FMT	Fractionated montmorillonite
HAP	Hydroxyapatite
HDPE	High-density polyethylene
HEC	Hydroxyethyl cellulose
hMSC	Human mesenchymal stem cell
HNT	Halloysite nanotube
HRR	Heat release rate
HTCC	N-(2-hydroxy) propyl-3-trimethylammonium chitosan chloride
MMT	Montmorillonite
PCL	Poly(ε-caprolactone)
PEO	Poly(ethylene oxide)
PLA	Poly(lactic acid)
PMMA	Poly(methyl methacrylate)
PNCs	Polymer nanocomposites
PP-MA	Poly(n-propyl methacrylate)
PUNC	Polyurethane nanocomposite
SAXS	Small-angle X-ray scattering
SBF	Simulated body fluid
SEM	Scanning electron microscopy
Tg	Glass-transition temperature
TGA	Thermogravimetric analysis
TPU	Thermoplastic polyurethane
UMTA	11-(Methacryloyloxy)undecyltrimethylammonium bromide
WPI	Whey protein isolate
XRD	X-ray diffraction

Two-Dimensional Nanostructures for Biomedical Technology. https://doi.org/10.1016/B978-0-12-817650-4.00005-X

5.1 Introduction

Over the past two decades, a variety of advanced materials with enhanced material properties have been developed through nanoscience and technology. Various types of nanofillers such as clays, carbon nanotubes, graphene, polyhedral oligomeric silsesquioxanes, polymeric nanofillers, metals, and metal-oxide ceramics have been used for the development of nanocomposites [1—5]. However, nanocomposites based on nanoclay have great potential in today's green materials research because a small weight percentage incorporation of clay nanoplatelets in matrix can enhance the mechanical and material properties of a matrix without losing inherent processability. Other than these, clays are naturally abundant, cheap, and benign to the environment and humans.

The commonly used clays for the preparation of clay-containing polymer nanocomposites (PNCs) belong to the family of 2:1 layered silicates or phyllosilicates. In their crystal structure, two tetrahedrally coordinated silicon atoms are fused to an edge-shared octahedral sheet of either aluminum or magnesium hydroxide (Fig. 5.1) [6]. Depending on the variation of clay and its source, the thickness of layers and their lateral dimension may vary from 30 nm to several micrometers or larger. As a result of stacking of clay layers, a regular van der Waals gap between

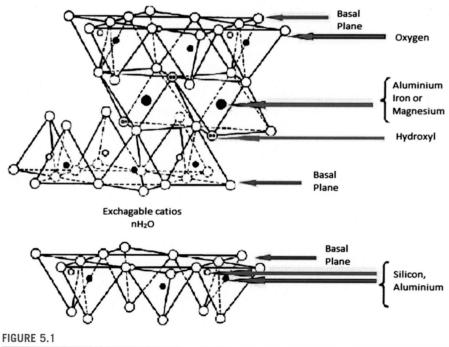

FIGURE 5.1

Smectite crystal structure [6].

Copyright 2019. Reproduced with permission from Senanayake et al.

the layers is formed, which is known as the interlayer or gallery. Isomorphic substitution within the layers (e.g., Al^{3+} replaced by Mg^{2+} or Fe^{2+} or Mg^{2+} replaced by Li^{1+}) generates negative charges that are counterbalanced by alkali and alkaline earth cations situated inside the galleries.

Montmorillonite (MMT) is the most commonly used clay for the preparation of polymer/clay nanocomposites. MMT has two types of structures: tetrahedral substituted and octahedral substituted MMTs. In tetrahedrally substituted clays the negative charge is located on the surface of the silicate layers and, hence, the polymer matrices can interact more readily with these clays than with octahedrally substituted materials. Owing to the hydrophilic nature of natural MMT, the clay is not able to bind to poly(lactic acid) (PLA) macromolecules in two-dimensional silicate galleries. The hydrophilic silicate surface of MMT is converted to an organophilic one for the successful intercalation of PLA chains into MMT. Generally, ion-exchange reactions with cationic surfactants, including primary, secondary, tertiary, and quaternary alkylammonium or alkylphosphonium cations, are performed for this purpose. Alkylammonium or alkylphosphonium cations in the organosilicates lower the surface energy of the inorganic host and improve the wetting characteristics of the polymer matrix, resulting in a larger interlayer spacing. Additionally, the alkylammonium and alkylphosphonium cations can provide the functional groups that can react with polymer matrix or, in some cases, initiate the polymerization of monomers to improve the strength of the interface between the inorganic and the polymer matrix [7a,b].

The thermal and mechanical properties of PNCs depend on the extent to which the clay nanoparticles are dispersed in the polymer matrix, and the extent is again dependent on two factors: one is physical and chemical interactions between the clay and the polymer and the other is the technique how they were prepared. Nanoclays have been incorporated to hydrogels such as polysaccharides (i.e., chitosan, gellan gum) to support adhesion and proliferation of cells [8,9]. The addition of nanoclay fillers enhances the physical and mechanical properties according to the desired application [10–15].

Thus according to the level of interaction ability between the clay nanoparticles and polymer matrix, three types of clay/polymer nanocomposites are possible (Fig. 5.2).

As they are nontoxic, lots of research interests have been focused on the study of various applications of nanoclays and their composites. This chapter highlights the basic aspects of clay nanostructures and the state of the art of biomedical applications of nanoclays and nanoclay-polymer composites.

5.2 Features of two-dimensional nanoclay

5.2.1 Nanoclay surface properties

Nanoclay polymer multilayers and clay oxide multilayers prepared through layer-by-layer self-assembly shows high bonding strength between interlayers. For example, $(ZrO_2\text{-}MMT)_{30}$, $(ZrO_2\text{-}MMT)_{60}$, $(SnO_2\text{-}MMT)_{30}$, and $(ZrO_2\text{-}MMT\text{-}SnO_2\text{-}MMT)_{15}$ are the multilayered films prepared through the layer-by-layer

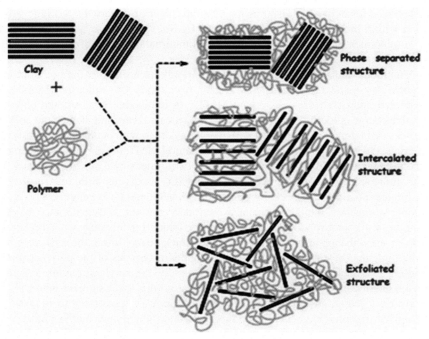

FIGURE 5.2

Possible types of polymer clay nanocomposite structures [7b].

(1) In intercalated PNCs, the polymer chains are inserted into the clay layered structure in a crystallographically regular fashion. The interlayers are composed of a few molecular layers of polymer, where the polymer chains are only partially intercalated between the clay layers.

(2) In an exfoliated nanocomposite, the individual clay layers are separated in a continuous polymer matrix by average distances. The clay layers are exfoliated completely in the polymer matrix.

(3) Flocculated PNCs are the same as intercalated PNCs. However, sometimes the silicate layers of clay are flocculated because of the hydroxylated edge-edge interactions of the platelets.

Copyright 2017. Reproduced with permission from Valapa et al.

deposition technique [16]. For the multilayers $(SnO_2\text{-}MMT)_{30}$ and $(ZrO_2\text{-}MMT\text{-}SnO_2\text{-}MMT)_{15}$, the total indention depth and residual depth are reduced with increase in annealing temperature, as shown in indentation load-displacement curves and indent impressions in Fig. 5.3. The film becomes stiffer with the increase in annealing temperature as a result of sink-in impressions.

Furthermore, the nonexfoliated clay particles present in the film reflected in the atomic force microscopic image as on the surface of impression a few irregular grains and a partial lateral crack. Again, this multilayer shows a smooth loading

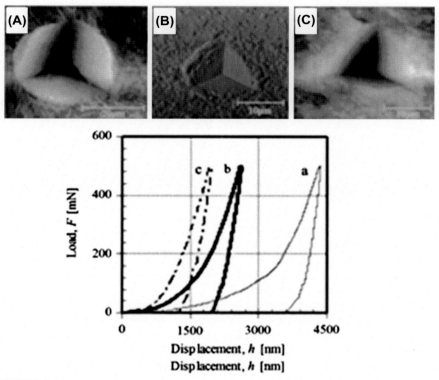

FIGURE 5.3

Atomic force microscopic images of typical indent impressions of $(SnO_2\text{-}MMT)_{30}$ multilayers with MMT 0.4 wt% that were (A) as-deposited, (B) annealed at 400°C, and (C) annealed at 600°C for 2 h, as well as the corresponding load-displacement curves [16].

curve indicating strong bonding between the clay layers or at the interface. The film has enhanced hardness due to the strong interaction between Zr^+ layers and clay layers expelling excess water during the annealing process. As the annealing temperature and concentration of clay suspension vary, the multilayers show variation in hardness and modulus. During the layer-by-layer deposition process, some clay particles were adsorbed in the multiple layer. Also, the structures of the annealed samples are highly amorphous as observed from X-ray diffraction (XRD) patterns. The mechanical behavior of these multilayers is affected by factors such as dipping control parameters, postdeposition treatment, and cationic and anionic precursors. The aggregation of unexfoliated clay particles or reaggregation of exfoliated clay particles results in the deposited clay layer having a rough surface. Also during the annealing process, oxide crystals grow resulting in the deformation of the clay layer.

FIGURE 5.4

Surface roughness of (A) as-deposited and (B) annealed (at 600°C) (ZrO$_2$-MMT)$_{30}$ with clay montmorillonite 0.4 wt% [16].

These issues give rise to the need for an effective exfoliating method for the preparation of a uniform clay layer (Fig. 5.4).

The interlayer nanostructured surface properties are modified by the intercalation reaction [17]. In designing functional nanostructure, smectite is used as the adsorbent of organic contaminates. To accommodate nonionic organic compounds, organophilic smectites are used as a support to the cationic and anionic guest species. Aromatic hydrocarbons are adsorbed very little on the hydrophilic surface of smectites, but are adsorbed by organoammonium-intercalated clays. The intercalated alkylammonium ions are arranged as a monolayer followed by their alkyl chains forming a parallel bilayer. Furthermore, the alkylammonium ions in $C_{18}3C_1N^+$-MMT, $C_{18}3C_1N^+$-fluorotetrasilicic mica, and $2C_{18}2C_1N^+$-saponite are arranged as a pseudotrimolecular layer. On introduction of retinal to these modified clay layers, color regulation and efficient isomerization of retinal occur through primary photochemical reactions. Similarly, amphiphilic dyes and *meso*-tetraphenylporphine are accommodated by clay-modified multilayers of alkylammonium ions.

Small organic cations form organically pillared clay with smectite clay. It can be further modified with organoammonium cations for the accommodation of guest species. Pillared clays show different adsorption properties for the adsorption of nonionic aromatic compounds compared to hydrophobic clays due to the interaction of nonionic aromatic compounds. In organically pillared clays, the adsorption is controlled by pore size and porosity. Thus different modified clays such as $[Ru(bpy)_3]^{2+}$ modified clays, where $[Ru(bpy)_3]^{2+}$ plays the role of pillar, and methylviologen (MV^{2+}) modified clays showed different adsorption behaviors toward phenolic compounds. However, the interaction between adsorbates and solid supports reflects from the release performance of the molecular species. For example, the slow release of alachlor from modified SWy-1 MMTs was studied by El-

Nahhal and his coworkers [18,19]. It was observed that the adsorbed alachlor amount was large in the benzyltrimethylammonium form. Similarly, hinokitiol-poly(ethylene terephthalate)-trimethylammonium (TMA)-MMT (Kunipia F) releases hinokitiol in a less amount, as TMA-MMT entraps hinokitiol.

Similarly, the hydrated cations present in the clay surface are replaced by organic cations that make the clay compatible with nonpolar species and more organophilic [20]. Ion-exchange reaction plays a role in this process, which facilitates replacement of inorganic cations by cationic surfactants on the surface. Thus polystyrene-clay nanocomposites (PSNCs) are found to have the ability to undergo ion-exchange reaction with organic modifiers. The clay particles are well dispersed in the PS matrix, indicating higher polymer chain interface area. The PSNCs are thermally more stable and have higher storage moduli than neat PS.

In a report by Balázs [21], the surface properties of exfoliated clay nanostructure belonging to kaolin group minerals were investigated with the nitrogen adsorption method and inverse gas chromatography method. After exfoliation, the surface energy of kaolinite nanostructures is decreased due to a change in the dispersion component of surface energy. The morphologies of exfoliated kaolinite and halloysite nanostructures are modified through the addition of organic minerals that bind strongly to their surfaces. Hydrogen peroxide removes this organic material, resulting in an irregular rearrangement of the exfoliated structure. The surface of the exfoliated halloysite nanostructure contains very strongly bound water molecules, and partial dehydroxylation cannot produce a smooth nanostructure surface without structural modification. The amount of inner and surface hydroxyls are decreased during the dehydroxylation process. As a result of the loss of these hydroxyl groups, possible functionalization in the surface is removed.

5.2.2 Nanoclay-based hybrids

Nanoclay-based hybrids have diverse applications in the field of materials science and biomedical technology. Nanoclay-reinforced polymer composites have excellent material characteristics including improved physical, thermal, and mechanical properties [22,23]. During the mixing process, higher clay loading results in an increase in the amount of air bubbles. Therefore the amount of nanoclay filler needs careful optimization for better material properties [24]. MMT nanoclay has been used extensively as the filler in hybrid polymeric composites owing to its excellent exfoliation/interaction chemistry, surface reactivity, high surface area, cost effectiveness, and easy availability [25]. The small addition of nanoclay to the polymer matrix showed variation in properties such as reduced gas permeability, improved solvent resistance, superior mechanical properties, increased thermal stability, and enhanced flame retardancy.

The effect of nanoclay as a filler on the mechanical properties of Kevlar fabric with S-glass fabric and polyester fabric was studied through improvement in tensile strength [26]. The tensile strengths of the composites increase with an increase in the weight percentage of the nanoclay. However, the strengths reach a maximum point

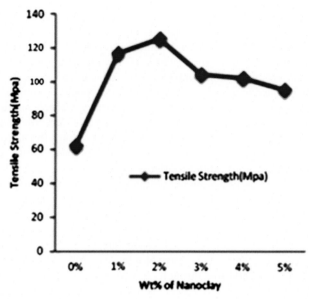

FIGURE 5.5

Tensile strength variations for different weight percentages of nanoclay.

with a certain weight percentage. Afterward, with increase in the weight percentage of nanoclay, the mixture becomes brittle and viscous forming voids (Figs. 5.5 and 5.6).

In a similar study [27] on the mechanical properties of epoxy/glass/nanoclay hybrid composites, it was shown that the interlaminar shear strength, tensile strength, flexural strength, Young's modulus, and microhardness of the composites increase as the nanoclay loading was increased up to 5 wt%. The tensile strength was increased due to the interfacial adhesion between the nanoclay and the epoxy matrix, which prevented the movement of matrix at the interface. Beyond 5 wt% of nanoclay addition, the viscosity increased. Also from a differential scanning calorimetry (DSC) study, it is observed that the glass-transition temperature of the hybrid nanoclay changes with the addition of weight percentage of nanoclay. The glass-transition temperature of the nanocomposites increases with the addition of nanoclay up to 2 wt%. This increase in T_g is because of the polymerization within the clay catalyzed by alkylammonium ions present in the nanoclay, which led to exfoliation and nanoclay dispersion. Further addition of nanoclay to polymer matrix resulted in the decrease in glass-transition temperature. This is because of the decrease in the cross-linking of nanocomposites. Similarly, in a cellulose nanofibrillar matrix, introduction of nanoclay improves the gas barrier properties of the materials to oxygen and water vapor. The highly ordered nanoplatelet-like structure of the mineral is retained in the hybrid clay/cellulose nanofibril, resulting in good gas barrier capability [28].

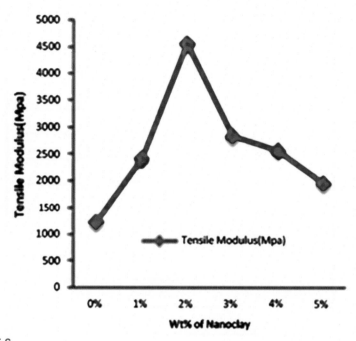

FIGURE 5.6

Tensile modulus variations for different weight percentages of nanoclay [26].

By comparing the melting behavior of neat Nylon-6 and hybrid Nylon-6-nanoclay with varying concentrations of MMT, it was found that both the neat and hybrid Nylon-6 show two sequences of melting depending on the crystallization temperature [29]. The experiments were carried out by Medellín-Rodríguez and his coworkers at several low and high isothermal crystallization temperatures. The DSC curve (Fig. 5.7) of neat Nylon-6 showed two sequences of melting: one at $80-190°C$ and the other at $195-200°C$. The change of the magnitude and position of melting endotherms is probably because of the molecular orientation of Nylon-6. However, in the hybrid structures NCH_2 and NCH_5, there are four sequences of melting and transition zones involving recrystallizations and double melting.

Absorption of water-insoluble and neutral Nile red dye onto disk-shaped nanoclay in the presence and absence of cationic surfactant shows some interesting spectroscopic properties [30]. In both the presence and absence of cationic surfactants such as cetyltrimethylammonium bromide (CTAB), the intensity of blue absorption band for the Nile red dye hybrid is more pronounced for higher dye loading. Spectroscopic analysis of the loading behavior of red-emissive dyes to modified Laponite suggested multi-site binding. The unmodified and CTAB-modified nanoclays have different orientation of dipole moments. The Nile red molecules are located between two clay disks at low molecules per disk (mpd). Thus the molecules are facing a less

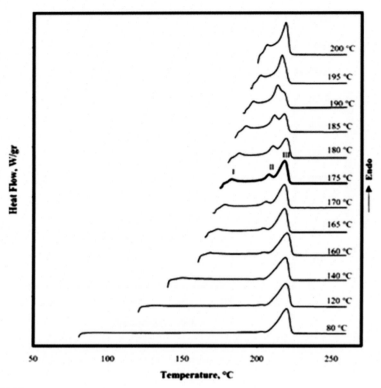

FIGURE 5.7

Linear heating curves differential scanning calorimetry traces of neat Nylon-6 after isothermal crystallization ($T_m°$) at 280°C, 3 min; t_c) 30 min). Crystallization temperatures are shown [29].

Copyright 2005. Reproduced with permission from Zapata-Espinosa et al.

polar environment and, as a result, the fluorescence spectrum is enhanced. With increasing concentration of mpd, the numbers of dye molecules present in the more polar environment are increased, resulting in reduced and moderate fluorescence quantum yield. The CTAB-modified clay is highly emissive reflected from the spectra, as in the red spectral region, broad and charge-transfer Stokes shifted emission band is observed.

The nacre-mimetic layered films (Fig. 5.8A), as a result of hybridization of ductile polymers such as poly(ethylene oxide) (PEO) and hydroxyethyl cellulose (HEC), and MMT have brittleness. The scanning electron microscopic image (Fig. 5.8B) of these hybrids revealed that the clay composites have layered structures with 100—200 nm thickness to the film surface. The MMT tactoids at higher magnifications (Fig. 5.8C) correspond to 10—20 layers of polymer, which are intercalated with MMT platelets. The PEO/MMT hybrid shows a more homogeneous

HEC /MTM   PEO /MTM

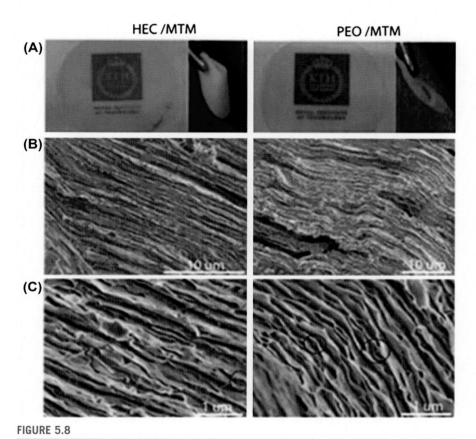

FIGURE 5.8

(A) Photographs of hybrid films on top of KTH logo showing optical translucency. Thickness of the films is ~50 μm. Image on the right shows the flexibility of hybrid films of ~4 × 4 cm². (B,C) Scanning electron micrographs showing cross sections of tensile fractured hybrid samples. Scale bars are 10 and 1 μm, respectively. Circles in (C) show montmorillonite (MTM) tactoids. MTM content in the hybrids is 80 wt% [31]. *HEC*, hydroxyethyl cellulose; *PEO*, poly(ethylene oxide).

Copyright 2013. Reproduced with permission from Sehaqui et al.

structure than HEC/MMT hybrid as confirmed by XRD studies (Fig. 5.9). The porosity present in the hybrid films is 17% and 25%, respectively, as observed from density measurement studies. In the tensile stress-strain curve of the hybrids (Fig. 5.10), a clear plastic deformation zone is observed that represented the ductility of the hybrids. From the thermogravimetric analysis curve (Fig. 5.11), it is concluded that the interaction of PEO with MMT clay is stronger than that of HEC. Furthermore, the hybrids have lower moisture uptake capacities than individual components because of the reduction in hydrophilicity of the hybrids as a result

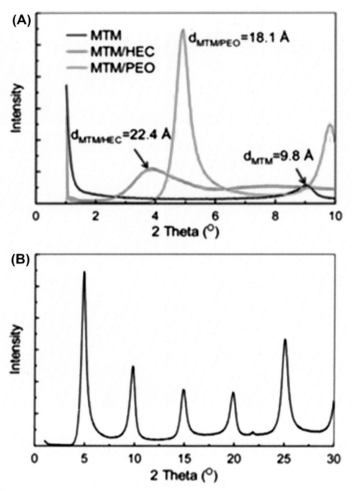

FIGURE 5.9

(A) X-ray diffractograms of pure montmorillonite (MTM) and poly(ethylene oxide)/ hydroxyethyl cellulose (PEO/HEC) hybrids with 80 and 60 wt% MTM content, respectively. (B) X-ray diffractogram of PEO hybrid between 0 and 30 degrees [31].

Copyright 2013. Reproduced with permission from Sehaqui et al.

of interaction of the polymer with the hydrophilic clay surface. Also the storage modulus of the hybrids with respect to relative humidity is quite higher than that of their constituents (Fig. 5.12A,B). PEO/MTM hybrids have oxygen permeability comparable to that of cellulose/MTM hybrids less affected by the relative humidities. However, the water vapor permeability of the PEO/MTM hybrid is lower than that of the cellulose/clay hybrid [31].

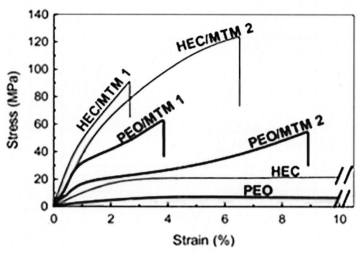

FIGURE 5.10

Uniaxial tensile stress-strain curves of montmorillonite (MTM) hybrids and neat polymer matrices [31]. *HEC*, hydroxyethyl cellulose; *PEO*, poly(ethylene oxide).

5.2.3 Polymer/clay nanocomposites

Nanocomposites are organic or inorganic multiphase solid materials with one of the phases having one dimension less than 100 nm. The interesting thing about the nanocomposites is that only a small amount of filler is required to enhance their properties. The synthesis of PNCs is an integral aspect of polymer nanotechnology. Various methods of processing such as melt intercalation process and in situ polymerization are applied for the preparation of PNCs. In the melt intercalation process, the polymer is melted at a high temperature and to this molten polymer, filler is added without the use of solvents (Fig. 5.13) [32].

In PNCs, the nanofillers are very fine and their total surface area is high when compared with the distance between their particles. The ability of nanocomposites filled with a low incorporation of nanoclay (<5%) to perform in an improved way depends on the modification of the surface of the clay and the quality of dispersion and exfoliation of the nanoparticles [33,34]. An intercalated or partially intercalated/partially exfoliated morphology is proved to be best exfoliated morphology for polar polymer/clay nanocomposites [35].

Smectite group of clays have been widely used in polymer clay nanocomposites. A tightly spaced clay stack is resulted from a smaller interlayer distance because of the presence of smaller ions such as Na^+, K^+, or Li^+ in pristine clay. However, the clay is intercalated as a result of larger interlayer spacing if an organic cationic group with a larger size replaces the metal counterion to make the clay more organophilic. To highlight the micromechanics of polymer-clay nanocomposites, Rao and

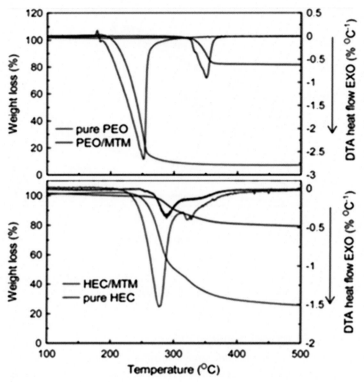

FIGURE 5.11

Thermogravimetric analysis plots showing weight loss and its derivative vs temperature for neat polymers and their hybrids with a montmorillonite (MTM) content of ~80 wt%. Downward peaks represent the differential thermal analysis (DTA) curve [31]. *HEC*, hydroxyethyl cellulose; *PEO*, poly(ethylene oxide).

coworkers [36] performed a series of experiments using a copolymer latex PBSMaSO$_3$ composed of butyl acrylate, styrene, methacrylamide, and 2-acrylamido-2-methylpropane sulfonic acid, sodium salt intercalated with MMT clay. It was observed that during the solution processing process, the dispersion and the well-ordered intercalated phase was formed uniformly.

In in situ polymerization (Fig. 5.14), the monomer is dispersed in the clay and some favorable conditions are set for the effective polymerization between its layers [37]. The clay particles attack the monomer units inside their galleries due to their greater surface energies. As a result of polymerization, the new polar species between layers are diffused.

In the solution method of synthesis, the clay is exfoliated in single layers in a solvent, where the polymer is soluble. Afterward, the polymer is embedded into

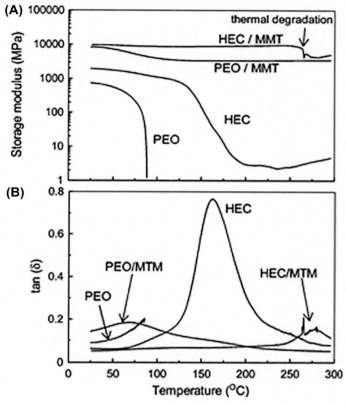

FIGURE 5.12

(A) Storage modulus and (B) tan δ as a function of temperature for neat polymers and their hybrids with 80 wt% montmorillonite (MTM) [31]. *HEC*, hydroxyethyl cellulose; *PEO*, poly(ethylene oxide).

Copyright 2013 Reproduced with permission from Sehaqui et al.

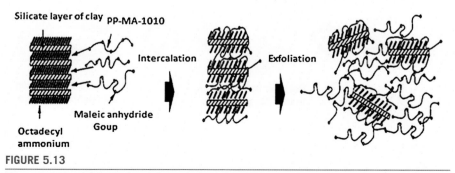

FIGURE 5.13

Representation of the intercalation process of PP-MA-1010 in the organized clay [32].

Copyright 2015. Reproduced with permission from Frontini et al.

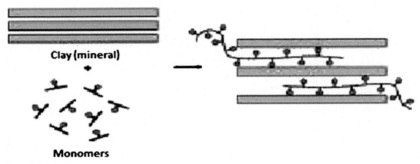

FIGURE 5.14

In situ polymerization of polymer nanocomposites (PCNs) [37].

the delaminated layers and the layers merge with the polymer when the solvent is evaporated (Fig. 5.15).

Again, intercalation in molten state is a technique that was developed by Vaia et al. [38]. In this method, the clay is mixed with a molten thermoplastic polymer matrix and the polymer is taken to the space between the layers forming a nanocomposite (Fig. 5.16). This method is the most flexible and environmentally benign method for the preparation of PNCs.

Katti and his coworkers [39] reported a multiscale approach to design polymer clay nanocomposites where the constant-force steered molecular dynamics method is used to study the mechanical properties of the constituents of the intercalated clay units in PCNs, whereas the finite element model is used to form a representative

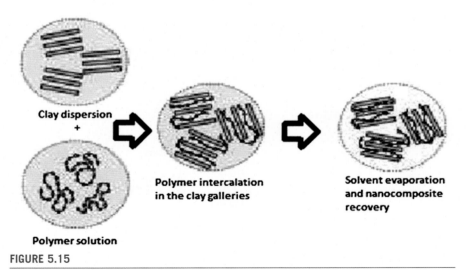

FIGURE 5.15

Synthesis of nanocomposites by dispersion in solution technique [37].

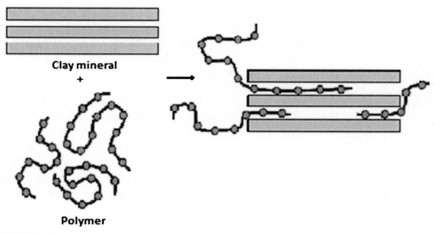

FIGURE 5.16

Nanocomposites synthesized by intercalation in the molten state [37].

Copyright 1993. Reproduced with permission from Vaia et al.

PCN model that stimulates the composite response of the intercalated clay and the polymer matrix. The study concluded that molecular interactions between the constituents of the nanocomposites should be taken into account to predict accurate mechanical response of PCNs. For nanocomposite formation, the organic modifier is required to access the clay galleries for occupation by the ions. However, the amount of organic modifier can be reduced by introducing inorganic exchange ions, which can co-occupy the clay gallery surfaces along with the organic onium ions. For example, Pinnavaia and his coworkers [40] reported a preparation of epoxy-clay nanocomposite by intercalation and exfoliation of homostructured mixed inorganic/organic cation-exchanged forms of MMT and a fluorohectorite clay. Depending on the fraction of onium ions in the mixed ions, intercalated and homostructures intercalated and exfoliated clay nanoclays were achieved. The mixed ion clay homostructures with onium ion-to−proton ratio in the range 25:75 to 65:35 showed improvement in mechanical properties. A long-chain diprotonated α,ω-diamine cation acting as the cross-linking agent enhances the prepolymer intercalation. In this method, up to 75% of the ion-exchange sites of naturally occurring MMT or fluorohectorite can be replaced with the same amount as in the fully exchanged organocation form.

Lafranche and his coworkers [41] developed compatibilized starch grafted polypropylene/organoclay nanocomposites by melt compounding. Polymer clay nanocomposites have also been prepared by radical photopolymerization and ion-exchange reaction of sodium MMT (Na-MMT) with N,N'-dimethylbenzenediazonium cation [42]. Trialkyl-imidazolium-treated MMT clays have been melt blended with PS to form PNCs, as reported by Trulove and his coworkers [43]. The polymer/clay nanocomposites can also be prepared by the spray casting technique [44]. This is a single experimental method for producing good layer oriented films of pure polymer, organoclay, or any variable composition.

The mechanical properties of polymer/clay nanocomposites are evaluated in terms of tensile and flexural tests. It is observed that the properties vary with the addition of nanoclay to polymer composites. A report by Cauvin [45] shows that addition of <3% nanoclay resulted in the increase in tensile modulus and strength, with reduced elongation at break. Similarly, dispersability and tensile strength of the clay are increased with loading of maleic anhydride—grafted polypropylene (MA-g-PP) as a compatibilizer. However, dispersion of clay particles in PP/clay is considered only at microscopic and nanoscopic scales [46]. The tensile and flexural properties of clay nanocomposites are influenced by a variety of factors such as interfacial interaction between clays and matrix, dispersability of the clay in the upper and core layers, degree of dispersion of the clay, interphase properties, and microstructure of upper and core structure of the clay.

Methacrylate monomer having a long alkyl chain with a quaternary ammonium salt group at the end can incorporate in hydrophilic clay platelets with parallel stack, leading to the highest aspect ratio. While investigating water absorption kinetics and mechanical properties, it was observed that the swelling ratio of both MMT-11-(methacryloyloxy)undecyltrimethylammonium bromide (pUMTA) and MMT-12-(methacryloyloxy)dodecyl 6-(hexanoyloxy)trimethylammonium bromide (pDHTA) increases with time, until after 24 h it saturates at 13 and 8 wt%, respectively. The swelling ratio for both the composites is determined by the hydrophilicity of the alkyl chain, which is again affected by the chain length [47].

Fracture toughness is another property of PNCs that is a measure of their ability to promptly resist fracture in the presence of a flow. The trends in the reported literature are contradictory, as there are fluctuations of toughness reported by several researchers. Ferreira et al. [48] reported that the toughness on PP/clay nanocomposites is decreased with reduction in failure displacement of the filled composites. Similarly, Chen et al. [49] reported the transition of fracture from ductile to brittle with reduced toughness, with increased clay loading. On the other hand, Saminathan et al. [50,51] reported an increase in the toughness of the compatibilized PP/nanoclay. The microscopic distribution of clay particles in the PP/clay nanocomposites affects the fracture toughness of the material [46]. As clay nanoparticles are acting as void nucleation sites, higher void nucleation and reduced void growth with extensive fibrillation result in significant reduction of toughness of these nanocomposites compared with neat PP. A study on the surface properties of the polymer clay nanocomposites reveals that the addition of even a small amount of nanoclay slightly increases the hardness of the skin layer of the PP/organoclay nanocomposite. The variation in wear resistance of the nanocomposite depends on the type and microstructure of the filler [52,53]. The specific surface area of nanofillers in a polymer matrix is measured through the small-angle X-ray scattering method. Thus a wide range of specific surfaces were determined in polymer/clay nanocomposites of natural rubber or styrene with different types of MMT clay. This method also results in the reduction of the tactoid size of the composite confirming that exfoliation occurs through the progressive peeling of the outer layer of the tactoids [54].

The influence of clay concentration on the interfacial interactions of the epoxy/clay nanocomposite and their free-volume properties has also been investigated [55]. Compared with pristine epoxy, the free volume size and the fractional free volume of the composite decrease in the nanocomposite. At lowest clay concentrations, the free positions are present in the epoxy-clay interfaces. A similar study on the effect of addition of nanoclays to the surface properties of polyethylene terephthalate/clay nanocomposite showed that the wettability of the clay nanocomposites depends on the hydrophilicity of the nanoclay particles [56].

The properties of nanoclay are affected by the surface modification of the clay. For example, surface modification of Na-MMT clay by plasma polymerization with methyl methacrylate (MMA) and styrene resulted in a change in the surface properties of the clay [57]. The PS/Na$^+$-MMT/nanocomposites exhibit better thermal properties than PS/Na$^+$-MMT/MMA nanocomposites (Figs. 5.17 and 5.18). Also the dispersion property of the modified composite depends on the polarity of the solvent.

Similar influence of organoclay of polymer/clay nanocomposite was studied by Pergal and his coworkers. The free —OH group present on the surface of the thermoplastic polyurethane nanocomposites (TPU-NCs) increases the hydrophilicity from

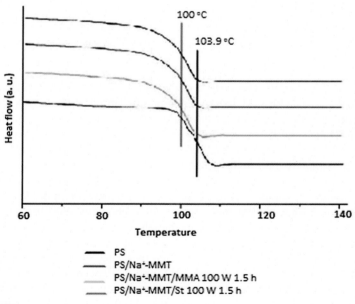

FIGURE 5.17

Curves obtained by differential scanning calorimetry of polystyrene (PS) and pristine PS/Na$^+$-montmorillonite (MMT), PS/Na$^+$-MMT/methyl methacrylate (MMA), and PS/Na$^+$-MMT/styrene (St) nanocomposites [57].

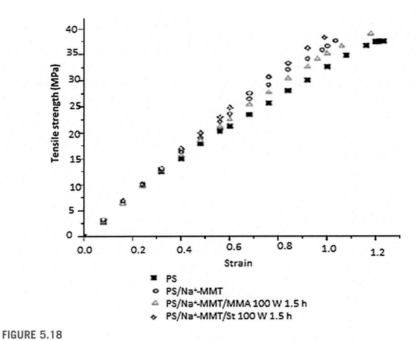

FIGURE 5.18

Tensile-strain diagram of polystyrene (PS) and pristine PS/Na$^+$-montmorillonite (MMT), PS/Na$^+$-MMT/methyl methacrylate (MMA), and PS/Na$^+$-MMT/styrene (St) nanocomposites [57].

TPU-NC1 to TPU-NC10. The surface of the TPU-NCs is heterogeneous and rough compared with pure TPU [58]. In a similar report by Pergal and his coworkers [59], a series of polyurethane nanocomposites was synthesized based on 4,4′-methylenedi-phenyldiisocyante and 1,4-butanediol as the comonomers of hard segments (HS) and poly(propylene oxide)-*b*-poly(dimethylsiloxane)-*b*-poly(propylene oxide) as a part of the soft segments (SS) and organically modified montmorillonite (OMt). The reinforcement of polyurethane matrix with OMt nanoparticles improved the thermal stability, storage modulus, Young's modulus, and tensile strength. Thus polyure-thane nanocomposites exhibit high surface hydrophobicity and they may have a good potentiality as a waterproof coating and elastomer in biomedical applications. The surface properties of thermally stable phosphonium-modified MMT clay on polymer/clay nanocomposite was investigated by Kamal et al. [60]. The quality of clay dispersion and the nanocomposite performance depend on the surface energy and interfacial interactions between the clay and the matrix. At room temperature, the surface energy of the polymer, compared with that of the organoclay surface, is enhanced by the presence and migration of lubricants, such as zinc stearate.

The shear stiff, micalike synthetic k-fluorohectorite clay/poly(methyl methacry-late) (PMMA) nanocomposites have significant mechanical properties compared with natural MMT clay [61]. The use of nanoplatelets resulted in increased fracture toughness in comparison to neat PMMA. Also, synthetic organoclay/PMMA has high dispersion quality. The enhanced fracture toughness is due to the dissipating energy through crack deflection and crack pinning, depending on the variation of stress state of the composite with different amounts of nanofillers.

Compared with those of traditional filled polymers, the thermal properties of nanocomposites are significantly improved. However, for polymers of low polarity, the compatibility between the clay and polyolefins is low, resulting in less significant improvements [62]. Albdiry et al [63] reported that good thermal properties of the nanocomposites can be achieved with different processing conditions, leading to good clay dispersion in the nanoclay composite. Other studies also refer to improvement in thermal conductivity by changing the processing pressure [64]. Costantino et al. [65] reported that under an oxidative atmosphere, nanoclay positively affects the thermal degradation of the injection-moulded intercalated PP/clay nanocomposite. It is also reported that the PS/clay nanocomposite synthesized by emulsion polymerization is found to have a higher thermal stability and higher storage modulus than neat PS [66,67].

To improve the rheological properties of polymer/clay nanocomposites, the native MMT clay is fractionated to exclude the large aggregates and modified by surfactant treatment, as shown in Fig. 5.5 [68].

Poly(3,4-ethylenedioxythiophene):polystyrene sulfonate (PEDOT:PPS) and polyionic liquid (PIL) polymers were self-assembled with synthetic nanoclay to study electric and ionic conductivities. Electrochemical impedance spectroscopy of Poly (ionic liquid) (PIL)/Synthetic saponite nanoclay, Sumecton (SUM) nacre mimetic revealed that the conductivities are much higher than the individual PIL and silica. It is probably due to increased ionic mobility at the particle-polymer interface, leading to the formation of ionic channels reducing the resistance [69].

5.3 Biomedical applications of nanoclay-based materials
5.3.1 In tissue engineering

Tissue engineering is a multidisciplinary field that applies the principles of biosciences and engineering in restoring and reconstructing imperfect tissues or organs. Tissue engineering has evolved as a versatile approach in human healthcare systems [70,71a,b].

Since the early 1990s, after the pioneering works of Langer ad Vacanti in tissue engineering [72], lots of studies have been done in this field. With the increasing bone-related surgeries and the limitations of current treatment methods, bone tissue engineering has contributed in various ways, including providing structural support to the body, protecting the vital internal organs, facilitating body movement, and acting as a reservoir of calcium and phosphate-based minerals [73–75]. The cells that can be used in bone tissue engineering are differentiated cells such as osteoblasts or undifferentiated cells such as mesenchymal stem cells. For the preparation of composite scaffolds, most of the bone tissue engineering studies use natural and synthetic polymers and fillers such as hydroxyapatite (HAP) and bioactive glass. Biocompatibility, biodegradability, high porosity, adequate pore size (100−500 μm), interconnected pores, and adequate mechanical properties are the basic requirements of a scaffold for tissue engineering applications. The choice of polymers is controlled by the biocompatibility and biodegradability of the scaffold, whereas the scaffold architecture can be controlled by the use of an appropriate processing technique [76].

The number of osteoblast cells in scaffolds based on chitosan/polygalacturonic acid (ChiPgA)-containing MMT clay modified with 5-aminovaleric acid was comparable to ChiPgA scaffolds containing HAP known for their osteoconductive properties [77]. According to photoacoustic Fourier transform infrared spectroscopic studies, the modified MMT clay was successfully incorporated in the ChiPgA-based scaffolds. Swelling studies on ChiPgA composite scaffolds showed that the cells and the nutrients have the ability to reach the interior parts of the scaffolds. Additionally, the ChiPgA scaffolds exhibiting porosity greater than 90% are appropriate for use in tissue engineering studies.

It is observed that nanoclays have remarkable compatibilities in in vivo and in vitro studies [10,12,15,78]. For example, as evaluated by Li et al. [79], the MMT nanoparticles have no effect on the mortality rate of Sprague-Dawley (SD) rats by oral feeding. The in vitro toxicity testing in highly phagocytic cultures showed good tolerance against two nanoclays, namely, clinoptilolite and sepiolite, [80] with lower toxic effects of clinoptilolite. Also, cytotoxicity studies of halloysite nanotubes (HNTs) having a fiberlike morphology have shown tolerance after 24 h of exposure to C6 glioma cell cultures with concentrations of 500 μg mL^{-1} [81]. Furthermore, toxicity measurements with neoplastic cell line models with change in concentration and incubation time showed that HNTs are safe for cells at concentrations of up to 75 μg mL^{-1} [82]. A hydrogel composed of general gum (GG), glycerol, and HNTs for soft tissue engineering applications such as for pancreas, liver, and skin regeneration was proposed by Bonifacio et al. [8]. The addition of glycerol to GG improved the material viscosity, while HNTs decreased water uptake by 30% −35%. The addition of 5% HNTs to the members of chitosan/HNTs prepared by solution casting resulted in improved mechanical properties as well as enhanced thermal stability, as reported by De Silva et al. [83]. Nitya et al. [15] performed in vitro studies of a fibrous polycaprolactone/HNT composite scaffold for bone tissue engineering and showed that these composites have enhanced mineralization, greater protein adsorption, and faster proliferation of human mesenchymal stem cells (hMSCs) seeded on these scaffolds. Moreover, a study by Zhou et al. [84] showed that the nanotopography and surface chemistry of poly(vinyl alcohol) (PVA) is changed if HNTs are added to PVA bionanocomposite films.

Aliabadi et al. [13] studied the antibacterial properties of the biocompatible chitosan ammonium salt *N*-(2-hydroxy) propyl-3-trimethylammonium chitosan chloride (HTCC)-modified MMT. The samples showed good antibacterial activities against both gram-negative and gram-positive bacteria because of the entrapment of bacteria between the intercalated structures of HTCC in MMT. These composites were further proposed for tissue engineering applications.

Other outstanding tissue engineering applications of clay nanomaterials have been further explored. For example, Payne and coworkers [14] modified Na-MMT with an amino acid to mineralize synthetic HAP, resembling the biogenic HAP in human bones. These HAP-clay materials are further incorporated into ChiPGA scaffolds and films for bone tissue engineering [85].

To perform osteogenic differentiation of hMSCs, a nanoclay-enriched electrospun poly(ε-caprolactone) (PCL) scaffold is developed by Gaharwar and his coworkers. This scaffold promotes in vitro mineralization when subjected to simulated body fluid. The alkaline phosphate activity of hMSCs and the deposition of mineralized extracellular matrix is increased as a result of hMSC adhesion and proliferation, indicating enhanced osteogenic differentiation of hMSCs on the electrospun scaffolds [86].

The osteogenic differentiation in hMSCs is directed by Laponite clay through an interaction by the ionic dissolution of the clay inside the cells and the subsequent increase in the intracellular concentration of ions and minerals [87−91]. The cell-regulating properties of Laponite have been extensively studied within polyethylene glycol (PEG) and gelatin-methacryloyl (GelMA) hydrogels [88,92,93] and PEG alone does not affect the cell binding. Subsequent studies with Laponite GelMA composites have revealed their profound impact over the proliferation and osteogenic differentiation of preosteoblast cells [88]. Again, chitosan-gelatin/nanohydroxyapatite-MMT (CS-Gel/nHA-MMT) composite is highly porous. The bioactivity studies of the scaffolds showed that the degradation rate is decreased and biomineralization is increased in the presence of nHA and MMT [94].

5.3.2 In drug delivery

The use of clay in drug delivery has been investigated from ancient times. Nanoclay minerals can act as efficient drug carriers with long-lasting drug concentrations. The clay nanocomposites have great potential as delivery hosts in the pharmaceutical field. For example, calcium-MMT is used extensively in the treatment of diarrhea, stomach ulcers, acne, anemia, pain, open wounds, colitis, hemorrhoids, intestinal problems, and a variety of other health issues. The investigation of clay-drug interaction and release mechanism is an essential contribution for the formulation of clay-based drug delivery system. Drug-clay interactions have been observed as a possible factor to modify drug release. Furthermore, drug stability is increased and drug delivery patterns are modified using clay minerals [95−103]. MMT provides sorption sites within its interlayer space and on the outer surface and edges. MMT has variable amounts of sodium and calcium ions, along with water for hydration. Na-MMT hydrates more than Ca-MMT and hence delamination is more prominent in Na-MMT than in Ca-MMT. Thus Ca-MMT is converted to Na-MMT for higher drug loading and drug release during pharmaceutical applications. Moreover, for effective higher drug loading, the improved aspect ratio, surface area, and pH and concentration of ionic species in the dispersing medium play a significant role [104−106].

There are several mechanisms that may be involved in the interaction between organoclay and drug molecules, including both strong and weak molecular bonding. Predominantly reported mechanisms are ion-pairing, ion-dipole interaction, covalent bonding, and cation exchange mechanisms (Fig. 5.19) [107–112]. Apart from molecular mechanisms, the processing factors also influence the incorporation of molecules in the MMT structure. The temperature also plays a role in drug adsorption on MMT.

HNTs are a good container for proteins and DNA for intracellular delivery [113]. They are also intercellular drug delivery vehicles, as they can travel easily through the cellular membranes without any affect. For the controlled release of drugs for hours, halloysite clay nanotubes are applied as potential nanomaterials. If the loaded HNTs are exposed to water, the drug gets released from the clay nanotubes. The nanotubes serve as a container to encapsulate and protect the drug molecules until the tubes are delivered into the cell. While halloysites are in the cytoplasm, they are decomposed by the intracellular enzymatic activity and the drug is released directly inside the cell [114]. A similar study was also done by Fakhrullin and his coworkers [115], in which halloysite clay nanotubes are specially focused for drug delivery purposes. Long and his coworkers studied a nanoclay-facilitated drug delivery system that could be effective for papillary thyroid cancer therapy. In this respect, the kaolinite nanoclay was expanded so that it can provide sufficient space for hosting doxorubicin molecules (Fig. 5.20) [116].

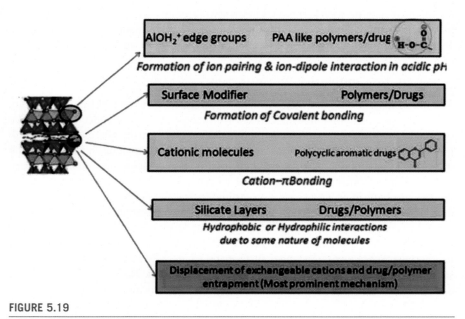

FIGURE 5.19

Various interaction mechanisms between clay and drug and/or polymer [112]. *PAA, polyacrylic acid.*

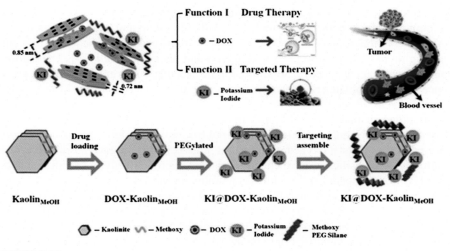

FIGURE 5.20

The procedure for synthesizing KI@DOX-Kaolin$_{MeOH}$ and its corresponding functions for tumor therapy [116]. *DOX*, doxorubicin; *PEG*, polyethylene glycol.

The in-depth biocompatibility of PCL/clay nanohybrids in terms of cell adhesion, genotoxicity, and nanocompatibility has been explored widely [117]. The biocomaptibility of the nanohybrids at the gene level showed the attuned nature of the particles within the subcellular localization in human epithelial cells. The biodegradation rate of pure PCL and its nanohybrid scaffolds showed an increase rate in the presence of nanoclays. Also, no genotoxic stress was observed in case of both pure and hybrid PCL as the surface for adhesion.

5.3.3 In antibacterial coating

The antimicrobial properties of clay nanocomposites and their applications as a surface coating material are the priority of researchers of material science in the recent days. For example [118,119], silver-based nanoclay showed strong antibacterial activity against both gram-positive and gram-negative bacteria. The mechanism of bacterial action of silver nanoparticles is still not well known. The silver nanoparticles bind strongly to electron donor groups in biological molecules containing sulfur, oxygen, or nitrogen causing defects in the bacterial cell wall so that the cell contents are lost [120]. The smaller particles with a larger surface-to-volume ratio provided a more efficient means of antibacterial activity than larger particles. As nanostructured antimicrobial materials have high surface area-to-volume ratio, compared with their higher scale counterparts, the antimicrobial nanocomposite packaging systems are particularly efficient for food packaging improving function and performance towards the food content [121].

The antimicrobial properties of *Gelidium corneum*/nanoclay composite film containing grapefruit extract or thymol was studied by Song and his coworkers [122]. Also, the antimicrobial film and coating applications of PCL nanocomposites containing thymol was investigated by Lagaron and his coworkers [123]. The natural antimicrobial agent uptake or solubility for the neat material is enhanced by the presence of the nanoplatelets, probably due to retention over the surface of the apolar biocide agent.

Bentonite nanoclay is hydrophilic in nature and can be dispersed in aqueous chitosan solution. Nanocomposite films and multilayer coating have improved the barrier properties against oxygen, water vapor, grease, and UV light transmission. The chitosan/layered silicates have enhanced the antimicrobial activity of chitosan acetate [124,125]. Biocompatible chitosan ammonium salt HTCC-modified MMT clay nanocomposites are evaluated for antibacterial efficiency and biocompatibility. The HTCC-modified MMT showed good antibacterial efficiency against both gram-positive (e.g., *Staphylococcus aureus*) and gram-negative bacteria (e.g., *Escherichia coli*). The enhanced antibacterial efficiency is probably due to the fact that the bacteria are entrapped between the intercalated structures (Figs. 5.21 and 5.22) [126a].

Furthermore, when incorporated with Cloisite 30B, the whey protein isolate/nanoclay composite films showed significant bacteriostatic effect against the gram-positive bacteria, *Listeria monocytogenes* [126b].

Unmodified nanoclays may enhance better interaction with chitosan by absorbing bacteria from the solution, although they are not antimicrobial by themselves [127]. Han and his coworkers [128] observed that compared with pure chitosan and Na-MMT, chitosan-MMT nanocomposites were significantly more effective against *S. aureus* and *E. coli*. Wang et al. [127,129−131] have carried out a series of studies on chitosan/rectorite intercalated nanocomposites with antimicrobial properties. The nanocomposites showed antimicrobial properties through a two-stage-mechanism: first, adsorbing bacteria from solution and immobilizing them on the clay surface and second, accumulating chitosan on the clay surface, inhibiting the bacterial growth. Based on further studies, the authors concluded that the antimicrobial activity is directly proportional to the amount or the interlayer distance of the nanoclay [132].

5.4 Commercial aspects

Clay nanocomposites play a vital role in various industries, and lots of work has been made toward the applications of clay nanocomposites. Numerous studies showed that clay nanocomposites have extensive potential applications such as in the development of various automotive fields, polymer automotive compounds, sporting goods, packaging, coating technology, building construction, and electrochemical and biomedical applications [133]. Organoclays blended with sorbents such as alum and activated carbon are used for the removal of contaminants of fresh water by treating industrialized and urban wastewater.

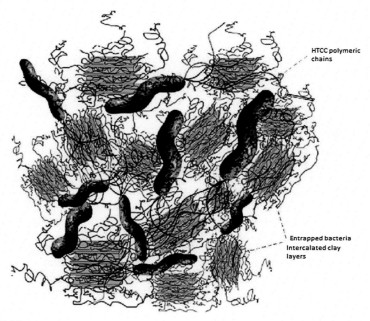

FIGURE 5.21

The synergistic antibacterial efficiency mechanism, entrapping bacteria between the intercalated HTCC/Mt structures [126a]. *HTCC, N*-(2-hydroxy) propyl-3-trimethylammonium chitosan chloride.

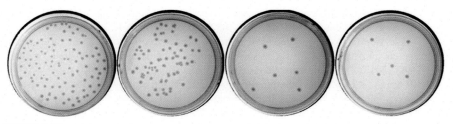

FIGURE 5.22

Results of the designed qualitative test for comparing the antibacterial efficiency of some samples against *Escherichia coli* [126a].

5.5 **Conclusion**

Nanocomposites introduced newer dimensions in different biomedical and industrial fields by giving enhanced properties to materials.

This chapter presents recent advances and biomedical applications of nanoclay materials. These natural aluminosilicate structures have been found to be cheap, biocompatible, and nontoxic, and lots of studies have been done on clay nanostructures for various applications such as tissue engineering, drug delivery, and antimicrobial activity. It can be concluded that clays and their intercalated nanocomposites/hybrids may pave newer ways for the development of novel biomaterials with tremendous possibility for commercialization.

References

[1] N.G. Sahoo, S. Rana, J.W. Cho, L. Li, S.W. Chan, Polymer nanocomposites based on functionalized carbon nanotubes, Progress in Polymer Science 35 (2010) 837—867.

[2] J.R. Potts, D.R. Dreyer, C.W. Bielawski, R.S. Ruoff, Graphene-based polymer nanocomposites, Polymer 52 (2011) 5—25.

[3] K.Y. Lee, Y. Aitomaki, L.A. Berglund, K. Oksman, A. Bismarck, On the use of nanocellulose as reinforcement in polymer matrix composites, Composites Science and Technology 105 (2014) 15—27.

[4] S. Barus, M. Zanetti, M. Lazzari, L. Costa, Preparation of polymeric hybrid nanocomposites based on PE and nanosilica, Polymer 50 (2009) 2595—2600.

[5] J. Zhang, E. Manias, C.A. Wikie, Polymerically modified layered silicates: an effective route to nanocomposites, Journal of Nanoscience and Nanotechnology 8 (2008) 1597—1615.

[6] G. Senanayake, G.K. Das, A.D. Lange, J. Li, D.J. Robinson, Reductive atmospheric acid leaching of lateritic smectite/nontronite ores in H_2SO_4/Cu(II)/SO_2 solutions, Hydrometallurgy 152 (2015) 44—54.

[7] [7a] S.S. Ray, M. Okamoto, Polymer/layered silicate nanocomposites: a review from preparation to processing, Progress in Polymer Science 28 (2003), 1539-15641.;
[7b] R.B. Valapa, S. Loganathan, G. Pugazhenthi, S. Thomas, T.O. Varghese, An overview of polymer/claynanocomposites, Clay/Polymer Nanocomposites Elsevier Inc. (2017). https://doi.org/10.1016/B978-0-323-46153-5.00002-1.

[8] M.A. Bonifacio, P. Gentile, A.M. Ferreira, Insight into halloysite nanotubes-loaded gellan gum hydrogels for soft tissue engineering applications, Carbohydrate Polymers 163 (2017) 280—291. https://doi.org/10.1016/j.carbpol.2017.01.064.

[9] R. Zafar, K.M. Zia, S. Tabasum, Polysaccharide based bionanocomposites, properties and applications: a review, International Journal of Biological Macromolecules 92 (2016) 1012—1024. https://doi.org/10.1016/j.ijbiomac.2016.07.102.

[10] K.S. Katti, D.R. Katti, R. Dash, Synthesis and characterization of a novel chitosan/montmorillonite/hydroxyapatite nanocomposite for bone tissue engineering, Biomedical Materials 3 (2008) 34122. https://doi.org/10.1088/1748-6041/3/3/034122.

[11] A. Olad, F.F. Azhar, F. Farshi Azhar, The synergetic effect of bioactive ceramic and nanoclay on the properties of chitosan—gelatin/nanohydroxyapatite—montmorillonite scaffold for bone tissue engineering, Ceramics International 40 (2014) 10061—10072. https://doi.org/10.1016/j.ceramint.2014.04.010.

[12] A.H. Ambre, K.S. Katti, D.R. Katti, Nanoclay based composite scaffolds for bone tissue engineering applications, Journal of Nanotechnology in Engineering and Medicine 1 (2010) 31013. https://doi.org/10.1115/1.4002149.

[13] M. Aliabadi, R. Dastjerdi, K. Kabiri, HTCC-modified nanoclay for tissue engineering applications: a synergistic cell growth and antibacterial efficiency, BioMed Research International (2013) 749240. https://doi.org/10.1155/2013/749240.

[14] S.A. Payne, D.R. Katti, K.S. Katti, Probing electronic structure of biomineralized hydroxyapatite inside nanoclay galleries, Micron 90 (2016) 78−86. https://doi.org/10.1016/j.micron.2016.09.001.

[15] G. Nitya, G.T. Nair, U. Mony, In vitro evaluation of electrospun PCL/nanoclay composite scaffold for bone tissue engineering, Journal of Materials Science: Materials in Medicine 23 (2012) 1749−1761. https://doi.org/10.1007/s10856-012-4647-x.

[16] D. Hou, G. Zhang, R.R. Pant, Z. Wei, S. Shen, Micromechanical properties of nanostructured clay-oxide multilayers synthesized by layer-by-layer self-assembly, Nanomaterials 6 (2016) 204, https://doi.org/10.3390/nano6110204.

[17] T. Okada, Y. Seki, Makoto, M. Ogawa, Designed nanostructures of clay for controlled adsorption of organic compounds, Journal of Nanoscience and Nanotechnology 14 (2014) 2121−2134.

[18] Y. El-Nahhal, S. Nir, T. Polubesova, L. Margulies, B.J. Rubin, Agric, Food Chemistry 46 (1998) 3305.

[19] Y. El-Nahhal, T. Undabeytia Polubesova, Y.G. Mishael, S. Nir, B. Rubin, Applied Clay Science 18 (2001) 309.

[20] N. Greesh, S.S. Ray, S. Bandyopadhyay, Role of nanoclay shape and surface characteristics on the morphology and thermal properties of polystyrene nanocomposites synthesized via emulsion polymerization, Industrial & Engineering Chemistry Research 52 (2013), 16220-1623.

[21] Z. Balázs, Synthesis, Structural and Surface Characterization of Clay Nanostructures Belonging to the Kaolin Group Minerals, PhD thesis, University of Pannonia, Doctoral School of Chemical Engineering and Material Science, 2017.

[22] P.M. Borba, A. Tedesco, M. Denise, D.M. Lenz, Effect of reinforcement nanoparticles addition on mechanical properties of SBS/Curaua Fiber composites, Materials Research 17 (2) (2014) 412−419.

[23] M.N. Gururaja, A.N. Hari Rao, A review on recent applications and future prospectus of hybrid composites, International Journal of Soft Computing and Engineering 1 (6) (2012) 352−355.

[24] N. Saba, M. Jawid, M. Asim, Recent Advances in Nanoclay/natural Fibers Hybrid Composites, Springer Science, 2016.

[25] M.F. Hossen, S. Hamdan, M.R. Rahman, M.M. Rahman, F.K. Liew, J.C. Lai, Effect of fiber treatment and nanoclay on the tensile properties of jute fiber reinforced polyethylene/clay nanocomposites, Fibers and Polymers 16 (2) (2015) 479−485.

[26] P. Neeraja, M.S.R.N. Kumar, M.S. Chowdary, Effect of nanoclay on tensile properties of hybrid polymer matrix composites, International Research Journal of Engineering and Technology (IRJET) 3 (6) (2016) 2810−2812.

[27] J.J. Karippal, H.N.M. Murthy, K.S. Rai, M. Sreejith, M. Krishna, Study of mechanical properties of epoxy/glass/nanoclay hybrid composites, Journal of Composite Materials (2014) 1−7.

[28] S. Mirmehdi, P.R.G. Hein, C.I.G.L. Sarantopoulos, M.V. Dias, G.H.D. Tonoli, Cellulose nanofibrils/nanoclay hybrid composite as a paper coating: effects of spray time, nanoclay content and corona discharge on barrier and mechanical properties of the coated papers, Food packaging and shelf life 15 (2018) 87−94.

[29] A. Zapata-Espinosa, F.J. Medellín-Rodríguez, N. Stribeck, Complex isothermal crystallization and melting behaviour of Nylon 6 nanoclay hybrids, Macromolecules 38 (2005) 4246−4253.

[30] T. Felbeck, T. Behnke, K. Hoffmann, M. Grabolle, M. Lezhnina, U.H. Kynast, U. Resch-Genger, Nile red-nanoclay hybrids: red emission optical probes for use in aqueous dispersion, Langmuir 29 (36) (2013) 11489−11497, https://doi.org/10.1021/la402165q.

[31] H. Sehaqui, J. Kochumalayil, A. Liu, T. Zimmermann, L.A. Berglund, Multifunctional nanoclay hybrids of high toughness, thermal, and barrier performance, ACS Applied Materials & Interfaces 5 (15) (2013) 7613−7620.

[32] P.M. Frontini, A.S. Pouzada, Trends in the multifunctional performance of polyolefin/clay nanocomposite injection moldings, Multifunctionality of Polymer Composites (2015) 213−244. ISBN: 978-0-323-26434-1, Elsevier.

[33] H.R. Dennis, D.L. Hunter, D. Chang, S. Kim, J.L. White, J.W. Cho, Effect of melt processing conditions on the extent of exfoliation in organoclay-based nanocomposites, Polymer 42 (23) (2001) 9513−9522. https://doi.org/10.1016/s0032-3861(01)00473-6.

[34] L. Chen, S.-C. Wong, T. Liu, X. Lu, C. He, Deformation mechanisms of nanoclay-reinforced maleic anhydride-modified polypropylene, Journal of Polymer Science Part B: Polymer Physics 42 (14) (2004) 2759−2768. https://doi.org/10.1002/polb.20108.

[35] K. Wang, S. Liang, Q. Zhang, R. Du, Q. Fu, An observation of accelerated exfoliation in iPP/organoclay nanocomposite as induced by repeated shear during melt solidification, Journal of Polymer Science Part B: Polymer Physics 43 (15) (2005) 2005−2012. https://doi.org/10.1002/polb.20487.

[36] Y.Q. Rao, J.M. Pochan, Mechanics of polymer-clay nanocomposites, Macromolecules 40 (2007) 290−296.

[37] K.A. Hernández-Hernández, J. Illescas, M.C. María del Carmen Díaz-Nava, C.R. Claudia Rosario Muro-Urista, S. Martínez-Gallegosand, R.E. Ortega-Aguilar, Polymer-clay nanocomposites and composites: structures, Characteristics, and their applications in the removal of organic compounds of environmental interest, Medicinal Chemistry 6 (2016) 3, https://doi.org/10.4172/2161-0444.100034.

[38] R.A. Vaia, H. Ishii, E.P. Giannelis, Synthesis and properties of two dimensional nanostructures by direct intercalation of polymer melts in layered silicates, Chemistry of Materials 5 (1993) 1694−1696.

[39] D. Sikdar, S.M. Pradhan, D.R. Katti, K.S. Katti, B. Mohanty, Altered phase model for polymer clay nanocomposites, Langmuir 24 (2008) 5599−5607.

[40] C.S. Triantafillidis, P.C. LeBaron, T.J. Pinnavaia, Homostructured mixed inorganic-organic ion clays: a new approach to epoxy polymer- exfoliated clay nanocomposites with a reduced organic modifier content, Chemistry of Materials 14 (2002) 4088−4095.

[41] R. Tessierm, E. Lafranche, P. Krawczak, Development of novel melt-compounded starch-grafted polypropylene/polypropylene-grafted maleicanhydride/organoclay ternary hybrids, Express Polymer Letters 6 (11) (2012) 937−952, https://doi.org/10.3144/expresspolymlett.2012.99.

[42] Z. Salmi, K. Benzarti, M.M. Chehimi, Diazonium cation-exchanged clay: an efficient, unfrequented route for making clay/polymer nanocomposites, Langmuir 29 (2013) 13323−13328, https://doi.org/10.1021/la402710r.

[43] P.C. Trulove, D.M. Fox, W.H. Awad, J.W. Gilman, C.D. Davis, T.E. Sutto, H.P. Maupin, H.C.D. Long, ACS the Application of Trialkyl-Imidazolium Ionic

Liquids and Salts for the Preparation of Polymer-Clay Nanocomposites, Symposium Series, American Chemical Society, Washington, 2007, https://doi.org/10.1021/bk-2007-0975.ch016.

[44] E. Dunkerley, D. Schmidt, Effects of composition, orientation and temperature on the O_2 permeability of model polymer/clay nanocomposites, Macromolecules 43 (2010) 10536−10544, https://doi.org/10.1021/ma1018846.

[45] L. Cauvin, D. Kondo, M. Brieu, N. Bhatnagar, Mechanical properties of polypropylene layered silicate nanocomposites: characterization and micro−macro modelling, Polymer Testing 29 (2) (2010) 245−250. https://doi.org/10.1016/j.polymertesting.2009.11.007.

[46] M.N. Bureau, F. Perrin-Sarazin, M.T. Ton-That, Polyolefin nanocomposites: essential work of fracture analysis, Polymer Engineering & Science 44 (6) (2004) 1142−1151. https://doi.org/10.1002/pen.20107.

[47] K. Sung, S. Nakagawa, N. Yoshie, Fabrication of water-resistant nacre-like polymer/clay nanocomposites via in situ polymerization, ACS Omega 2 (2017) 8475−8482.

[48] J.A.M. Ferreira, P.N.B. Reis, J.D.M. Costa, B.C.H. Richardson, M.O.W. Richardson, A study of the mechanical properties on polypropylene enhanced by surface treated nanoclays, CompositesPart B Eng 42 (6) (2011) 1366−1372. https://doi.org/10.1016/j.compositesb.2011.05.038.

[49] L. Chen, S.-C. Wong, S. Pisharath, Fracture properties of nanoclay-filled polypropylene, Journal of Applied Polymer Science 88 (14) (2003) 3298−3305. https://doi.org/10.1002/app.12153.

[50] K. Saminathan, P. Selvakumar, N. Bhatnagar, Fracture studies of polypropylene/nanoclay composite. Part I: effect of loading rates on essential work of fracture, Polymer Testing 27 (3) (2008) 296−307. https://doi.org/10.1016/j.polymertesting.2007.11.008.

[51] K. Saminathan, P. Selvakumar, N. Bhatnagar, Fracture studies of polypropylene/nanoclay composite. Part II: failure mechanism under fracture loads, Polymer Testing 27 (4) (2008) 453. https://doi.org/10.1016/j.polymertesting.2008.01.011.

[52] K. Friedrich, Z. Zhang, A.K. Schlarb, Effects of various fillers on the sliding wear of polymer composites, Composites Science and Technology 65 (15−16) (2005) 2329−2343. https://doi.org/10.1016/j.compscitech.2005.05.028.

[53] G. Malucelli, F. Marino, Abrasion Resistance of Polymer Nanocomposites—A Review, 2012.

[54] C. Marega, V. Causin, R. Saini, A. Marigo, A.P. Meera, S. Thomas, K.S.U. Devi, A direct SAXS approach for the determination of specific surface area of clay in polymer-layered silicate nanocomposites, Journal of Physical Chemistry B 116 (2012) 7596−7602, https://doi.org/10.1021/jp303685q.

[55] P.N. Patil, K. Sudarshan, S.K. Sharma, P. Maheshwari, S.K. Rath, M. Patri, P.K. Pujari, Investigation of nanoscopic free volume and Interfacial Interaction in an epoxy resin/modified Clay Nanocomposite using positron annihilation spectroscopy, ChemPhysChem 13 (2012) 3916−3922, https://doi.org/10.1002/cphc.201200593.

[56] M. Parvinzadeh, S. Moradian, A. Rashidi, M.-E. Yazdanshenas, Effect of the addition of modified nanoclays on the surface properties of the resultant polyethyleneterephthalate/clay nanocomposites, Polymer-Plastics Technology and Engineering (2010). Taylor& Francis, https://doi.org/10.1080/03602551003664628.

[57] R.I. Narro-Céspedes, M.G. Neira-Velázquez, L.F. Mora-Cortes, E. Hernández-Hernández, A.O. Castañeda-Facio, M.C. Ibarra-Alonso, Y.K. Reyes-Acosta, G. Soria-Arguello, J.J. Javier Borjas-Ramos, Surface modification of sodium Montmorillonite Nanoclay by plasma polymerization and its effect on the properties of

polystyrene nanocomposites, Hindawi Journal of Nanomaterials (2018). Article ID 2480798, https://doi.org/10.1155/2018/2480798.

[58] M.V. Pergal, I.S. Stefanovi, R. Poreba, M. Steinhart, P.M. Jovancic, S. Ostoji, M. Spirkova, Influence of organoclay content on the structure, morphology and surface related properties of novel poly(dimethylsiloxane)-based polyurethane/organoclay nanocomposites, Industrial & Engineering Chemistry Research 56 (17) (2017) 4970−4983.

[59] S. StefanovićI, M. Špírková, S. Ostojić, P. Stefanov, V.B. Pavlović, M.V. Pergal, Montmorillonite/poly(urethane-siloxane) nanocomposites: morphological, thermal, mechanical and surface properties, Applied Clay Science 149 (1) (2017) 136−146.

[60] M.R. Kamal, J.U. Calderon, B.R. Lennox, Surface energy of modified nanoclays and its effect on polymer/clay nanocomposites, Journal of Adhesion Science and Technology 23 (5) (2009) 663−688, https://doi.org/10.1163/156856108X379164.

[61] B. Fischer, M. Ziadeh, A. Pfaff, J. Breu, V. Altstädt, Impact of large aspect ratio, shear-stiff, mica-like clay on mechanical behaviour of PMMA/clay nanocomposites, Polymer 53 (2012) 3230−3237.

[62] A.M. Mazrouaa, Polypropylene nanocomposites, in: F. Dogan (Ed.), Polypropylene, InTech, Rijeka, Croatia, 2012.

[63] M. Albdiry, B. Yousif, H. Ku, K. Lau, A critical review on the manufacturing processes in relation to the properties of nanoclay/polymer composites, Journal of Composite Materials 47 (9) (2013) 1093−1115. https://doi.org/10.1177/0021998312445592.

[64] M.G. Battisti, W. Friesenbichler, Injection molding compounding of PP polymer nanocomposites, Strojniški vestnik Journal of Mechanical Engineering 59 (11) (2013) 662−668.

[65] A. Costantino, V. Pettarin, J. Viana, A. Pontes, A. Pouzada, P. Frontini, Morphology − performance relationship of polypropylene− nanoclay composites processed by shear controlled injection moulding, Polymer International 62 (11) (2013) 1589−1599. https://doi.org/10.1002/pi.4543.

[66] N. Greesh, S. Ray, J. Bandyopadhyay, Role of nanoclay shape and surface characteristics on the Morphology and thermal properties of polystyrene nanocomposites, synthesized via emulsion polymerization, Industrial & Engineering Chemistry Research 52 (2013) 16220−16231, https://doi.org/10.1021/ie4024929.

[67] H. Qin, S. Zhang, C. Zhao, G. Hu, M. Yang, Flame retardant mechanism of polymer/clay nanocomposites based on polypropylene, Polymer 46 (19) (2005) 8386−8395. https://doi.org/10.1016/j.polymer.2005.07.019.

[68] B.H. Cipriano, T. Kashiwagi, X. Zhang, S.R. Raghavan, A simple method to improve the clarity and Rheological properties of polymer/clay nanocomposites by using fractionated clay particles, ACS Apllied Materials and Interfaces 1 (1) (2009) 130−135.

[69] R.O. Mäkiniemi, P. Das, D. Hönders, K. Grygiel, D. Cordella, C. Detrembleur, J. Yuan, A. Walther, Conducting, self-assembled, nacre-mimetic polymer/clay nanocomposites, ACS Applied Materials & Interfaces 7 (29) (2015) 15681−15685.

[70] R. Langer, J.P. Vacanti, Tissue engineering, Science 260 (1993) 920.

[71] [71a] M. Murugan, S. Ramakrishna, Nano-featured scaffolds for tissue engineering: A review of spinning methodologies tissue engineering, Tissue Engineering 12 (3) (2006);

[71b] L. Peña-Parás, J.A. Sánchez-Fernández, R. Vidaltamayo, Nanoclays for Biomedical Applications.Handbook of Ecomaterials, Springer, 2018. https://doi.org/10.1007/978-3-319-48281-1_50-1.

[72] R. Langer, J.P. Vacanti, Tissue engineering, Science 260 (5110) (1993) 920–926.

[73] R. Murugan, S. Ramakrishna, Development of nanocompositesfor bone grafting, Composites Science and Technology 65 (15–16) (2005) 2385–2406.

[74] S.V. Dorozhkin, Calcium orthophosphates, Journal of Materials Science 42 (4) (2007) 1061–1095.

[75] A.J. Salgado, O.P. Coutinho, R.L. Reis, Bone tissue engineering:state of the art and future trends, Macromolecular Bioscience 4 (8) (2004) 743–765.

[76] C. Viseras, C. Aguzzi, P. Cerezo, A. Lopez-Galindo, Uses of clay minerals in semisolid health care and therapeutic products, Applied Clay Science 36 (1–3) (2007) 37–50.

[77] A.H. Ambre, K.S. Katti, D.R. Katti, Nanoclay based composite scaffolds for bone tissue engineering applications, Journal of Nanotechnology in Engineering and Medicine 1 (2010) 1–9, 031013.

[78] M.S. Nazir, M. Haafiz, M. Kassim, Characteristic properties of nanoclays and characterization of nanoparticulates and nanocomposites, in: H. Essabir, M. Raji, R. Bouh (Eds.), Nanoclay Reinforced Polymer Composites, Springer, 2016, pp. 35–55.

[79] P.R. Li, J.C. Wei, Y.F. Chiu, Evaluation on cytotoxicity and genotoxicity of the exfoliated silicate nanoclay, ACS Applied Materials & Interfaces 2 (2010) 1608–1613. https://doi.org/10.1021/am1001162.

[80] Y. Toledano-Magaña, L. Flores-Santos, G. Montes De Oca, Effect of clinoptilolite and sepiolite nanoclays on human and parasitic highly phagocytic cells, BioMed Research International (2015) 164980. https://doi.org/10.1155/2015/164980.

[81] A. Sánchez-Fernández, L. Peña-Parás, R. Vidaltamayo, Synthesization, characterization, and in vitro evaluation of cytotoxicity of biomaterials based on halloysite nanotubes, Materials 7 (2014) 7770–7780. https://doi.org/10.3390/ma7127770.

[82] V. Vergaro, E. Abdullayev, Y.M. Lvov, Cytocompatibility and uptake of halloysite clay nanotubes, Biomacromolecules 11 (2010) 820–826. https://doi.org/10.1021/bm9014446.

[83] R.T. De Silva, P. Pasbakhsh, K.L. Goh, Physico-chemical characterisation of chitosan/halloysite composite membranes, Polymer Testing 32 (2013) 265–271. https://doi.org/10.1016/j.polymertesting.2012.11.006.

[84] W.Y. Zhou, B. Guo, M. Liu, Poly(vinyl alcohol)/halloysite nanotubes bio-nanocomposite films: properties and in vitro osteoblasts and fibroblasts response, Journal of Biomedical Materials Research Part A 93 (2010) 1574–1587. https://doi.org/10.1002/jbm.a.32656.

[85] A.H. Ambre, D.R. Katti, K.S. Katti, Nanoclays mediate stem cell differentiation and mineralized ECM formation on biopolymer scaffolds, Journal of Biomedical Materials Research Part A 101A (2013) 2644–2660. https://doi.org/10.1002/jbm.a.34561.

[86] A.K. Gaharwar, S. Mukundan, E. Karaca, A. Dolatshahi-Pirouz, A. Patel, K. Rangarajan, S.M. Mihaila, G. Iviglia, H. Zhang, A. Khademhosseini, Nanoclay-enriched poly(e-caprolactone) electrospun scaffolds for osteogenic differentiation of human mesenchymal stem cells, Tissue Engineering Part A 19 (S1) (2014), https://doi.org/10.1089/ten.tea.2013.0281.

[87] M. Mehrali, A. Thakur, C.P. Pennisi, S. Talebian, A. Arpanaei, M. Nikkhah, A. Dolatshahi-Pirouz, Nanoreinforced hydrogels for tissue engineering:biomaterials that are compatible with load-bearing and electroactive tissues, Advances in Materials (2017) 1603612.

[88] J.R. Xavier, T. Thakur, P. Desai, M.K. Jaiswal, N. Sears, E. Cosgriff-Hernandez, R. Kaunas, A.K. Gaharwar, ACS Nano 9 (2015) 3109.

[89] J.I. Dawson, J.M. Kanczler, X.B. Yang, G.S. Attard, R.O. Oreffo, Advances in Materials 23 (2011) 3304.

[90] A.K. Gaharwar, S.M. Mihaila, A. Swami, A. Patel, S. Sant, R.L. Reis, A.P. Marques, M.E. Gomes, A. Khademhosseini, Advances in Materials 25 (2013) 3329.

[91] S.M. Mihaila, A.K. Gaharwar, R.L. Reis, A. Khademhosseini, A.P. Marques, M.E. Gomes, Biomaterials 35 (2014) 9087.

[92] A.K. Gaharwar, V. Kishore, C. Rivera, W. Bullock, C.-J. Wu, O. Akkus, G. Schmidt, Macromolecular Bioscience 12 (2012) 779.

[93] A.K. Gaharwar, P.J. Schexnailder, B.P. Kline, G. Schmidt, ActaBiomaterials 7 (2011) 568.

[94] A. Oladn, F.F. Azhar, The synergetic effect of bioactive ceramic and nanoclay on the properties of chitosan—gelatin/nanohydroxyapatite—montmorillonite scaffold for bone tissue engineering, Ceramics International 40 (2014) 10061—10072.

[95] M. Massaro, S. Riela, C. Baiamonte, Dual drug-loaded halloysite hybrid-based glycocluster for sustained release of hydrophobic molecules, RSC Advances 6 (91) (2016) 87935—87944.

[96] E. Ruiz-Hitzky, P. Aranda, M. Darder, G. Rytwo, Hybrid materials based on clays for environmental and biomedical applications, Journal of Materials Chemistry 20 (42) (2010) 9306—9321.

[97] W.O. Yah, A. Takahara, Y.M. Lvov, Selective modification of halloysite lumen with octadecylphosphonic acid: new inorganic tubular micelle, Journal of the American Chemical Society 134 (3) (2012) 1853—1859.

[98] Y. Zhang, M. Long, P. Huang, Emerging integrated nanoclay-facilitated drug-delivery system for papillary thyroid cancer therapy, Scientific Reports 6 (2016) 33335.

[99] L. Li, H. Fan, L. Wang, Z. Jin, Does halloysite behave like an inert carrier for doxorubicin? RSC Advances 6 (59) (2016) 54193—54201.

[100] M. Gibaldi, S. Feldman, Establishment of sink conditions in dissolution rate determinations. Theoretical considerations and application to nondisintegrating dosage forms, Journal of Pharmaceutical Sciences 56 (10) (1967) 1238—1242.

[101] T. Higuchi, Mechanism of sustained-action medication theoretical analysis of rate of release of solid drugs dispersed in solid matrices, Journal of Pharmaceutical Sciences 52 (12) (1963) 1145—1149.

[102] P.L. Ritger, N.A. Peppas, A simple equation for description of solute release I. Fickian and non-Fickian release from non-swellable devices in the form of slabs, spheres, cylinders or discs, Journal of Controlled Release 5 (1) (1987) 23—36.

[103] A. Wilczak, T.M. Keinath, Kinetics of sorption and desorption of copper (II) and lead (II) on activated carbon, Water Environment Research 65 (3) (1993) 238—244.

[104] G. Lazzara, S. Riela, R.F. Fakhrullin, Clay-based drug-delivery systems: what does the future hold? Therapeutic Delivery 8 (8) (2017) 633—636.

[105] E. Tombácz, M. Szekeres, Colloidal behavior of aqueous montmorillonite suspensions: the specific role of pH in the presence of indifferent electrolytes, Applied Clay Science 27 (2004) 75—94.

[106] L. Sciascia, M.L. Turco Liveri, M. Merli, Kinetic and equilibrium studies for the adsorption of acid nucleic bases onto K10 montmorillonite, Applied Clay Science 53 (2011) 657.

[107] I. Calabrese, G. Cavallaro, G. Lazzara, M. Merli, L. Sciascia, M.L. Turco Liveri, Preparation and characterization of bio-organoclays using nonionic surfactant, Adsorption 22 (2016) 105—116.

[108] L. Aristilde, C. Marichal, J. Miéhé-Brendlé, B. Lanson, L. Charlet, Interactions of oxytetracycline with a smectite clay: a spectroscopic study with molecular simulations, Environmental Science and Technology 44 (2010) 7839−7845.

[109] L.S. Porubcan, C.J. Serna, J.L. White, S.L. Hem, Mechanism of adsorption of clindamycin and tetracycline by montmorillonite, Journal of Pharmaceutical Sciences 67 (1978) 1081−1087.

[110] R. Suresh, S.N. Borkar, V.A. Sawant, V.S. Shende, S.K. Dimble, Nanoclay drug delivery system, Int J Pharm SciNanotechnol 3 (2010) 901−905.

[111] D. Vasudevan, T.A. Arey, D.R. Dickstein, M.H. Newman, T.Y. Zhang, H.M. Kinnear, M.M. Bader, Nonlinearity of cationic aromatic amine sorption to aluminosilicates and soils: role of intermolecular cation− π interactions, Environmental Science and Technology 47 (2013) 14119.

[112] S. Jayrajsinh, G. Shankar, Y.K. Agrawal, L. Bakre, Montmorillonite nanoclay as a multifaceted drug-delivery carrier: a review, Journal of Drug Delivery Science and Technology (2017), https://doi.org/10.1016/j.jddst.2017.03.023.

[113] Y.M. Lvov, M.M. DeVilliers, R.F. Fakhrullin, The application of halloysite tubule nanoclay in drug delivery, Expert Opinion on Drug Delivery 13 (7) (2016) 977−986, https://doi.org/10.1517/17425247.2016.1169271, 2016.

[114] M. Dzamukova, E. Naumenko, Y.M. Lvov, Enzyme-activated intracellular drug delivery with tubule clay nanoformulation, Scientific Reports 5 (2015) 10560−10566.

[115] G. Lazzara, S. Riela, R.F. Fakhrullin, Clay-based drug-delivery systems: what does the future hold? Therapeutic Delivery 8 (8) (2017) 633−646.

[116] A. Tang, Y. Zhang, M. Long, P. Huang, H. Yang, S. Chang, Y. Hu, L. Mao, Emerging integrated nanoclay facilitated drug delivery system for papillary thyroid cancer therapy, Scientific Reports 6 (2016) 33335, https://doi.org/10.1038/srep33335.

[117] N.K. Singh, S.K. Singh, D. Dash, B.P. Purkayastha, J.K.R.P. Maiti, Nanostructure controlled anti-cancer drug delivery using poly(3-caprolactone) based nanohybrids, Journal of Materials Chemistry (2012), https://doi.org/10.1039/c2jm32340k.

[118] M.A. Busolo, P. Fernandez, M.J. Ocio, J. Lagaron, Novel silver-based nanoclay as an antimicrobial in polylactic acid food packaging coatings, Food Additives & Contaminants: Part A 27 (11) (2010) 1617−1626.

[119] U. Konwar, N. Karak, M. Mandal, Vegetable oil based highly branched polyester/clay silver nanocomposites as antimicrobial surface coating materials, Progress in Organic Coatings 68 (2010) 265−273.

[120] E. Weir, A. Lawlor, A. Whelan, F. Regan, Analyst 133 (2008) 835−845.

[121] S.M.A. Tawakkal, M.J. Cran, J. Miltz, S.W. Bigger, A review of poly(lactic acid)-based materials for antimicrobial packaging, Journal of Food Science 79 (8) (2014) 1477−1490.

[122] G.-O. Lim, S.-A. Jang, K.B. Song, Physical and antimicrobial properties of Gelidium corneum/nano-clay compositefilm containing grapefruit seed extract or thymol, Journal of Food Engineering 98 (2010) 415−420.

[123] M.D. Sanchez-Garcia, M.J. Ocio, E. Gimenez, J.M. Lagaron, Novel polycaprolactone nanocomposites containing thymol of interest in Antimicrobial film and coating applications, Journal of Plastic Film and Sheeting 24 (2008) 239.

[124] J.M. Lagaron, P. Fernandez-Saiz, M.J.J. Ocio, Agricultural and Food Chemistry 55 (2007) 2554.

[125] J. Vartianen, M. Mikko Tuominen, K. Nättinen, Bio-hybrid nanocomposite coatings from sonicated chitosan and nanoclay, J. Applied Polymer 116 (6) (2010) 3638−3647.

[126] [126a] M. Aliabadi, R. Dastjerdi, K. Kabiri, HTCC-modified nanoclay for tissue engineering applications:A synergistic cell growth and antibacterial efficiency, Hindawi Publishing Corporation BioMed Research International (2013). Article ID 749240, https://doi.org/10.1155/2013/749240;
[126b] R. Sothornvit, J.-W. Rhim, S.-I. Hong, Effect of nano-clay type on the physical and antimicrobial properties of whey protein isolate/clay composite films, Journal of Food Engineering 91 (2009) 468−473.

[127] X. Wang, Y. Du, J. Yang, X. Wang, X. Shi, Y. Hu, Preparation, characterization and antimicrobial activity of chitosan/layered silicate nanocomposites, Polymer 47 (2006) 6738−6744.

[128] Y.S. Han, S.H. Lee, K.H. Choi, I. Park, Preparation and characterization of chitosan-clay nanocomposites with antimicrobial activity, Journal of Physics and Chemistry of Solids 71 (2010) 464−467.

[129] X. Wang, Y. Du, J. Luo, B. Lin, J.F. Kennedy, Chitosan/organic rectorite nanocomposite films: structure, characteristic and drug delivery behaviour, Carbohydrate Polymers 69 (2007) 41−49.

[130] X. Wang, Y. Du, J. Luo, J. Yang, W. Wang, J.F. Kennedy, A novel biopolymer/rectorite nanocomposite with antimicrobial activity, Carbohydrate Polymers 77 (2009) 449−456.

[131] Y. Wang, Q. Zhang, C. Zhang, P. Li, Characterisation and cooperative antimicrobial properties of chitosan/nano-ZnO composite nanofibrous membranes, Food Chemistry 132 (2012) 419−427.

[132] M.C. de A. Henriette, Antimicrobial nanostructures in food packaging, Trends in Food Science & Technology XX (2012) 1−14.

[133] F. Ali, H. Ullah, Z. Ali, F. Rahim, F. Khan, Z.U. Rehman, Polymer-clay nanocomposites, preparations and current applications: a review, Current Nanomaterials 1 (2016) 83−95.

Metal-organic frameworks for biomedical applications

Ezgi Gulcay[1], Ilknur Erucar[2]

[1]*Master Student, Department of Mechanical Engineering, Faculty of Engineering, Ozyegin University, Cekmekoy, Istanbul, Turkey;* [2]*Assistant Professor, Department of Natural and Mathematical Sciences, Faculty of Engineering, Ozyegin University, Cekmekoy, Istanbul, Turkey*

6.1 Background

Metal-organic frameworks (MOFs) are a unique class of porous materials due to their large pore sizes (from micro- to mesoporosity), tunable chemical and physical properties, and adequate structural stability. MOFs are crystalline materials consisting of metal complexes linked by organic ligands to create permanent porous networks [1]. Alkaline earth metals, transition metals, and lanthanides are used as metal sources, whereas carboxylates, phosphonates, and polyamines, which are mainly derived from benzene, imidazole, and oxalic acid, are commonly used as organic ligands in MOF synthesis [2]. The rapid development of MOF research is based on the reticular design concept, which was first introduced by Yaghi et al. [3]. In this systematic approach, the same building units can be used to construct predetermined frameworks with different physical properties. For example, Yaghi and coworkers [4] used the same topology of IRMOF-1 (isoreticular MOF-1, also known as MOF-5), which has Zn-O-C clusters and benzene links to construct 16 different series of IRMOFs. The schematic design of IRMOF series is shown in Fig. 6.1. All these frameworks have the same three-dimensional cubic topology; however, their pore sizes range from 3.8 to 28.8 Å due to the various types of the organic linkers [5]. MOFs are generally synthesized from molecular building blocks by using hydro-/solvothermal procedures [6]. Details of MOF synthesis can be found in the literature [7—9].

Since there are numerous types of organic ligands and metal cations, an infinite number of MOFs can be theoretically synthesized. Once an MOF is synthesized, its crystal structure is deposited with a unique "refcode" (entry ID) consisting of six letters in the Cambridge Structural Database (CSD) [10]. CSD is a database that includes experimentally reported atomic coordinates of the organic and inorganic molecules. Moghadam et al. [11] identified all MOFs in this database and created the most complete and updated MOF database within the CSD. Currently, there are 90947 MOFs available in the CSD, November 2018. The high number of synthesized MOFs provides an opportunity for both experimental and computational studies by creating various types of interesting materials with a range of pore sizes,

Two-Dimensional Nanostructures for Biomedical Technology. https://doi.org/10.1016/B978-0-12-817650-4.00006-1

173

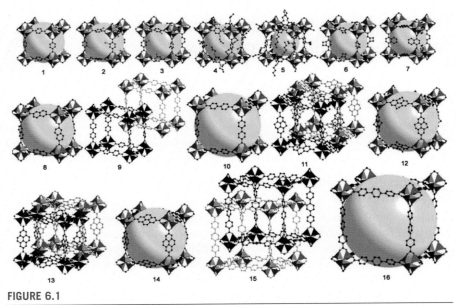

FIGURE 6.1

A large series of isoreticular metal-organic frameworks (IRMOFs)-*n* (*n* = 1–16). The voids are represented by yellow spheres.

Reprinted from Rowsell JLC, Yaghi, OM. Metal–organic frameworks: a new class of porous materials. Microporous and Mesoporous Materials 2004;73(1):3–14 with permission of Elsevier.

surface areas, and different physical and chemical properties. A great deal of attention has been paid in the recent years toward the synthesis and characterization of MOFs for various applications, including gas storage [12], drug storage [13], gas separation [14], catalysis [15], chemical sensing [16], luminescence [17], biomedical imaging [18], and drug delivery [19]. Among these applications, biomedical applications of MOFs have recently started to grow rapidly in the literature. This was illustrated in Fig. 6.2 by the increase in the number of published papers. Fig. 6.2 shows that within the past few years, there has been a growing interest in the biomedical applications of MOFs, including drug delivery (total number of publications from 2006 to 2018: 1227), bioimaging (total number of publications from 2012 to 2018: 50), and biosensing (total number of publications from 2011 to 2018: 59).

As Fig. 6.2 shows, MOFs have been recently investigated especially for drug delivery in biomedicine. Drug encapsulation or drug delivery is used to enhance drug biodistribution, the biological half-life of active species, and their therapeutic effects. Nanoparticles such as lipids, polymers, metal clusters, and carbon structures have been used for drug encapsulation so far [20,21]. However, these systems have several drawbacks including low drug loading capacities (<5 wt%), rapid release, and toxicity [22]. As an alternative to traditional drug carriers, MOFs

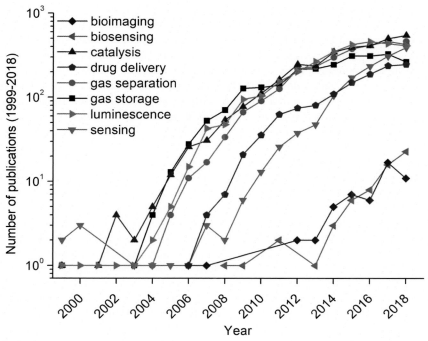

FIGURE 6.2

The number of publications featuring the terms "metal organic framework" and "bioimaging", "biosensing", "catalysis", "drug delivery", "gas separation", "gas storage", "luminescence" and "sensing" in their titles. Accessed: 2018-11-07 at Web of Science.

have been tested as potential drug carriers due to their porous structures and functional polar or nonpolar groups. The first biomedical application of MOFs was based on drug delivery, which was carried out by Férey et al. [23]. His group investigated the Materials of Institut Lavoisier (MIL) family, including MIL-100, MIL-101, and MIL-53, as an analgesic and antiinflammatory drug, ibuprofen carrier. Drug loading capacities of MIL-100 (0.35 g g^{-1}) and MIL-101 (1.38 g g^{-1}) were much higher than those of traditional materials, such as mesoporous silica (0.34 g g^{-1}), due to their large pores (25−34 Å) and high surface areas (3100−5900 m^2 g^{-1}). The formation of specific interactions between the deprotonated ibuprofen molecules and the Lewis acid metal sites and the organic moieties of MOFs also enhanced their ibuprofen loading capacities. The same group [24] later examined the potential of different MIL series for the encapsulation of several anticancer and antiviral drugs, including busulfan (BU, [1,4-butanediol-dimethylsulfonate]), doxorubicin (DOX), azidothymidine triphosphate (AZT-TP), and cidofovir (CDV). These pioneering studies have motivated researchers to study MOFs in biomedicine.

It is important to note that to use MOFs in biomedicine, organic ligands and metal cations should be nontoxic, biocompatible, and stable in physiologic conditions. Due to tunable characteristics of MOFs, biocompatible MOFs have been synthesized by using endogenous linkers that are biomolecules composed of constitutive properties of body composition and nontoxic metal cations. Toxicity is detected by oral lethal dose 50 (LD_{50}) [22], which is the amount of a compound that kills half the members of a population after a predetermined test duration. Calcium (Ca), magnesium (Mg), zinc (Zn), iron (Fe), titanium (Ti), and zirconium (Zr) have been commonly preferred as metal cations, whose toxicity range from 0.025 g kg^{-1} to 30 g kg^{-1}. Cations, which have high LD_{50} values and high daily requirements in the body, are required for biomedical applications. Bioactive metals, such as antiarthritic gold (Au^+), antibacterial Zn^{2+}, antidiabetic vanadium ($V^{5+,4+}$), and antiulcer bismuth (Bi^{3+}), can be also used to fabricate biocompatible MOFs [25]. Nucleobases, citrates, formates, aspartates, amino acids, peptides, gallates, fumarates, muconates, and cyclodextrins (CDs) have been used as endogenous linkers [22]. The first bio-MOF (bio-MOF-1) was synthesized by Rosi and coworkers [26] using Zn^{2+} ions and adenine to create an anionic framework for cationic drug (procainamide) encapsulation. In 2010, bio-MOF-11 was synthesized by the same group [27] using two cobalt ions (Co^{2+}) bridged by two adenines as a paddle-wheel cluster. Bio-MOF-11 analogues, bio-MOF-n ($n = 12-14$), were also synthesized by using different aliphatic monocarboxylates, and the water stability of these bio-MOFs was reported [28]. Rosi and coworkers [29,30] later synthesized another analogue series of bio-MOFs-n ($n = 100-103$) by using Zn^{2+} ions. Among all bio-MOF series, bio-MOFs-n ($n = 100-103$) have the highest surface areas ($2704-4410$ m^2 g^{-1}) due to their highly porous structures. Bio-MIL series (bio-MIL-1, 3, and 5) are also alternative biocompatible MOFs due to their nontoxic and biodegradable porous carboxylate structures. For example, Serre and coworkers [31] synthesized Ca-based bio-MIL-3 [31] and Zn-based bio-MIL-5 [32] and reported that bio-MIL-5 can be used for the treatment of skin disorders. Another biocompatible MOF series are edible MOFs (also known as sugar MOFs), CD-MOFs, which were synthesized using γ-CD building units and potassium ions (K^+) by Stoddart and coworkers [33]. They noted that CD-MOFs (1, 2 and 3) can be prepared with suitable food ingredients and these MOFs can be good alternatives for renewable natural products due to their high stability (up to around 473 K) [33]. Lan and coworkers [34] also synthesized two different CD-MOFs using β-CD ligand and cesium (Cs^+) ions and tested them for anticancer drug encapsulation such as 5-fluorouracil (5-FU) and methotrexate (MTX). These studies showed that MOFs with endogenous linkers and metal sites, such as bio-MOFs, bio-MILs, and CD-MOFs, have potential in biomedical applications.

Besides MOFs with endogenous linkers, exogenous linkers-based MOFs, including MIL series as mentioned earlier, and some widely studied MOFs such as Zr-based MOFs, zeolitic imidazolate framework (ZIF)-8, and Cu-BTC (copper benzene-1,3,5-tricarboxylate) have also been studied for biomedicine. For example, a series of Zr-based MOFs with dicarboxylate bridging ligands, especially UiO-66

(UiO stands for University of Oslo) and UiO-67, have been widely studied for drug encapsulation due to their nontoxicity and high structural stability [35,36]. These materials were studied for the encapsulation of anticancer drugs, such as cisplatin and paclitaxel (another name is taxol), by Filippousi and coworkers [37] and a slow drug release profile, which is desired for the controlled delivery of drug molecules, was reported. Cu-BTC was tested for 5-FU encapsulation by Lucena and coworkers [38] and a slow 5-FU release profile was observed. In another study, ZIF-8 having tetrahedral units covalently joined by Zn^{2+} metal ions was investigated by Sun and coworkers [39] for the efficient encapsulation of 5-FU. ZIF-8 has been widely studied in the literature for its high stability and high surface area ($1630 \, m^2 \, g^{-1}$) [40]. Similarly, Zheng and coworkers [41] tested ZIF-8 for the controlled delivery of DOX. These pioneering experimental studies have been influential in the development of research on biomedical applications of MOFs.

6.2 Potential applications of metal-organic frameworks in biomedicine

The biomedical applications of MOFs include various research areas including drug storage and drug delivery, biomedical gas storage, biosensing, and molecular imaging. In this part, we mainly discuss and summarize the recent experimental and computational studies on MOFs for drug storage, drug delivery, and biomedical gas storage applications. We refer the reader to excellent reviews about sensing, imaging, and various therapeutic applications of MOFs [42−48].

6.2.1 Drug storage and drug delivery

The wide range of MOFs with endogenous and exogenous ligands provides an opportunity to test them for different types of therapeutic activities. Drug storage and drug delivery processes include encapsulation of drug molecules within the pores and/or surface of a material and releasing the drug molecules at a specific rate to a specific site. Efficient drug delivery systems should have a controlled drug releasing mechanism and high drug loading capacities. There are four specific features of MOFs to use them for drug storage and drug delivery applications: (1) they are the most porous structures with large surface areas and high pore volumes to adsorb different molecules, (2) they have tunable cavities for strong host-guest interactions and molecular sieving behavior, (3) they have adequate thermal stability for human body conditions, and (4) they can be synthesized with various methods considering their toxicity and biocompatibility. Additionally, surface modification, which is used for enhancing stability, water solubility, drug loading, and surface morphology for different biomedical applications, can also be performed with MOFs. For example, Horcajada and coworkers [24] used polyethylene glycol (PEG) to modify the surfaces of nanoparticles and to decrease particle aggregation.

Considering these features, MOFs can be regarded as great alternatives for the delivery of drug molecules.

Drug loading in MOFs can be mainly investigated with two different mechanisms: (1) covalent and (2) noncovalent binding, based on the drug-MOF interactions. In Fig. 6.3, three different strategies for MOF-based drug delivery systems are demonstrated [49]. In the encapsulation strategy (noncovalent binding), drug molecules are located within the pores of MOFs via π-π stacking, van der Waals forces, and electrostatic interactions [50]. In the direct self-assembly method (covalent binding), cargo molecules are used as ligands when the framework is constructed via coordination bonds. In the postsynthesis strategy, drug molecules are located on the surface and/or metal sites, as well as the linkers of presynthesized MOFs via coordination and covalent bonds based on "click chemistry". All these strategies can be used together for an MOF-based drug delivery system [49]. Table 6.1 summarizes drug delivery applications of MOFs based on experimental methods. As shown in Table 6.1, MOFs have been commonly used as carriers for antiinflammatory and anticancer drug molecules.

In 2006, the first study on drug delivery applications of MOFs was performed by Férey's group [23] using MIL-100 and MIL-101, two chromium (Cr)-based MOFs with different organic components, BTC and benzenedicarboxylic acid (BDC), respectively. They reported that MIL-100 has lower ibuprofen uptake ($0.35 \ g \ g^{-1}$) than MIL-101 ($1.38 \ g \ g^{-1}$) due to narrower pore apertures of MIL-100 (25 and 29 Å) than those of MIL-101 (29 and 34 Å). They also investigated their drug release profile in simulated body fluid (SBF, pH = 7.4) at 310 K and reported that the complete release of ibuprofen from MIL-100 (MIL-101) occurs within 3 days (within

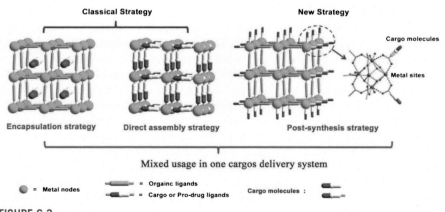

FIGURE 6.3

Drug loading strategies of metal-organic frameworks (MOFs).

Reprinted from Wang L, Zheng M, Xie Z. Nanoscale metal—organic frameworks for drug delivery: a conventional platform with new promise. Journal of Materials Chemistry B 2018;6(5):707—717 with permission of the Royal Society of Chemistry.

Table 6.1 Summary of experimental studies on MOFs used for drug storage and drug delivery.

MOFs	Building components	Cargos	Reference
MIL-100 MIL-101	Terephthalate, Cr	Ibuprofen	[23]
MIL-53	Terephthalate, Cr, Fe	Ibuprofen	[51]
Bio-MOF-1	Adenine, Zn	Procainamide	[26]
Bio-MIL-1	Acetate, nicotinate, Fe	Nicotinic acid	[52]
MIL-100 MIL-101_NH_2 MIL-53 MIL-88A MIL-88Bt MIL-89	Trimesate, Fe Aminoterephthalate, Fe Terephthalate, Fe Fumarate, Fe Tetramethylterephthalate, Fe Muconate, Fe	BU, AZT-TP, CDV, DOX, ibuprofen, caffeine, urea, benzophenone-3, and benzophenone-4	[24]
CPO-27-Ni	Terephthalate, Ni	RAPTA-C	[53]
ICODIP	TATAT, Zn	5-FU	[54]
ZIF-8	Methylimidazole, Zn	5-FU	[39]
ADUWIH (MOF-1)	Formadide, Zn	5-FU	[55]
Cu-BTC	Tricarboxylate, Cu	5-FU	[38]
MIL-88B(Fe)-X X = 4H, $2CF_3$, $2CH_3$, OH, $4CH_3$, 4F, CH_3, NH_2, NO_2, Br	Terephthalate, Fe	Caffeine	[56]
MIL-53 MIL-100 MIL-127 UiO-66	Tetracarboxylate, Fe Dicarboxylate, Zr	Caffeine	[35]
MIL-100	Trimesate, Fe	AZT, AZT-MP, and AZT-TP	[57]
MOF-74	DOT, Fe	Ibuprofen	[58]
PORLAL	Carboxylate, Gd	DOX	[59]
UVUFEX (UiO-Cis)	Dicarboxylate, Zr	Cisplatin and pooled siRNA	[60]
Bio-MIL-5	Azelate, Zn	Azelaic acid	[32]
Medi-MOF-1	Methoxyphenolate, Zn	Ibuprofen	[61]
PCN-333-Al	Porphyrin, Al	Enzyme capsulation	[62]
ZIF-7 ZIF-8	Methylimidazole, Zn	DOX	[63]
PCN-221-Cu	Porphyrin, Cu	MTX	[64]
ZJU-64 ZJU-64-CH_3	Adenine, Zn	MTX	[65]

Continued

Table 6.1 Summary of experimental studies on MOFs used for drug storage and drug delivery.—*cont'd*

MOFs	Building components	Cargos	Reference
MIL-100	Trimesate, Fe	RAPTA-C	[66]
MIL-100	Trimesate, Fe	Aspirin	[67]
NAKREZ	Carbazole, Zn	5-FU	[68]
PI-2-COF PI-3-COF	Amino	5-FU, captopril, and ibuprofen	[69]
ZIF-8	Methylimidazole, Zn	DOX	[41]
ZJU-101	BPYDC, Zr	Diclofenac sodium	[70]
CD-MOF-1 CD-MOF-2	CD, Cs	5-FU and MTX	[34]
mesoMOF	Terephthalate, Fe	DOX	[71]
MIL-100 UiO-66_COOH	Trimesate, Fe Dicarboxylate, Zr	Gentamicin	[72]
ZIF-8	Methylimidazole, Zn	Insulin	[73]

5-FU, 5-fluorouracil; AZT-MP, azidothymidine monophosphate; AZT-TP, azidothymidine triphosphate; BTC, benzene-1,3,5-tricarboxylate; BU, busulfan; CD, cyclodextrin; CDV, cidofovir; COF, covalent organic framework; CPO, coordination polymer of Oslo; DOT, dioxidoterephthalate; DOX, doxorubicin; MIL, Materials of Institut Lavoisier; MOF, metal-organic framework; MTX, methotrexate; PCN, porous coordination network; PI, polyimide; RAPTA-C, Ru(p-cymene)Cl₂(1,3,5-triaza-7-phosphadamantane); siRNA, small interfering RNA; TATAT, 5,5′,5″-(1,3,5-triazine-2,4,6-triyl)tris(azanediyl)triisophthalate; UiO, University of Oslo; ZIF, zeolitic imidazolate framework; ZJU, Zhejiang University.

6 days). The drug delivery process was governed by the interactions between drug molecules and the Lewis acid metal sites of these two MOFs. Similarly, the same research group [51] investigated the ibuprofen encapsulation performance of flexible Cr-based and Fe-based MIL-53 in 2008. Both Cr-based and Fe-based MIL-53 series gave similar ibuprofen uptakes as 0.2 g g^{-1} ($\sim$17.4 wt%), indicating that metal sites do not have a significant effect on ibuprofen adsorption. The adsorption mechanism can be explained by the hydrogen bonding between the carboxylic group of ibuprofen molecules and the hydroxyl group of the frameworks. Drug uptake capacities of these MOFs were found to be lower than those of MIL-100 and MIL-101 due to the narrow pores ($\sim$8 Å) of MIL-53 series. The flexibility of MIL-53 series also provided a breathing phenomenon for structural transition under various stimuli such as pressure, temperature, and pH and a long-term ibuprofen releasing profile (within 3 weeks in SBF at 310 K) was observed. This work mainly pointed out that pore openings of flexible MOFs can be optimized for drug delivery applications.

Besides MIL series, ibuprofen encapsulation in different types of MOFs has been extensively studied in the literature. For example, in 2014, MOF-74-Fe (CPO-27-Fe(II), CPO stands for coordination polymer of Oslo), which was constructed by dioxidoterephthalate ligand, was tested for ibuprofen storage at 333 K by Yang and coworkers [58]. Following the procedure developed by Long's group [74],

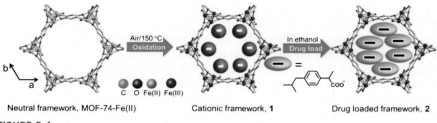

Neutral framework, MOF-74-Fe(II)   Cationic framework, **1**   Drug loaded framework, **2**

FIGURE 6.4

The oxidation process of MOF-74-Fe(II) and anionic ibuprofen loading. *MOF*, metal-organic framework.

Reprinted from Hu Q, Yu J, Liu M, Liu A, Dou Z, Yang Y. A low cytotoxic cationic metal–organic framework carrier for controllable drug release. Journal of Medicinal Chemistry 2014;57(13):5679–5685 with permission of ACS.

they added positive charges to the host by oxidation reaction of the Fe(II) nodes in the air at 423 K for 20 min. Fig. 6.4 illustrates the neutral framework of MOF-74 and its postoxidation process. As shown in Fig. 6.4, MOF-74-Fe(II) has one-dimensional hexagonal channels, and these large accessible channels are suitable for encapsulation of large drug molecules. The anionic ibuprofen loading in the cationic MOF-74-Fe was reported as ~ 15.9 wt% and the complete delivery of ibuprofen took around 100 h in phosphate-buffered saline (PBS, pH = 7.4) at 310 K.

Zhu and coworkers [61] also investigated ibuprofen uptake in a Zn-based medi-MOF-1 that was built with curcumin, a pharmaceutical ingredient. They reported that almost complete drug release from medi-MOF-1 (97%) occurs within 80 h in PBS (pH = 7.4) at 310 K. Qian's group [70] synthesized a cationic MOF, ZJU (Zhejiang University)-101 and tested it for delivery of an anionic antiinflammatory drug, diclofenac sodium, in inflamed tissues (pH = 5.4) and normal tissues (pH = 7.4). ZJU-101, which was constructed by Zr^{4+} ions and 2,2′-bipyridine-5,5′-dicarboxylate (BPYDC) ligands, gave high diclofenac sodium uptake of 0.546 g g^{-1}. The release of diclofenac sodium in acidic environment (inflamed tissues, pH = 5.4) was faster than that in normal tissues (pH = 7.4), which was attributed to the rapid ion exchange mechanism between the cationic framework and the anionic drug molecule at acidic conditions. Lin's group [67] studied encapsulation of another antiinflammatory drug, aspirin, in MIL-100(Fe) by an immersion method under stirring and investigated drug release profile in PBS (pH = 7.4) and 0.1 M HCl (hydrochloric acid, pH = 1.2) at 310 K. They chose MIL-100(Fe) for aspirin (molecular size, ~ 4.1 Å) encapsulation due to large pore sizes of MIL-100 (18 × 32 Å) and reported the maximum aspirin uptake in MIL-100(Fe) as 1806 μmol mg^{-1}. They also tested the release profile for aspirin in HCl solution to mimic blood and gastric pH and almost no drug release was observed in this medium. However, in PBS a complete release of aspirin was observed within 14 days. All these experimental studies showed that both rigid and flexible MOFs can be great candidates for antiinflammatory drug encapsulation.

As we mentioned in the previous section, Rosi and coworkers [26] synthesized the first bio-MOF (as bio-MOF-1) using adenine ligand and Zn^{2+} ions for storage and release of an antiarrhythmic drug, procainamide. Procainamide encapsulation in bio-MOF-1 was performed by a cation exchange mechanism due to interactions between the anionic structure of bio-MOF-1 and the cationic drug molecule. After 15 days the maximum drug loading (0.22 g g^{-1}, 18 wt%) in this framework was achieved and the complete release of drug took around almost 72 h. However, only a small amount of drug (20%) was released in pure water. In a different study, Serre's group [52] synthesized the first bio-MIL structure, bio-MIL-1. In bio-MIL-1, the cargo molecule was directly used as a ligand (nicotinic acid) and nicotinic acid was released completely after 2 h. Degradation of bio-MIL-1 was also tested under the same condition (in PBS pH $= 7.4$ at 310 K), and after a few hours, this structure degraded completely. The same research group [32] synthesized bio-MIL-5 with Zn^{2+} ions and azelaic acid (AzA) used in dermatologic treatments of acne vulgaris and rosacea. Release experiments were carried out in two different media, MilliQ water (pH $= 6.0$) and Mueller Hinton cation (MHCA, pH $= 7.4$), under stirring at 310 K. In water, it took over for 70 days to release about 57% Zn and 51% AzA. In the MHCA medium including many organic and inorganic nutrients, complete release of Zn and AzA occurred in around 70 days. Antibacterial properties of bio-MIL-5 against *Staphylococcus epidermidis* were also examined, and the number of bacteria was found to be reduced significantly after 2 days of incubation. These studies showed that therapeutic organic linkers can be directly used in the synthesis of biocompatible MOFs. A controlled delivery system with a reasonable release time can be developed by using MOFs as carrier materials.

MOFs were also studied for anticancer and antiviral drugs, as well as some cosmetic molecules. In 2010, Horcajada and coworkers [24] investigated efficient drug delivery systems using nontoxic porous Fe(III)-based MOFs. Besides MIL-100, they built new MOFs with carboxylate organic linkers (MIL-53, MIL-88A, MIL-88Bt, MIL-89, and MIL-101_NH$_2$). They tested *in vitro* degradation of fumarate-based MIL-88A and trimesate-based MIL-100 under physiologic conditions. After 7 days of incubation at 310 K, almost complete degradation of those MOFs occurred. However, they reported that MIL-88A showed lower toxicity than MIL-100 due to the higher LD$_{50}$ of fumaric acid (10.7 g kg^{-1}). They carried out encapsulation experiments of BU, an amphiphilic and antitumoral drug, by soaking it in saturated drug solutions. The maximum BU uptakes in MIL-89, MIL-88A, MIL-100, and MIL-53 were found as 9.8 wt%, 8.0 wt%, 25.5 wt%, and 14.3 wt%, respectively. Among these four MOFs, MIL-100 was the only rigid structure and gave the highest BU uptake due to its large pore apertures (25×29 Å). Encapsulation of DOX, used for breast cancer treatment, within MIL-100 was also tested and the maximum DOX loading was found as 9.1 wt%. They also investigated the encapsulation of two antiviral drugs, AZT-TP and CDV, for HIV (human immunodeficiency virus) in four MOFs. For both drug molecules, the highest drug uptakes (42 wt% for AZT-TP and 41.9 wt% for CDV) were reported for rigid MIL-101_NH$_2$, which was attributed to its large pore sizes (29×34 Å) and the amino-functional group that enhanced the host-drug interactions. Release of three drug molecules (AZT-TP, CDV, and DOX) from MIL-100 showed no burst effect, and the slow release mechanism was explained with the diffusion of drug molecules from the pores and favorable drug-host interactions.

Adsorption of different cosmetic molecules such as caffeine, urea, benzophenone-3, and benzophenone-4 in MIL-53 and MIL-100 was also investigated by Horcajada and coworkers [24]. Results showed that these MOFs are promising for encapsulation of cosmetic molecules. Similarly, Maurin and coworkers [56] tested a series of 10 different MIL-88B(Fe)-X series (X = -4H, -2CF$_3$, -2CH$_3$, -OH, -4CH$_3$, -4F, -CH$_3$, -NH$_2$, -NO$_2$, -Br) for caffeine encapsulation using the impregnation method at room temperature under stirring. Among these MOFs, MIL-88B(Fe)-CH$_3$, which has the nonpolar group -CH$_3$, gave the lowest caffeine uptake (9 wt%), whereas MIL-88B(Fe)-OH gave the highest caffeine uptake (22 wt%). They explained that MOFs with nonpolar groups do not make hydrogen bonds with the guest molecules, resulting in lower caffeine uptake. The same research group [35] later used the same method to investigate caffeine uptake in seven MOFs including MIL-100(Fe), MIL-127(Fe), MIL-53(Fe), MIL-53(Fe)_Br, UiO-66(Zr), UiO-66(Zr)_Br, and UiO-66(Zr)_NH$_2$. Among these MOFs, MIL-100(Fe), MIL-127(Fe), MIL-53(Fe), and UiO-66(Zr) exhibited promising performances for caffeine adsorption, storing 50 wt%, 16 wt%, 30 wt%, and 24 wt% caffeine, respectively. In comparison with MIL-88B series, these MOFs showed high caffeine uptakes, except MIL-127(Fe). Interestingly, the flexible MIL-53 gave high performance for caffeine adsorption due to its full pore opening mechanism in the presence of ethanol-caffeine solution. Fast caffeine release profile from these MOFs was observed in PBS medium, indicating the degradation of the frameworks. On the other hand, in distilled water, there was no release of ligands, as no degradation of structures was observed. In a different study, Gref and coworkers [57] tested the encapsulation of antiviral drugs, AZT, AZT-TP, and azidothymidine monophosphate (AZT-MP) in MIL-100(Fe) with incubating the nanoparticles from 30 min to 72 h. AZT encapsulation in MIL-100(Fe) was not efficient ($\sim$1.4 wt%) due to poor ionic interactions between the framework and the drug molecule. They also reported that AZT-MP loading in MIL-100 ($\sim$36 wt%) was higher than AZT-TP loading in this MOF ($\sim$25 wt%) because small AZT-MP molecules (263 $\mathring{A}^3$) could find more space within the pores of MIL-100 than large AZT-TP (356 $\mathring{A}^3$) molecules. AZT-MP molecules ($\sim$60%) were released more quickly than AZT-TP molecules ($\sim$45%) after 8 h of incubation. This study emphasized that MOFs can be efficiently used for the delivery of drug molecules that contain phosphorylates. With these pioneering studies, the research on drug and/or cosmetic molecule encapsulation in MOFs has started to gain momentum with the interest in developing new cargo delivery applications of MOFs.

In 2011, nonconventional anticancer metallodrug Ru(p-cymene)Cl$_2$(1,3,5-triaza-7-phosphadamantane (RAPTA-C), which is used to suppress lung metastases, was tested for desorption from CPO-27-Ni (Ni, nickel) by Barea and coworkers [53]. RAPTA-C uptake in this MOF was reported as 1.1 g g^{-1}. They estimated that

10 mg of RAPTA-C was desorbed from the MOF after 12 h in SBF (pH = 8.05) at 310 K. This study shows that MOFs can be also used for the delivery of nonconventional metallodrugs. In 2016, Navarro and coworkers [66] tested RAPTA-C encapsulation in MIL-100(Fe) using the impregnation method and reported 0.42 g RAPTA-C per gram of MIL-100(Fe) after 4 h. They then investigated RAPTA-C delivery from MIL-100(Fe) in two different media, i.e., (1) SBF and (2) SBF with bovine serum albumin (BSA, 5.4% w/v), at 310 K for 6 days. After 3 days, RAPTA-C release from MIL-100(Fe) in SBF with BSA ($\sim$80%) was higher than that in pure SBF ($\sim$50%). Presence of BSA proteins led to a faster diffusion profile due to better binding between protein molecules and metallodrugs. Wang and coworkers [54] studied another anticancer drug, 5-FU, used for various cancer treatments, such as skin, anal, pancreatic, and breast cancers, using an MOF, ICODIP. This MOF was built with triazine (5,5',5''-(1,3,5-triazine-2,4,6-triyl)tris(azanediyl) triisophthalate [TATAT]) derivatives, and the maximum 5-FU loading was reported as 0.5 g g^{-1}. To evaluate the 5-FU delivery profile in this MOF, they used phosphate PBS (pH = 7.4) as a buffer solution at room temperature. After a week, 86.5% of the drug was released from the framework with no burst effect. They suggested that MOFs with TATAT organic linkers can be used for both high drug loadings and slow drug release profiles. The same research group [39] investigated 5-FU delivery from ZIF-8 in two different pH media to examine the pH effect on drug delivery. Experiments for 5-FU loading in ZIF-8 (0.66 g g^{-1}) were carried out by the impregnation method. They reported that the release of 5-FU from ZIF-8 in PBS (pH = 7.4) is slower than that in acetate buffer (pH = 5.0) at 310 K, indicating that 5-FU delivery is faster in acidic environments. This study shows that drug delivery to tumors can be studied by MOFs that have pH sensitivity. Similarly, Ma and coworkers [55] studied 5-FU encapsulation within an MOF, ADUWIH (MOF-1), and the 5-FU loading in MOF-1 was reported as 0.37 g g^{-1}. Nascimento and coworkers [38] also examined 5-FU encapsulation in Cu-BTC and showed a slow drug release profile within this MOF. Drug loading in Cu-BTC was reported as 0.82 g g^{-1} after 7 days under stirring conditions. Cao and coworkers [68] synthesized a water-stable MOF, NAKREZ, with carbazole organic ligands and Zn^{2+} anions. They reported the maximum 5-FU loading in NAKREZ as 53.3 wt% ($\sim$0.6 g g^{-1}). After 3 days, one-step drug release profile occurred in PBS ($\sim$65%) and water (82%) environments due to hydrogen bonds and π bond interactions between the drug molecule and the organic skeleton part.

Covalent organic frameworks (COFs), which have strong covalent bonds between carbon (C), hydrogen (H), and oxygen (O) atoms, have also been tested for encapsulation of drug molecules [75]. Two COFs, namely, polyimide (PI)-2-COF and PI-3-COF, were synthesized by Zhao and coworkers [69] and tested for encapsulation of three drugs, 5-FU, captopril, and ibuprofen. 5-FU delivery from these structures occurred completely after 3 days in PBS at 310 K. Lan and coworkers [34] examined the storage performances of two sugar MOFs with β-CD ligands, CD-MOF-1, and CD-MOF-2, for two different anticancer drugs, 5-FU and MTX, which have been widely used to decelerate the growth of breast, leukemia, and lung cancer cells. Drug loading experiments were carried out in drug/ethanol solution for 6 days under stirring, and 5-FU (MTX) loadings in CD-MOF-1 and CD-MOF-2 were found as 1.379 g g^{-1} and 1.510 g g^{-1} (0.689 g g^{-1} and 1.217 g g^{-1}), respectively. Similarly, Qian and coworkers [65] tested two novel Zn-based porous MOFs with adenine linkers, ZJU-64 and ZJU-64-CH$_3$, for MTX delivery at pH 7.4 and two different temperatures (310 and 333 K). ZJU-64-CH$_3$ (10.63 wt%) gave lower MTX uptake than ZJU-64 (13.45 wt%) due to the presence of methyl groups that block its channels. In another study performed by Qian and coworkers [64], a porphyrin-based MOF, porous coordination network (PCN)-221(Cu), was also tested for MTX encapsulation. The maximum MTX loading was reported as 0.40 g g^{-1} at 333 K for 72 h. Drug release experiments were performed in PBS solution with two different pH values (7.4 and 2.0) at 310 K. After 72 h, the amount of drug at pH 2.0 (40%) was lower than that at pH 7.4 (100%) due to the electrostatic interactions between porphyrin ligands and drug molecules at gastric pH environment. They also tested cell cytotoxicity of PCN-221 using PC12 (rat pheochromocytoma) cells. As the concentrations (5, 20, 50, and 100 μg mL^{-1}) increased, PC12 cell viability gave a constant dosage as around 90%. The maximum dosage of MOF (100 μg mL^{-1}) led to higher cytotoxicity, whereas cell viability was found around 59%, indicating that PCN-221 can be a low-toxic carrier material. This is the first study in the literature to show that MOFs can be efficient nanocarriers for oral anticancer drug delivery.

In 2014, an MOF (PORLAL) with a fluorescent linker pDBI (1,4-bis(5-carboxy-1H-benzimidazole-2-yl)benzene) and gadolinium (Gd^{3+}) ions was tested for delivery of the anticancer drug DOX by Banerjee and coworkers [59]. The maximum DOX loading in PORLAL was reported as 12 wt% due to its large pore apertures (19 × 12 Å). They also examined DOX release profile in PBS (pH = 7.4) and citric acid (pH = 5.0), and after 5 h, DOX release in acidic environment (44%) occurred more quickly than in PBS (22%). In a follow-up study, Chakraborty and coworkers [63] tested DOX encapsulation in two MOFs, ZIF-7 and ZIF-8, and reported the maximum DOX uptakes in these MOFs as 40 wt% and 52 wt%, respectively. Zou and coworkers [41] also investigated DOX encapsulation in ZIF-8 using the one-pot synthesis method to improve the drug loading efficiency. They reported that DOX uptake in this material can be achieved up to 20 wt%, one of the highest DOX uptakes among the MOF carriers. They also tested its delivery profile at a wide range of pH (5, 5.5, 6, 6.5, and 7.4) at 310 K, and after almost 8 days, complete drug release occurred at low pH values (pH = 5.0−6.0) due to the framework decomposition under acidic conditions. In 2017, Li and coworkers [71] synthesized a mesoporous metal-organic framework (mesoMOF) with Fe^{3+} ions using the double-template method, including cetyltrimethylammonium bromide and citric acid as double-template agents. These double-template molecules were chosen to create the pores and to formulate a biocompatible MOF. They reported that DOX can be encapsulated in this mesoMOF up to 55 wt% without burst effect at 310 K due to its large surface area (2964 m^2 g^{-1}) and large pore diameters (22 Å).

MOFs can be also used for encapsulation of biomacromolecules such as proteins, enzymes, and DNA, which requires several considerations such as chemical affinity, controlled release, and efficiency. For example, Feng's group [62] tested a porphyrin-based MOF, PCN-333, for encapsulation of three different enzymes, horseradish peroxidase (HRP), cytochrome (Cyt) *c*, and microperoxidase (MP)-11. The maximum loadings of HRP, Cyt *c*, and MP-11 in PCN-333 were found as 22.7, 77.0, and 478 $\mu mol\ g^{-1}$, respectively. As MP-11 ($11 \times 17 \times 33$ Å) is smaller than HRP ($40 \times 44 \times 68$ Å) and Cyt *c* ($26 \times 32 \times 33$ Å), MP-11 uptake was found to be the highest. In another study, Hoop and coworkers [73] tested ZIF-8 for the encapsulation of a hormone, insulin, due to its high stability and remarkable insulin uptake capacity (20.73 wt%). They elucidated that this amount of insulin is adequate for type I and II diabetes based on their daily required amount of insulin (0.5 IU/kg/day). They also investigated the toxicology of ZIF-8 in six different cell lines, including kidney, skin, breast, blood, bone, and connective tissue cell lines, and reported that ZIF-8 exhibits biocompatible properties in these cell lines up to $30\ \mu g\ mL^{-1}$. Lin and coworkers [60] investigated the co-delivery of an ovarian cancer drug, cisplatin, and small interfering RNAs (siRNAs) from a Zr-based MOF (UVUFEX) to achieve efficient therapy for silencing genes. The analysis of nuclear magnetic resonance showed that cisplatin is located within the pores of UVUFEX by noncovalent encapsulation and the phosphates in siRNA are located on Zr metal sites of the structure. The uptakes of cisplatin and siRNA in UVUFEX were found as 12.3 wt% and 81.6 wt%, respectively. This study shows that co-delivery of drug molecules using MOFs can be used for the treatment of cancer. Blanco-Prieto and coworkers [72] tested the release of an antibiotic, gentamicin, from MIL-100 (Fe) and UiO-66_COOH. They reported that UiO-66_COOH cannot adsorb large gentamicin molecules ($16.7 \times 7.4 \times 6.5$ Å) due to its narrow openings (~ 6 Å). They also used two solvents, distilled water and dichloromethane (DCM), to investigate the effect of the solvent used in the impregnation method on drug loading efficiency. Results showed that gentamicin uptake in DCM (from 400 to 607 $\mu g\ mg^{-1}$) is higher than that in water (207 $\mu g\ mg^{-1}$), indicating that choosing an adequate solvent in the impregnation method is important to achieve a high drug loading efficiency.

So far, we have discussed the examples of MOF-based drug delivery systems. In all these studies, MOFs were synthesized as crystalline structures. Amorphous MOFs (a_mMOFs) have been examined in drug delivery applications due to their highly robust structures [76]. Fairen-Jimenez's group [76] investigated the encapsulation of a hydrophilic drug, calcein, in UiO-66 and its amorphous structure (a_mUiO-66), which was achieved by the ball-milling process. Almost complete calcein release from UiO-66 occurred within 2 days, whereas calcein delivery from a_mUiO-66 increased up to 30 days ($\sim 80\%$) in PBS solution (pH = 7.4) at 310 K. This study shows that the drug release time can be increased using the amorphization technique for all other MOFs and crystalline materials. The same research group [77] used a family of Zr-based MOFs, having different organic ligand forms of BDC, and tested them for calcein and α-cyano-4-hydroxycinnamic acid (α-CHC) delivery. The highest α-CHC loading was reported as 15.2 wt% (31 wt%) in Zr-based MOFs. After encapsulation of drug molecules, the ball-milling process was similarly performed for the amorphization of crystalline Zr-based MOFs. Results showed that a complete drug release profile is observed within 2—3 days for the crystalline Zr-based MOFs, whereas this time can be increased up to 15 days for the a_mMOFs. These pioneering studies showed that the amorphization technique can be used to control the drug delivery mechanism within MOFs.

We discussed the recent experimental studies on storage of drugs, enzymes, hormones, and cosmetic molecules and their delivery in MOFs. All these studies are very important in the scientific community and provide new opportunities for synthesizing new functional structures and developing novel techniques for biomedical applications of MOFs.

6.2.2 Biomedical gas storage

MOFs have been great candidates for storage of biomedical gases such as carbon monoxide (CO), nitric oxide (NO), hydrogen sulfide (H_2S), and O_2 due to their highly porous structures and tunable chemical and physical properties. CO, H_2S, and NO are biological signaling molecules that can freely permeate through cell membranes [22]. All these gases are produced endogenously in the body and vital for life. O_2 storage is also very important for medical devices. Table 6.2 shows both experimental and computational studies of MOFs on biomedical gas storage.

The first biological signaling molecule, also known as a gasotransmitter molecule, is NO in Table 6.2. NO is a colorless, diatomic molecule and has been used in different *in vivo* and *in vitro* applications for antibacterial, antithrombotic, and wound healing purposes [89]. The concentration of NO in the body is important because a very high NO concentration can cause hypotension, bleeding, and inflammation, whereas a very low NO concentration can cause hypertension and weakness in fighting against bacteria [22]. The first study on NO delivery in various MOFs including CPO-27-Ni, CPO-27-Co, Cu-BTC, MIL-53(Al), and MIL-53(Cr) at 310 K was reported by Morris and coworkers [78]. They emphasized that porous materials with open metal sites have enhanced performance for the adsorption and release of NO due to strong binding to the metal sites of MOFs.

For example, CPO-27-m (m = Ni or Co and ~ 7 mmol NO/g MOFs) and Cu-BTC (~ 3 mmol g^{-1} MOF) exhibited higher NO uptakes than MIL-53-n (n = Al or Cr and ~ 1 mmol g^{-1} MOFs) at 1 bar and 310 K. The same research group [79] also investigated NO adsorption in Cu-sulfoisophthalate (SIP)-3 at 298 K up to 1 bar due to its controllable phase transformation and adequate thermal stability. NO molecules were adsorbed above 0.3 bar and the highest NO uptake capacity of Cu-SIP-3 was reported as 1.1 mmol g^{-1} at 1 bar. McKinlay and coworkers [80] investigated the release profile of NO from two different MOFs, CPO-27-Ni and Cu-BTC. They reported that complete delivery of NO from CPO-27-Ni (~ 2 mmol g^{-1} after ~ 30min) occurrs quicker than that from Cu-BTC (~ 2 h). Bio-MIL-3 (an MOF having azobenzene derivatives and Ca^{2+} cations) was also studied for NO adsorption at 298 K up to 20 bar by Serre and coworkers [31].

Table 6.2 Several examples of MOFs used for biomedical gas storage.

MOFs	Components	Gases	Reference
CPO-27-Co CPO-27-Ni Cu-BTC MIL-53(Al) MIL-53(Cr)	Terephthalate, Co, Ni Tricarboxylate, Cu Terephthalate, Al, Cr	NO	[78]
Cu-SIP-3	Phthalate, Cu	NO	[79]
CPO-27-Ni Cu-BTC	Terephthalate, Ni Tricarboxylate, Cu	NO	[80]
Bio-MIL-3	Tetracarboxylate, Ca	NO	[31]
CPO-27-Ni	Terephthalate, Ni	CO	[81]
MIL-88B NH$_2$-MIL-88B	Terephthalate, Fe	CO	[82]
MIL-53(n) n = Al, Cr, Fe MIL-47(V) MIL-100(Cr) MIL-101(Cr)	Terephthalate	H$_2$S	[83]
MIL-47 MIL-53	Terephthalate, V Cr	H$_2$S	[84]
CPO-27-n n = Ni, Zn	Terephthalate	H$_2$S	[85]
NU-125 Cu-BTC UiO-66	Dicarboxyphenyl, Cu Tricarboxylate, Cu Dicarboxylate, Zr	O$_2$	[86]
UiO-66	Dicarboxylate, Zr	O$_2$	[87]
UMCM-152	Dicarboxylate, Cu	O$_2$	[88]

BTC, *benzene-1,3,5-tricarboxylate*; CPO, *coordination polymer of Oslo*; MIL, *Materials of Institut Lavoisier*; MOF, *metal-organic framework*; SIP, *sulfoisophthalate*; NU, *Northwestern University*; UiO, *University of Oslo*; UMCM, *University of Michigan Crystalline Material*.

Although bio-MIL-3 adsorbed lower NO ($\sim$0.8 mmol g^{-1} at 1 bar) than CPO-27 families ($\sim$7 mmol g^{-1}), this MOF exhibited slow NO release without any burst effect. Therefore bio-MIL-3 can be a promising material for effective NO delivery. According to these recent studies, MOFs can be the potential carriers for controlled delivery of NO to biological systems.

Another gasotransmitter molecule is CO, an odorless and colorless gas. It can irreversibly bind to hemoglobin (Fe metal center) and affect the O$_2$ delivery process in the body, resulting in tissue hypoxia and poisoning due to this strong binding. However, CO-releasing molecules have been receiving attention due to their potential for therapeutic applications including antiinflammatory, antithrombotic, and antiproliferative activities [90]. However, developing an effective and controlled-delivery mechanism for CO is highly required due to its high toxicity. MOFs as potential carriers have also been tested for CO delivery. For example, Chavan et al. [81] examined CO adsorption in CPO-27-Ni at 303 K and reported that CO can strongly bind to Ni^{2+} cations due to strong electrostatic interactions. Metzler-Nolte and coworkers [82] investigated the release of CO from flexible Fe-based MIL-88B and NH_2-MIL-88B in PBS (pH = 7.4) at 310 K and reported that NH_2-MIL-88B delivers more CO (0.69 µmol/1 mg of framework within 76 min) than MIL-88B (0.36 µmol/1 mg of framework within 38 min). Results showed that CO release kinetics depend on the degradability of these two MOFs in PBS. These two highly flexible MOFs can be great candidates for CO storage and delivery due to their controllable CO delivery profiles. There are only a few studies on CO uptake of MOFs; however, MOFs have the potential for CO storage due to favorable electrostatic interactions between CO molecules and the frameworks' atoms, which provide a controlled-delivery mechanism for CO molecules.

The third gasotransmitter gas is H_2S, which is produced in different cells and tissues of humans by enzymatic synthesis. H_2S has also been used for the treatment of cardiovascular diseases, diabetes, and other conditions because of its antiinflammatory effects [91]. Weireld and coworkers [83] investigated H_2S uptake in six different MOFs, MIL-53(Al, Cr, Fe), MIL-47(V), MIL-100(Cr), and MIL-101(Cr) at 303 K. They reported that MIL-100 and MIL-101 have strong interactions with H_2S, exhibiting high H_2S uptakes as 16.7 mmol g^{-1} and 38.4 mmol g^{-1} at $\sim$20 bar, respectively. These MOFs adsorbed much higher H_2S than the MIL-53-n series (n = Al, Cr, Fe; 13.12, 11.77, and 8.53 mmol g^{-1} at 16 bar, respectively) and MIL-47 (14.6 mmol g^{-1} at 16 bar) due to their larger pore apertures. Since the MIL-53 series and MIL-47 have approximately similar pore sizes ($\sim$8.5 Å), they exhibited almost similar H_2S uptakes. Maurin and coworkers [84] also studied H_2S uptake in MIL-47(V) and MIL-53(Cr) under the same conditions (at 303 K and up to $\sim$20 bar) by using both experimental and simulation methods. They reported that MIL-47-V ($\sim$14 mmol g^{-1}) has a slightly higher H_2S uptake than MIL-53-Cr ($\sim$12 mmol g^{-1}) at 18 bar. This was attributed to the flexible structure of MIL-53(Cr). When the pressure was increased, the framework of MIL-47(V) remained rigid, whereas the pore sizes of MIL-53(Cr) became narrower, resulting in a lower gas uptake. Morris and coworkers [85] reported the maximum H_2S uptake in CPO-27-Ni (Zn) as $\sim$12 mmol g^{-1} ($\sim$10 mmol g^{-1}) at 303.15 K and 1 bar. They also investigated H_2S release profile from these MOFs and reported that the amount of H_2S released from Ni-CPO (1.8 mmol g^{-1}) is much higher than Zn-CPO (0.5 mmol g^{-1}) after 30 min. Zn-CPO-27 was chosen for *in vitro* experiments, which was tested in Krebs buffer solution (pig coronary artery bath) at 310 K, because of its lower toxicity than Ni-CPO. Zn-CPO-27 stayed in this solution only 5 min because of artery relaxation. H_2S released from Zn-CPO-27 was found to be biologically active related to vascular relaxation. These studies show that MOFs can also be used for the adsorption and delivery of H_2S molecules.

O_2 has been treated as a medical gas like CO, NO, and H_2S. When human cells, tissues, and organs require more O_2, it is supplied by inhalation mask [92]. High-pressure O_2 storage is significant for the treatment of respiratory insufficiency and the hyperbaric O_2 treatment for CO poisoning [93]. High-pressure cylinder tanks (140 bar as adsorption and 5 bar as desorption pressures) have been used for medical

O$_2$ storage. Smaller and efficient storage systems can be designed using MOFs owing to their large surface areas and high pore volumes. To the best of our knowledge, there are a few studies on O$_2$ storage of MOFs in the literature. Piscopo et al. [87] synthesized two fluorine-containing UiO-66 series and O$_2$ adsorption capacity of these MOFs was enhanced with the favorable O-fluorine interactions. Farha and coworkers [86] first used a computational method to screen 10000 hypothetical MOFs to identify the best performing material for O$_2$ storage and then performed experiments for the selected materials. They reported that Cu-BTC (13.2 mol kg^{-1}) and NU-125 (17.4 mol kg^{-1}) gave high O$_2$ uptakes at 140 bar and room temperature, indicating that these two MOFs can be promising adsorbents for O$_2$ storage. Similarly, Fairen-Jimenez et al. [88] performed molecular simulations to predict O$_2$ storage in 2932 existing MOFs and validated their predictions by experiments for the top material, UMCM-152 (UMCM stands for University of Michigan Crystalline Material). They reported that UMCM-152 can deliver 22.5% more O$_2$ than the best material known to date (NU-125) and can enhance the O$_2$ deliverable capacity by 90% compared with the capacity of an empty tank. This computational study shows that molecular simulations can be used to identify promising candidates for biomedical gas storage. Details of molecular simulations and recent computational studies are discussed in the following section.

6.3 Molecular simulations of metal-organic frameworks for drug storage and drug delivery

6.3.1 Computational methods

Molecular simulations can provide insights into the drug adsorption and release mechanisms in MOFs at the atomistic level and guide experiments for the design of novel drug carrier systems. Quantum mechanics (QM) and classical mechanics (CM) are used to identify favorable drug-host configurations and compute molecular interactions, respectively.

In QM calculations, the time-dependent and nonrelativistic Schrödinger equation is solved as follows:

$$H\Psi = E\Psi \tag{6.1}$$

where E is the total energy of the system, Ψ is the n-electron wave function based on the identities and positions of the nuclei and electrons, and H is the Hamiltonian operator showing the kinetic and potential energies for each particle in the system. Hamiltonian operator can be expressed as follows[94]:

$$H = -\frac{\hbar^2}{2m_e} \sum_{i}^{\text{electrons}} \nabla_i^2 - \frac{\hbar^2}{2} \sum_{A}^{\text{nuclei}} \frac{1}{M_A} \nabla_A^2 - \frac{e^2}{4\pi\varepsilon_0} \sum_{i}^{\text{electrons}} \sum_{A}^{\text{nuclei}} \frac{Z_A}{r_{iA}}$$

$$+ \frac{e^2}{4\pi\varepsilon_0} \sum_{i(i>j)}^{\text{electrons}} \sum_{j}^{\text{electrons}} \frac{1}{r_{ij}} + \frac{e^2}{4\pi\varepsilon_0} \sum_{A(A>B)}^{\text{nuclei}} \sum_{B}^{\text{nuclei}} \frac{Z_A Z_B}{R_{AB}} \tag{6.2}$$

where Z_A is the nuclear charge, M_A is the mass of nucleus A, m_e is the mass of the electron (e), R_{AB} is the distance between nuclei A and B, r_{ij} is the distance between electrons i and j, r_{iA} is the distance between electron i and nucleus A, ε_0 is the permittivity of free space, and $\hbar$ is the Planck constant divided by 2π. Several assumptions, including Born-Oppenheimer and Hartree-Fock approximations, are done to solve Eq. (6.2). Born-Oppenheimer approximation assumes that the nuclei move much more slowly than electrons, resulting a zero-kinetic energy term and a constant coulombic term for the nuclei in Eq. (6.2). On the other hand, Hartree-Fock approximation assumes that electrons move independently, and the coordinates of a single electron can be expressed as a set of differential equations to be solved numerically. This method also assumes that molecular orbitals can be written as linear combinations of defined functions, which are known as basis functions [94]. Another widely used QM method is density functional theory (DFT), which depends on the electron density of the system rather than the many-body wave function [94]. DFT method is the preferred QM method in the literature to examine the electronic structure of molecular systems due to lower computational cost than Hartree-Fock models. The favorable drug-host configurations can be obtained by using DFT simulations. To compute the partial charges of atoms in drug molecules, electrostatic potential, which is the energy of the interaction of a unit positive charge at some point in space with the nuclei and the electrons of a molecule, is commonly used.

In CM calculations, the microscopic state of a system can be defined by using the positions and momenta of a set of particles. The Hamiltonian of the system is written as a function of coordinates (q) and momenta (p) of the particles:

$$H(q,p) = K(p) + U(q) \tag{6.3}$$

where K is the kinetic energy of the system and U is the potential energy, which is based on the coordinates of individual atoms and pairs. The molecular energy can be expressed by using Taylor expansion in bonds, bends, and torsions as follows [95]:

$$U = \sum_{\text{bonds}} u_b(r) + \sum_{\text{bends}} u_\theta(\theta) + \sum_{\text{torsions}} u_\varphi(\varphi) + \sum_{\text{nonbonded}} u_{nb}(r) + \dots \tag{6.4}$$

where u_b is the bond stretching potential related to the bond length (r), u_θ is the bending energy that depends on the angle (θ) between two particles, u_φ is the torsional potential and it depends on the torsional angle φ, and u_{nb} is the nonbonded energy (long-range intermolecular interactions) between particles. Three-body and higher terms are not generally considered in molecular simulations due to high computational cost.

The intermolecular interactions (nonbonded) between two particles are commonly considered as van der Waals forces and long-range coulombic interactions. Lennard-Jones (LJ) 12-6 potential is (shown in Eq. 6.5) widely used to compute nonbonded interactions in molecular simulations[96].

$$U_{LJ} = 4\varepsilon_{ij} \left[\left(\frac{\sigma_{ij}}{r_{ij}} \right)^{12} - \left(\frac{\sigma_{ij}}{r_{ij}} \right)^{6} \right] \tag{6.5}$$

In this equation, U_{LJ} is the intermolecular potential between two particles (i and j), r is the distance of separation from the center of one particle to the center of the other particle, ε_{ij} is the well depth, and σ_{ij} represents a molecular length scale based on the particle diameter. The parameters, σ and ε, in the LJ potential are obtained from a force field that includes a set of equations to calculate the potential energy between interacting atoms with specific parameters [97]. The universal force field [98] and DREIDING [99] are generally used in molecular simulations of MOFs [100,101]. The atom-based summation method is used for van der Waals terms, and a cut-off radius is defined to neglect the interactions between all pairs of atoms, which are further beyond this distance. In molecular simulations, periodic boundary conditions were applied to diminish the surface effects. For the cross-interactions between two different LJ sites, Lorentz-Berthelot mixing rules are used [102,103]. The electrostatic potential energy (U_{ij}) can be calculated by the Coulomb potential, as shown in Eq. (6.6), where ε_0, q_i, and q_j represent the electric constant and partial atomic charges of i and j, respectively.

$$U_{ij} = \frac{1}{4\pi\varepsilon_0} \frac{q_i q_j}{r_{ij}} \tag{6.6}$$

The partial charges of MOFs' atoms can be calculated by using different methods including the connectivity-based atomic contribution (CBAC) method [104], the extended charge equilibration method (EQeq) [105], and DFT-based methods [106]. The CBAC and EQeq methods are less accurate than DFT-based methods, but due to their low computational cost, these methods are widely used in molecular simulations including a large number of atoms.

In order to compute the amount of adsorbed drug molecules in MOFs, Monte Carlo simulations including grand-canonical Monte Carlo (GCMC) and configurational bias Monte Carlo (CBMC) methods are used. These methods are based on the ensemble averaging in statistical mechanics. GCMC simulations assume a grand-canonical ensemble (μ, V, T) to predict the average number of particles in the system at equilibrium. In molecular simulations, "Widom test particle insertion method" is performed to compute the chemical potential. The chemical potential of the system is based on the energy change due to the insertion of this test particle. An appropriate equation of state can also be used to convert the chemical potential to the corresponding fugacity. In GCMC simulations, molecules are added into the system with trial moves, such as particle addition, particle deletion, particle rotation, and particle displacement. Based on an acceptance rule, the energy of the system is computed for these different moves. In CBMC simulations, molecules are added into the system part by part biasing a growth process based on the energetically favorable configurations [95]. Similar to conventional GCMC, the bias is removed based on an acceptance rule. This method is suitable for long-chain molecules. The output of GCMC and CBMC simulations is the adsorbed number of drug molecules in an MOF.

Isosteric heat of adsorption (Q_{st}) is also computed in molecular simulations to identify MOF-drug binding interactions using the ensemble average fluctuations [107].

$$Q_{st} = RT - \left(\frac{\langle U_{ads} \times N_{ads} \rangle - \langle U_{ads} \rangle \times \langle N_{ads} \rangle}{\langle N_{ads}^2 \rangle - \langle N_{ads} \rangle \times \langle N_{ads} \rangle} \right) \tag{6.7}$$

where R is the ideal gas constant, T is the temperature, $\langle U_{ads} \rangle$ is the average potential energy of adsorbed phase, and $\langle N_{ads} \rangle$ is the average number of adsorbed drug molecules in MOFs.

Molecular dynamics (MD) simulations are used to compute diffusion coefficients of drug molecules in MOFs [107]. In MD simulations, Newton's equations of motion are solved numerically. Intermolecular and intramolecular interactions are considered to compute the forces on the molecules and the positions and velocities of drug molecules are computed until the equilibrium is reached [108]. The self-diffusivities of drug molecules in the pores of MOFs are computed from the MD trajectories after a predefined simulation time (generally in nanoseconds). The self-diffusivity ($D_{i,\text{self}}$) is calculated from the mean-squared displacement of the individual tagged particles by the Einstein relation [109].

$$D_{i,\text{self}} = \lim_{t \to \infty} \frac{1}{dNt} \left\langle \sum_{i=1}^{N} [\vec{r}_i(t) - \vec{r}_i(0)]^2 \right\rangle \tag{6.8}$$

where $\vec{r}(t)$ is the position of the tagged particle at time t, N is the number of drug molecules, d is the number of spatial dimension for diffusion (d = 2, 4, or 6 for one, two, or three dimensions, respectively), and the angular brackets represent the ensemble average.

6.3.2 Computational studies on metal-organic frameworks for drug storage and drug delivery

Table 6.3 shows the summary of computational studies on MOFs for drug storage and drug delivery applications. The first computational study on a drug delivery application of MOFs was performed in 2009 by Babarao and Jiang [110]. They examined the adsorption and diffusion behaviors of ibuprofen, an analgesic and anti-inflammatory drug, in MIL-101 and UMCM-1 using molecular simulations and DFT calculations. The saturated loading of ibuprofen in MIL-101 (1.11 g g^{-1}) was found

Table 6.3 Summary of computational studies on MOFs for drug storage and drug delivery.

MOFs	Drugs	Methods	Reference
MIL-101 (Cr) UMCM-1	Ibuprofen	DFT, GCMC, MD	[110]
MIL-53(Fe) MIL-53(Fe)-NH$_2$	BU	DFT	[111]
MUWRIH MUWRON MUWRUT	5-FU	GCMC	[112]
RONSIY RONSOE	5-FU	Molecular docking	[113]
Bio-MOF-n (n = 1, 11, 100) UMCM-1	Ibuprofen	DFT, GCMC, MD	[114]
Bio-MOF-100 CD-MOF-1 MIL-53, MIL-100, MIL-101 MOF-74	Ibuprofen	GCMC	[115]
HMOF-1 MIL-47(V), MIL-53(Cr)	Ibuprofen and lysine	GCMC	[116]
IRMOF-n (n = 14, 16)	Tamoxifen	DFT	[117]
CUQDUP	BU	DFT, GCMC	[118]
NTU-Z11 GDMU	5-FU	DFT, GCMC	[119]
CD-MOF-n (n = 1−3) Bio-MOF-n (n = 1, 11, 12, 100 −102) IZUMUM NUDKON MIL-53, MIL-100, MIL-101 MOF-74 (VOGTIV) 9 MOF-74 series (RAVVUH-RAVWAO, RAVWES, RAVWIW) RAVWOC, RAVWUI, RAVXAP RAVXET, RAVXIX)	Ibuprofen, caffeine, and urea	DFT, CBMC, MD	[13]
UiO-n (n = 66, 67)	Paclitaxel and cisplatin	MD	[37]
IRMOF-16 OH-functionalized IRMOF-16	Gemcitabine	DFT, GCMC	[120]
IRMOF-74-III	Gemcitabine	DFT, GCMC, MD	[121]
MOF-74 9 MOF-74 series (RAVVUH-RAVWAO, RAVWES, RAVWIW RAVWOC, RAVWUI, RAVXAP, RAVXET, RAVXIX)	MTX and 5-FU	CBMC, MD	[19]
MAPNOJ	5-FU	Semiempirical AM1 method	[122]

5-FU, 5-fluorouracil; AM1, *Austin Model 1*; BU, *busulfan*; CBMC, *configurational-bias Monte Carlo*; CD, *cyclodextrin*; DFT, *density functional theory*; DMF, *dimethylformamide*; GCMC, *grand-canonical Monte Carlo*; GDMU, ([Zn$_3$(µ$_3$-O)(BTC)$_2$(DMF)]. 2NH$_2$(CH$_3$)$_2$.4(H$_2$O)}$_n$); HMOF, *heterometal-organic framework*; IRMOF, *isoreticular MOF*; MD, *molecular dynamics*; MIL, *Materials of Institut Lavoisier*; MOF, *metal-organic framework*; MTX, *methotrexate*; NTU-Z11, [Zn$_3$(µ$_3$-O)(BTC)$_2$(H$_2$O)]$_n$; UiO, *University of Oslo*; UMCM, *University of Michigan Crystalline Material*.

to be consistent with the experimental measurements (1.37 g g^{-1}). The small difference in ibuprofen uptakes was attributed to the remaining solvent molecules in experimental studies. Diffusion of ibuprofen in MIL-101 was found to be slower than that in UMCM-1 because of the smaller pore volume of MIL-101 than that of UMCM-1 and the coordination bond formation between the carboxylic O of ibuprofen and the chromium site of MIL-101. Similarly, Bei et al. [114] studied ibuprofen adsorption in bio-MOFs-1, -11, and -100 and UMCM-1 using molecular simulations. Among these materials, bio-MOF-100 gave the highest ibuprofen uptake (2.03 g g^{-1}) at 1 bar and 298 K due to its large pore sizes (15 × 20 Å) and high pore volume (2.9 cm^3 g^{-1}). In a follow-up study, Bernini et al. [115] examined the ibuprofen storage performances of six different MOFs, including bio-MOF-100, CD-MOF-1, MIL-53, MIL-100, MIL-101, and MOF-74. They too reported the highest ibuprofen uptake in bio-MOF-100 (1.97 g g^{-1}). They also showed a high potential energy of adsorbed ibuprofen in CD-MOF-1, bio-MOF-100, and MOF-74, indicating a long ibuprofen release time in these materials. A different approach in which a liquid medium in GCMC simulations was used to investigate the ibuprofen adsorption in HMOF-1 (heterometal-organic framework), MIL-47, and MIL-53 was performed by Bueno-Perez et al. [116]. Ibuprofen uptake in MIL-53 was found to be similar to that reported by Bernini et al. [115], indicating that water molecules do not have a significant effect on saturated ibuprofen loading in MIL-53. They also showed that MOFs can separate mixtures of lysine enantiomers (amino acids). In our previous work [13], we also computationally studied ibuprofen adsorption in 24 different MOFs. Results (given in Fig. 6.5) showed that bio-MOF-100 and MOF-74 series are promising candidates for ibuprofen storage and they can outperform a widely studied MOF, MIL-101 (Cr). We also observed that there is a linear relationship between ibuprofen uptake of MOFs having similar chemical topologies and their pore volumes and pore sizes.

In this computational study, uptakes of two cosmetic molecules, caffeine (liporeducer) and urea (hydrating agent), in these 24 MOFs were also predicted. Similarly, the bio-MOF-100 series and MOF-74 series (RAVVUH-RAVXIX) were found to be promising for the storage of caffeine and urea molecules. The transport of ibuprofen, caffeine, and urea within the pores of bio-MOF-100, bio-MOF-102, and RAVXIX was then studied by performing MD simulations and slow diffusion of these molecules was reported owing to strong host-guest interactions. All these computational studies showed that MOFs are potential materials for ibuprofen encapsulation and strong host-drug interactions can lead a slow drug release process, which is desired for efficient drug delivery systems.

MOFs were also tested for anticancer drug delivery by performing both DFT calculations and molecular simulations. For example, Chalati et al. [111] both experimentally and computationally studied BU, an alkylating agent used in chemotherapy adsorption in MIL-53 (Fe) and MIL-53 (Fe)-NH$_2$. DFT calculations were performed for BU-loaded MIL-53. Strong hydrogen bonds between BU molecule and the matrix and the weak van der Waals and/or CH-π interactions between sulfonate and methyl groups of BU and the organic linker of MIL-53 were observed. The presence

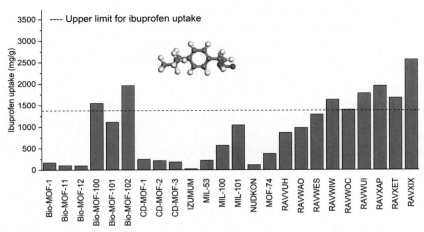

FIGURE 6.5

Ibuprofen adsorption in 24 different MOFs. Dashed line shows the current experimental limit of ibuprofen uptake of MIL-101(Cr). *MIL*, Materials of Institut Lavoisier; *MOF*, metal-organic framework.

Reprinted from Erucar I, Keskin S. Efficient storage of drug and cosmetic molecules in biocompatible metal organic frameworks: a molecular simulation study. Industrial and Engineering Chemistry Research 2016;55(7): 1929–1939 with permission of the ACS.

of amino group in MIL-53(Fe)-NH$_2$ enhanced the interaction energy between BU molecule and the framework, indicating a more controlled drug delivery process. Similarly, Ma et al. [118] performed DFT and GCMC simulations to compute optimized geometries of a BU-loaded MOF and the Q$_{st}$ of BU within the pores of the framework. Results showed that the methyl and nitryl hydrophobic groups of the framework are close to the BU molecules and the Q$_{st}$ of BU in this MOF at 298 K and infinite dilution was found as 167.95 kJ mol^{-1}. This high energy was attributed to the strong adsorption of BU molecule in this MOF.

Another widely computationally studied anticancer drug is 5-FU, as shown in Table 6.3. Liu et al. [112] performed GCMC simulations to compute the saturated uptake amount of 5-FU in three different MOFs. The uptake amounts were found to be consistent with the experimental measurements. The same research group [113] later performed molecular docking calculations to examine the favorable conformations of 5-FU within the two isoreticular Cu-MOFs. Molecular docking simulations are commonly used to predict the drug binding mechanism and compute the minimum binding energy of drug-host complexes. These calculations showed that 5-FU confines within the pores of MOFs and interacts with the metal and carboxylate groups of the ligand. In another study, Wang et al. [119] performed GCMC simulations to study the adsorption isotherms of 5-FU in two isostructural MOFs, namely, NTU-Z11 ([Zn$_3$(μ_3-O)(BTC)$_2$(H$_2$O)]$_n$) and GDMU ({[Zn$_3$(μ_3-O)(BTC)$_2$(DMF:dimethylformamide)]·2NH$_2$(CH$_3$)$_2$·4(H$_2$O)}$_n$). They initially performed DFT simulations to compute partial atomic charges of MOFs and then predicted the adsorbed amount of 5-FU in these materials using GCMC simulations. They found that NTU-Z11 (0.4 g g^{-1}) exhibits higher saturated 5-FU uptake capacity than GDMU (0.22 g g^{-1}). Results also showed that GDMU shows a steep adsorption behavior for 5-FU at low fugacities because of the presence of cations within the framework, which enhances host-guest interactions. In a follow-up study, Wang et al. [122] performed simulations to investigate 5-FU adsorption in a Co-based MOF using the semiempirical AM1 (Austin Model 1) method. They showed that 5-FU can interact with the carboxylate O atoms via hydrogen bonding.

Storage and diffusion of various anticancer drugs, including taxol, cisplatin, tamoxifen, and gemcitabine, have also been tested in the literature. For example, Filippousi and coworkers [37] studied the adsorption of two drugs, hydrophobic paclitaxel and hydrophilic cisplatin, in two Zr-based MOFs, UiO-66 and UiO-67. They used MD simulations to examine the mobility of drug molecules and their favorable locations within these MOFs having some linker defects. They reported that since paclitaxel (taxol) molecules are larger than cisplatin molecules, linker defects exist in these MOFs to provide the required porosity for the adsorption of taxol. MD simulations showed that taxol molecules are located within the middle of the created cage, whereas cisplatin molecules are close to the accessible basic site Zr-O-Zr of the inorganic units in these materials. Koukaras et al. [117] performed DFT calculations to examine the binding of tamoxifen, which is commonly used for the treatment of breast cancer in two functionalized MOFs, IRMOF-14 and IRMOF-16. They incorporated an acidic and aromatic hydroxyl group to increase the binding sites and enhance acid-base and hydrogen bonding interactions with the -N(CH$_3$)$_2$ group and the O atom of tamoxifen. The binding interaction energies were calculated as 16.16 kcal mol^{-1} for IRMOF-14 and 8.32 kcal mol^{-1} for IRMOF-16. These energies were reported to be reasonable for drug adsorption and delivery. In another study, Kotzabasaki et al. [120] performed both DFT and GCMC simulations to examine the adsorption behavior of an anticancer drug, i.e., gemcitabine, in IRMOF-16 and OH-functionalized IRMOF-16. Similar to the studies that we mentioned earlier, the presence of functionalized (hydroxyl) group in the linker enhanced the binding energy due to the increase in the number of active binding sites. They also showed that the performance of these materials (4.3 g g^{-1}, 4.7 g g^{-1}) can outperform the conventional drug delivery systems such as lipid-coated mesoporous silica nanoparticles (0.4 g g^{-1}) in terms of saturated gemcitabine-loading capacity. The same research group [121] studied the adsorption and diffusion behavior of the same drug in IRMOF-74-III and OH-functionalized IRMOF-74-III. DFT calculations showed that the interaction energy of gemcitabine with both MOFs is similar, although the hydroxyl group in the organic linker generates additional interaction sites. The saturated gemcitabine-loading capacity in these materials was found to be high (1.1 g g^{-1}), indicating a promising storage system for gemcitabine. MD simulations in which MOFs and drug molecules were considered as flexible were then performed to examine the mobility of drug molecules in these MOFs. In the simulated system shown in Fig. 6.6, 32 wt% gemcitabine

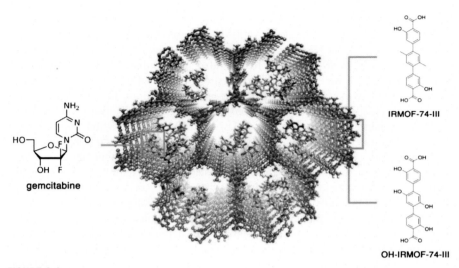

FIGURE 6.6

Representation of 32 wt% gemcitabine loaded in a unit cell of metal-organic framework (MOF) (IRMOF-74-III and OH-functionalized IRMOF-74-III) MOF: C, gray; Mg, green; O, red; H, white. Gemcitabine: C, cyan; O, red; N, blue; H, white; F, light blue. *IRMOF,* isoreticular MOF.

Reprinted from Kotzabasaki M, Galdadas I, Tylianakis E, Klontzas E, Cournia Z, Froudakis, GE. Multiscale simulations reveal IRMOF-74-III as a potent drug carrier for gemcitabine delivery. Journal of Materials Chemistry B 2017;5(18):277–3282 with permission of the Royal Society of Chemistry.

was incorporated in both MOFs. MD simulations showed that the diffusion of gemcitabine in IRMOF-74-III is slightly slower than that in OH-functionalized IRMOF-74-III because the -CH_3 group of IRMOF-74-III is bulkier than the -OH group of IRMOF-74-III. Density profiles also showed that drug aggregates within the pores of IRMOF-74-III hinder the diffusion of drug molecules in IRMOF-74-III. This study also showed that the free volume of an MOF is much correlated with the saturated drug loading capacity compared with the functionalization of the MOF surface.

All these computational studies showed that MOFs are promising candidates for the storage of anticancer drugs. In all these studies, the adsorption of a single molecule in an MOF was investigated. Recently, our group performed molecular simulations to examine the potential of 10 IRMOF-74 materials for concurrent adsorption and diffusion of two anticancer drug molecules, MTX and 5-FU [19]. Among the MOF-74 series, RAVXIX exhibited the highest MTX (2.8 g g^{-1}) and 5-FU (4.2 g g^{-1}) uptakes owing to its extremely high pore volume (3.74 cm^3 g^{-1}). Slow diffusion of drug molecules was also reported in these materials, indicating a controlled drug delivery mechanism. This study showed that MOFs can be alternative carriers for multidrug delivery systems, which are significant for combination therapies in cancer research.

We discussed the recent research on computational modeling of MOFs for drug delivery applications. The computational methods used in all these works are highly useful to identify the promising MOFs before performing experimental studies. They can also complement experimental research by providing fundamental insights into understanding the adsorption and diffusion mechanism of drug molecules within the pores of MOFs at the atomistic level.

6.4 Critical considerations in the biomedical use of metal-organic frameworks

MOFs have great potential in biomedical applications, especially in drug delivery, as we discussed earlier. However, to use MOFs in biomedicine, several issues including biodegradability, stability, toxicity, shape, pore size, and biodistribution should be considered. Physicochemical properties of MOFs such as size, shape, colloidal stability, and surface charge can affect the behavior of drug-loaded MOFs in body fluid. Particle size and shape are also important for the administration routes. For example, for the parenteral route the particles should be <200 nm, indicating that size is important for cellular uptake and degradation [123]. Therefore, to fabricate nano-MOFs in various sizes several synthesis methods, including hydrosolvothermal, reverse-phase microemulsions, sonochemical, and microwave-assisted approaches, were developed in the literature[22].

Stability is one of the critical issues that should be addressed for biomedical applications of MOFs. Some MOFs such as MIL-100(Fe), UiO-66, and ZIF-8 were found to have hydrothermal stability, which depends on the charge and coordination number of the cations, the chemical functionality of the organic linker, the framework dimensionality, and the interpenetration [22]. However, the information about the stability of MOFs in body fluid is highly required to understand the biodegradable character of MOFs. There are only a few studies on the biodegradability of MOFs. For example, bio-MIL-1, MIL-101(Fe), MIL-88A, and MIL-100(Fe) degrade rapidly in PBS (pH, 7.4) [24,27,52]. The degradation of MOFs can take a few days to many weeks based on the crystallinity, composition, and particle sizes of MOFs [124]. To increase the stability of MOFs and enhance their biocompatibility, surface modification has been applied by introducing different functional groups or molecules. For example, PEG has been used to prevent the aggregation of nanoparticles and stabilize them [123]. However, all these approaches are still in their infancy and more studies are required to examine the stability of MOFs in different SBFs.

Another important issue is the toxicity of MOFs, which is commonly evaluated by their metal ions and organic linkers. LD_{50} value is mainly used to determine the toxicity of MOFs. However, the required dose, frequency, and the duration of treatment and the administration route should also be considered to evaluate the administration, distribution, metabolization, and excretion processes. To evaluate the toxicity of MOFs, *in vitro* and *in vivo* experiments should be performed [50]. *In vitro* experiments follow a procedure in a controlled environment outside living organisms or cells, whereas *in vivo* experiments use a whole living organism. The toxicity of MOFs is evaluated by performing *in vitro* experiments in cell lines. For example, human promyelocytic leukemia (HL-60) cells, cervical cancer HeLa cells, and human lung NCI-H446 cancer cells are used for different types of MOFs in the literature [124]. On the other hand, animal studies and clinical trials can be given as examples of *in vivo* studies. To date, there are a few toxicologic experiments for various types of MOFs. The first study on the identification of toxicity of MOFs using *in vivo* approach was performed by Horcajada et al. in 2010 [24]. They examined the toxicity of Fe-based nanoMOFs (MIL-88A) using rats as animal models. The biodistribution of Fe-based nanoMOFs after intravenous administration was shown in Fig. 6.7.

Fe concentration in different tissues was evaluated and found to be increased in liver and spleen 1 day after particle administration. It was stabilized after 7 days. No

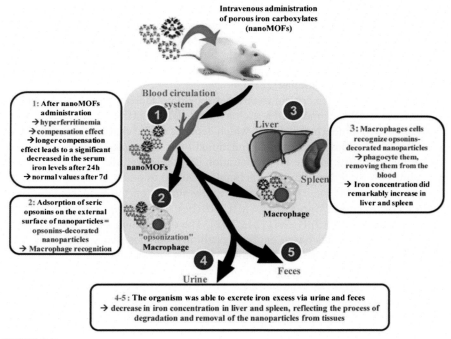

FIGURE 6.7

Biodistribution of Fe-based nano metal organic frameworks (nanoMOFs).

Reprinted from Horcajada P, Gref R, Baati T, Allan PK, Maurin G, Couvreur P, et al. Metal–organic frameworks in biomedicine. Chemical Reviews 2011;112(2):1232–1268 with permission of the ACS.

immune or inflammatory reactions were observed. In 2014, Horcajada and co-workers [20,21] carried out *in vitro* and *in vivo* toxicologic studies for MOFs with different metal sites and organic linkers, such as Fe, Zn, or Zr and carboxylate or imidazolate linkers. They reported that Fe-based carboxylate structures are less toxic than Zn-based imidazolate structures and 14 different MOFs were found to have low cytotoxicity for HeLa and J774 cell lines. In a review, Simon-Yarza et al. [50] summarized all *in vivo* studies of various MOFs, including CD-MOF-1, Cu-BTC, MIL-88, MIL-100, MIL-101, UiO-66, and ZIF-8. Owing to the tunable characteristics of MOFs, endogenous linkers and metals provide an opportunity to synthesize nontoxic MOFs, but the synthesis procedure and solvents can also affect the toxicity of drug-loaded MOFs. Therefore toxicity studies are still critical for biomedical applications of MOFs.

6.5 Conclusions and future perspectives

In the past decade, studies on biomedical applications of MOFs have rapidly grown and the recent developments and encouraging results facilitate the development of MOF research in biomedicine. MOFs can be regarded as the promising candidates especially for drug storage and drug delivery systems and they outperform traditional drug carrier materials in terms of saturated drug uptakes, as we discussed earlier. However, several issues must be addressed to use MOFs in biomedical applications:

- In order to use MOFs in biomedicine, synthesis conditions should be modified to control some of the important properties including particle size, shape, purity, and colloidal stability of MOFs. Currently, MOFs are commonly synthesized by heating a mixture of ligands and metal salts under hydrothermal conditions. The solvents that are used in the synthesis of MOFs have generally toxic properties, which limit their potential in biomedical applications. To use MOFs in biomedicine, green, low-cost, and scalable synthesis methods should be developed.
- Molecular simulations and DFT calculations are highly useful to provide valuable information about the host-drug systems. However, the system size and the timescale of the simulated environment should be modified to represent a more realistic system. Additionally, as drug molecules are large, efficient sampling methods should be developed in atomistic simulations. Parallelization for Monte Carlo methods should be done to speed up calculations to predict the adsorption behavior of drug molecules within the pores of MOFs. Considering the rapid growth in computational power and sophisticated methods, multiscale simulations can be used to study longer timescales to mimic more realistic physiologic systems. It is important to note that current molecular simulations do not provide any information about the toxicity, release kinetics, and/or

degradation mechanisms of MOFs in physiologic environment. All these issues are considered in experimental studies.

- Another important issue is the understanding of co-adsorption and co-delivery mechanisms of drugs using MOFs. Investigation of the synergistic effects of drugs within the pores of MOFs will be highly useful for the development of combination therapies. In the literature, there are only a few studies on multi-drug delivery of a limited number of MOFs. More experimental and computational studies are highly required for the development of various therapeutic agents for the application of interest.
- Encapsulation of large biomolecules such as genetic materials and proteins into MOFs has also been started. Considering the tunable properties of MOFs and their large numbers, both computational and experimental studies are required to investigate the encapsulation mechanism of large molecules in MOFs.
- Commercial applications of MOFs are still in their infancy. The market potential of MOFs should be considered and compared with the conventional materials used in healthcare systems. However, before investigation of the market potential, more studies including both *in vitro* and *in vivo* experiments are required for preclinical applications. The biodegradation, toxicity, long-term biosafety, and stability of MOFs should be investigated in detail by performing experiments. All these issues should be addressed by scientists from various research areas ranging from natural sciences and engineering to medicine.

References

[1] H.-C. Zhou, J.R. Long, O.M. Yaghi, Introduction to metal-organic frameworks, Chemical Reviews 112 (2) (2012) 673—674.

[2] P. Silva, S.M. Vilela, J.P. Tomé, F.A.A. Paz, Multifunctional metal—organic frameworks: from academia to industrial applications, Chemical Society Reviews 44 (19) (2015) 6774—6803.

[3] H. Li, M. Eddaoudi, M. O'Keeffe, O.M. Yaghi, Design and synthesis of an exceptionally stable and highly porous metal-organic framework, Nature 402 (6759) (1999) 276—279.

[4] M. Eddaoudi, J. Kim, N. Rosi, D. Vodak, J. Wachter, M. O'keeffe, et al., Systematic design of pore size and functionality in isoreticular MOFs and their application in methane storage, Science 295 (5554) (2002) 469—472.

[5] J.L.C. Rowsell, O.M. Yaghi, Metal—organic frameworks: a new class of porous materials, Microporous and Mesoporous Materials 73 (1) (2004) 3—14.

[6] C. Janiak, J.K. Vieth, MOFs, MILs and more: concepts, properties and applications for porous coordination networks (PCNs), New Journal of Chemistry 34 (11) (2010) 2366—2388.

[7] S. Furukawa, J. Reboul, S. Diring, K. Sumida, S. Kitagawa, Structuring of metal—organic frameworks at the mesoscopic/macroscopic scale, Chemical Society Reviews 43 (16) (2014) 5700—5734.

[8] Y.-R. Lee, J. Kim, W.-S. Ahn, Synthesis of metal-organic frameworks: a mini review, Korean Journal of Chemical Engineering 30 (9) (2013) 1667–1680.

[9] N. Stock, S. Biswas, Synthesis of metal-organic frameworks (MOFs): routes to various MOF topologies, morphologies, and composites, Chemical Reviews 112 (2) (2011) 933–969.

[10] F.H. Allen, The Cambridge Structural Database: a quarter of a million crystal structures and rising, Acta Crystallographica Section B Structural Science 58 (3) (2002) 380–388.

[11] P.Z. Moghadam, A. Li, S.B. Wiggin, A. Tao, A.G. Maloney, P.A. Wood, et al., Development of a Cambridge structural database subset: a collection of metal–organic frameworks for past, present, and future, Chemistry of Materials 29 (7) (2017) 2618–2625.

[12] E. Atci, I. Erucar, S. Keskin, Adsorption and transport of CH_4, CO_2, H_2 mixtures in a bio-MOF material from molecular simulations, Journal of Physical Chemistry C 115 (14) (2011) 6833–6840.

[13] I. Erucar, S. Keskin, Efficient storage of drug and cosmetic molecules in biocompatible metal organic frameworks: a molecular simulation study, Industrial and Engineering Chemistry Research 55 (7) (2016) 1929–1939.

[14] C. Altintas, I. Erucar, S. Keskin, High-throughput computational screening of the metal organic framework database for CH_4/H_2 separations, ACS Applied Materials and Interfaces 10 (4) (2018) 3668–3679.

[15] S.M. Vilela, P. Horcajada, MOFs as supports of enzymes in biocatalysis, in: Metal-Organic Frameworks: Applications in Separations and Catalysis, Wiley-VCH, 2018, pp. 447–476.

[16] L.E. Kreno, K. Leong, O.K. Farha, M. Allendorf, R.P. Van Duyne, J.T. Hupp, Metal–organic framework materials as chemical sensors, Chemical Reviews 112 (2) (2011) 1105–1125.

[17] S.E. Miller, M.H. Teplensky, P.Z. Moghadam, D. Fairen-Jimenez, Metal-organic frameworks as biosensors for luminescence-based detection and imaging, Interface Focus 6 (4) (2016) 20160027–20160041.

[18] H.-S. Wang, Metal–organic frameworks for biosensing and bioimaging applications, Coordination Chemistry Reviews 349 (2017) 139–155.

[19] I. Erucar, S. Keskin, Computational investigation of metal organic frameworks for storage and delivery of anticancer drugs, Journal of Materials Chemistry B 5 (35) (2017) 7342–7351.

[20] C. Tamames-Tabar, D. Cunha, E. Imbuluzqueta, F. Ragon, C. Serre, M.J. Blanco-Prieto, et al., Cytotoxicity of nanoscaled metal–organic frameworks, Journal of Materials Chemistry B 2 (3) (2014) 262–271.

[21] C. Tamames-Tabar, A. García-Márquez, M. Blanco-Prieto, C. Serre, P. Horcajada, MOFs in pharmaceutical technology, in: Bio-and Bioinspired Nanomaterials, John Wiley & Sons, 2014, pp. 83–112.

[22] P. Horcajada, R. Gref, T. Baati, P.K. Allan, G. Maurin, P. Couvreur, et al., Metal–organic frameworks in biomedicine, Chemical Reviews 112 (2) (2011) 1232–1268.

[23] P. Horcajada, C. Serre, M. Vallet-Regí, M. Sebban, F. Taulelle, G. Férey, Metal-organic frameworks as efficient materials for drug delivery, Angewandte Chemie 118 (36) (2006) 6120–6124.

[24] P. Horcajada, T. Chalati, C. Serre, B. Gillet, C. Sebrie, T. Baati, et al., Porous metal−organic framework nanoscale carriers as a potential platform for drug delivery and imaging, Nature Materials 9 (2) (2010) 172−178.

[25] T. Simon-Yarza, S. Rojas, P. Horcajada, C. Serre, The situation of metal-organic frameworks in biomedicine, in: Comprehensive Biomaterials II, Elsevier, 2017, pp. 720−749.

[26] J. An, S.J. Geib, N.L. Rosi, Cation-triggered drug release from a porous zinc−adeninate metal−organic framework, Journal of the American Chemical Society 131 (24) (2009) 8376−8377.

[27] J. An, S.J. Geib, N.L. Rosi, High and selective CO_2 uptake in a cobalt adeninate metal−organic framework exhibiting pyrimidine- and amino-decorated pores, Journal of the American Chemical Society 132 (1) (2010) 38−39.

[28] T. Li, D.-L. Chen, J.E. Sullivan, M.T. Kozlowski, J.K. Johnson, N.L. Rosi, Systematic modulation and enhancement of CO_2:N_2 selectivity and water stability in an isoreticular series of bio-MOF-11 analogues, Chemical Science 4 (4) (2013) 1746−1755.

[29] J. An, O.K. Farha, J.T. Hupp, E. Pohl, J.I. Yeh, N.L. Rosi, Metal-adeninate vertices for the construction of an exceptionally porous metal-organic framework, Nature Communications 3 (2012) 604−610.

[30] T. Li, M.T. Kozlowski, E.A. Doud, M.N. Blakely, N.L. Rosi, Stepwise ligand exchange for the preparation of a family of mesoporous MOFs, Journal of the American Chemical Society 135 (32) (2013) 11688−11691.

[31] S.R. Miller, E. Alvarez, L. Fradcourt, T. Devic, S. Wuttke, P.S. Wheatley, et al., A rare example of a porous Ca-MOF for the controlled release of biologically active NO, Chemical Communications 49 (71) (2013) 7773−7775.

[32] C. Tamames-Tabar, E. Imbuluzqueta, N. Guillou, C. Serre, S. Miller, E. Elkaïm, et al., A Zn azelate MOF: combining antibacterial effect, CrystEngComm 17 (2) (2015) 456−462.

[33] R.A. Smaldone, R.S. Forgan, H. Furukawa, J.J. Gassensmith, A.M. Slawin, O.M. Yaghi, et al., Metal−organic frameworks from edible natural products, Angewandte Chemie International Edition 49 (46) (2010) 8630−8634.

[34] J. Liu, T.-Y. Bao, X.-Y. Yang, P.-P. Zhu, L.-H. Wu, J.-Q. Sha, et al., Controllable porosity conversion of metal-organic frameworks composed of natural ingredients for drug delivery, Chemical Communications 53 (55) (2017) 7804−7807.

[35] D. Cunha, M. Ben Yahia, S. Hall, S.R. Miller, H. Chevreau, E. Elkaïm, et al., Rationale of drug encapsulation and release from biocompatible porous metal−organic frameworks, Chemistry of Materials 25 (14) (2013) 2767−2776.

[36] M.J. Katz, Z.J. Brown, Y.J. Colón, P.W. Siu, K.A. Scheidt, R.Q. Snurr, et al., A facile synthesis of UiO-66, UiO-67 and their derivatives, Chemical Communications 49 (82) (2013) 9449−9451.

[37] M. Filippousi, S. Turner, K. Leus, P.I. Siafaka, E.D. Tseligka, M. Vandichel, et al., Biocompatible Zr-based nanoscale MOFs coated with modified poly (ε-caprolactone) as anticancer drug carriers, International Journal of Pharmaceutics 509 (1−2) (2016) 208−218.

[38] F.R.S. Lucena, L.C. de Araújo, M.d.D. Rodrigues, T.G. da Silva, V.R. Pereira, G.C. Militão, et al., Induction of cancer cell death by apoptosis and slow release of 5-fluoracil from metal-organic frameworks Cu-BTC, Biomedicine and Pharmacotherapy 67 (8) (2013) 707−713.

[39] C.-Y. Sun, C. Qin, X.-L. Wang, G.-S. Yang, K.-Z. Shao, Y.-Q. Lan, et al., Zeolitic imidazolate framework-8 as efficient pH-sensitive drug delivery vehicle, Dalton Transactions 41 (23) (2012) 6906–6909.

[40] K.S. Park, Z. Ni, A.P. Côté, J.Y. Choi, R. Huang, F.J. Uribe-Romo, et al., Exceptional chemical and thermal stability of zeolitic imidazolate frameworks, Proceedings of the National Academy of Sciences of the United States of America 103 (27) (2006) 10186–10191.

[41] H. Zheng, Y. Zhang, L. Liu, W. Wan, P. Guo, A.M. Nyström, et al., One-pot synthesis of metal–organic frameworks with encapsulated target molecules and their applications for controlled drug delivery, Journal of the American Chemical Society 138 (3) (2016) 962–968.

[42] C. Doonan, R. Ricco, K. Liang, D. Bradshaw, P. Falcaro, Metal–organic frameworks at the biointerface: synthetic strategies and applications, Accounts of Chemical Research 50 (6) (2017) 1423–1432.

[43] M. Giménez-Marqués, T. Hidalgo, C. Serre, P. Horcajada, Nanostructured metal–organic frameworks and their bio-related applications, Coordination Chemistry Reviews 307 (2) (2016) 342–360.

[44] S. Kempahanumakkagari, V. Kumar, P. Samaddar, P. Kumar, T. Ramakrishnappa, K.-H. Kim, Biomolecule-embedded metal-organic frameworks as an innovative sensing platform, Biotechnology Advances 36 (2) (2018) 467–481.

[45] S. Keskin, S. Kızılel, Biomedical applications of metal organic frameworks, Industrial and Engineering Chemistry Research 50 (4) (2011) 1799–1812.

[46] K. Lu, T. Aung, N. Guo, R. Weichselbaum, W. Lin, Nanoscale metal–organic frameworks for therapeutic, imaging, and sensing applications, Advanced Materials 30 (37) (2018) 1707634–1707654.

[47] H.Y. Yang, Y. Li, D.S. Lee, Multifunctional and stimuli-responsive magnetic nanoparticle-based delivery systems for biomedical applications, Advanced Therapeutics 1 (2) (2018) 1800011–1800028.

[48] J. Zhou, G. Tian, L. Zeng, X. Song, X.W. Bian, Nanoscaled metal-organic frameworks for biosensing, imaging, and cancer therapy, Advanced Healthcare Materials 7 (10) (2018) 1800022–1800043.

[49] L. Wang, M. Zheng, Z. Xie, Nanoscale metal–organic frameworks for drug delivery: a conventional platform with new promise, Journal of Materials Chemistry B 6 (5) (2018) 707–717.

[50] T. Simon-Yarza, A. Mielcarek, P. Couvreur, C. Serre, Nanoparticles of metal–organic frameworks: on the road to in vivo efficacy in biomedicine, Advanced Materials 30 (37) (2018) 1707365–1707380.

[51] P. Horcajada, C. Serre, G. Maurin, N.A. Ramsahye, F. Balas, M. Vallet-Regi, et al., Flexible porous metal-organic frameworks for a controlled drug delivery, Journal of the American Chemical Society 130 (21) (2008) 6774–6780.

[52] S.R. Miller, D. Heurtaux, T. Baati, P. Horcajada, J.-M. Grenèche, C. Serre, Biodegradable therapeutic MOFs for the delivery of bioactive molecules, Chemical Communications 46 (25) (2010) 4526–4528.

[53] E.Q. Procopio, S. Rojas, N.M. Padial, S. Galli, N. Masciocchi, F. Linares, et al., Study of the incorporation and release of the non-conventional half-sandwich ruthenium (II) metallodrug RAPTA-C on a robust MOF, Chemical Communications 47 (42) (2011) 11751–11753.

[54] C.Y. Sun, C. Qin, C.G. Wang, Z.M. Su, S. Wang, X.L. Wang, et al., Chiral nanoporous metal-organic frameworks with high porosity as materials for drug delivery, Advanced Materials 23 (47) (2011) 5629–5632.

[55] Y. Wang, J. Yang, Y.Y. Liu, J.F. Ma, Controllable syntheses of porous metal−organic frameworks: encapsulation of Ln^{III} cations for tunable luminescence and small drug molecules for efficient delivery, Chemistry − A European Journal 19 (43) (2013) 14591–14599.

[56] C. Gaudin, D. Cunha, E. Ivanoff, P. Horcajada, G. Chevé, A. Yasri, et al., A quantitative structure activity relationship approach to probe the influence of the functionalization on the drug encapsulation of porous metal-organic frameworks, Microporous and Mesoporous Materials 157 (2012) 124–130.

[57] V. Agostoni, R. Anand, S. Monti, S. Hall, G. Maurin, P. Horcajada, et al., Impact of phosphorylation on the encapsulation of nucleoside analogues within porous iron (III) metal−organic framework MIL-100(Fe) nanoparticles, Journal of Materials Chemistry B 1 (34) (2013) 4231–4242.

[58] Q. Hu, J. Yu, M. Liu, A. Liu, Z. Dou, Y. Yang, A low cytotoxic cationic metal−organic framework carrier for controllable drug release, Journal of Medicinal Chemistry 57 (13) (2014) 5679–5685.

[59] T. Kundu, S. Mitra, P. Patra, A. Goswami, D. Díaz Díaz, R. Banerjee, Mechanical downsizing of a gadolinium (III)-based metal−organic framework for anticancer drug delivery, Chemistry - A European Journal 20 (33) (2014) 10514–10518.

[60] C. He, K. Lu, D. Liu, W. Lin, Nanoscale metal−organic frameworks for the Co-delivery of cisplatin and pooled siRNAs to enhance therapeutic efficacy in drug-resistant ovarian cancer cells, Journal of the American Chemical Society 136 (14) (2014) 5181–5184.

[61] H. Su, F. Sun, J. Jia, H. He, A. Wang, G. Zhu, A highly porous medical metal−organic framework constructed from bioactive curcumin, Chemical Communications 51 (26) (2015) 5774–5777.

[62] D. Feng, T.-F. Liu, J. Su, M. Bosch, Z. Wei, W. Wan, et al., Stable metal-organic frameworks containing single-molecule traps for enzyme encapsulation, Nature Communications 6 (2015) 5979–5987.

[63] C. Adhikari, A. Das, A. Chakraborty, Zeolitic imidazole framework (ZIF) nanospheres for easy encapsulation and controlled release of an anticancer drug doxorubicin under different external stimuli: a way toward smart drug delivery system, Molecular Pharmaceutics 12 (9) (2015) 3158–3166.

[64] W. Lin, Q. Hu, K. Jiang, Y. Yang, Y. Yang, Y. Cui, et al., A porphyrin-based metal−organic framework as a pH-responsive drug carrier, Journal of Solid State Chemistry 237 (2016) 307–312.

[65] W. Lin, Q. Hu, J. Yu, K. Jiang, Y. Yang, S. Xiang, et al., Low cytotoxic metal−organic frameworks as temperature-responsive drug carriers, ChemPlusChem 81 (8) (2016) 804–810.

[66] S. Rojas, F.J. Carmona, C.R. Maldonado, E. Barea, J.A. Navarro, RAPTA-C incorporation and controlled delivery from MIL-100(Fe) nanoparticles, New Journal of Chemistry 40 (7) (2016) 5690–5694.

[67] B. Singco, L.-H. Liu, Y.-T. Chen, Y.-H. Shih, H.-Y. Huang, C.-H. Lin, Approaches to drug delivery: confinement of aspirin in MIL-100(Fe) and aspirin in the de novo synthesis of metal−organic frameworks, Microporous and Mesoporous Materials 223 (2016) 254–260.

[68] P.P. Bag, D. Wang, Z. Chen, R. Cao, Outstanding drug loading capacity by water stable microporous MOF: a potential drug carrier, Chemical Communications 52 (18) (2016) 3669−3672.

[69] L. Bai, S.Z.F. Phua, W.Q. Lim, A. Jana, Z. Luo, H.P. Tham, et al., Nanoscale covalent organic frameworks as smart carriers for drug delivery, Chemical Communications 52 (22) (2016) 4128−4131.

[70] Y. Yang, Q. Hu, Q. Zhang, K. Jiang, W. Lin, Y. Yang, et al., A large capacity cationic metal−organic framework nanocarrier for physiological pH responsive drug delivery, Molecular Pharmaceutics 13 (8) (2016) 2782−2786.

[71] Z. Wu, N. Hao, H. Zhang, Z. Guo, R. Liu, B. He, et al., Mesoporous iron-carboxylate metal−organic frameworks synthesized by the double-template method as a nanocarrier platform for intratumoral drug delivery, Biomaterials Science 5 (5) (2017) 1032−1040.

[72] X. Unamuno, E. Imbuluzqueta, F. Salles, P. Horcajada, M. Blanco-Prieto, Biocompatible porous metal-organic framework nanoparticles based on Fe or Zr for gentamicin vectorization, European Journal of Pharmaceutics and Biopharmaceutics 132 (2018) 11−18.

[73] M. Hoop, C.F. Walde, R. Riccò, F. Mushtaq, A. Terzopoulou, X.-Z. Chen, et al., Biocompatibility characteristics of the metal organic framework ZIF-8 for therapeutical applications, Applied Materials Today 11 (2018) 13−21.

[74] E.D. Bloch, L.J. Murray, W.L. Queen, S. Chavan, S.N. Maximoff, J.P. Bigi, et al., Selective binding of O_2 over N_2 in a redox−active metal−organic framework with open iron(II) coordination sites, Journal of the American Chemical Society 133 (37) (2011) 14814−14822.

[75] A.P. Cote, A.I. Benin, N.W. Ockwig, M. O'keeffe, A.J. Matzger, O.M. Yaghi, Porous, crystalline, covalent organic frameworks, Science 310 (5751) (2005) 1166−1170.

[76] C. Orellana-Tavra, E.F. Baxter, T. Tian, T.D. Bennett, N.K. Slater, A.K. Cheetham, et al., Amorphous metal−organic frameworks for drug delivery, Chemical Communications 51 (73) (2015) 13878−13881.

[77] C. Orellana-Tavra, R.J. Marshall, E.F. Baxter, I.A. Lázaro, A. Tao, A.K. Cheetham, et al., Drug delivery and controlled release from biocompatible metal−organic frameworks using mechanical amorphization, Journal of Materials Chemistry B 4 (47) (2016) 7697−7707.

[78] N.J. Hinks, A.C. McKinlay, B. Xiao, P.S. Wheatley, R.E. Morris, Metal organic frameworks as NO delivery materials for biological applications, Microporous and Mesoporous Materials 129 (3) (2010) 330−334.

[79] B. Xiao, P.J. Byrne, P.S. Wheatley, D.S. Wragg, X. Zhao, A.J. Fletcher, et al., Chemically blockable transformation and ultraselective low-pressure gas adsorption in a non-porous metal organic framework, Nature Chemistry 1 (4) (2009) 289−294.

[80] A.C. McKinlay, P.K. Allan, C.L. Renouf, M.J. Duncan, P.S. Wheatley, S.J. Warrender, et al., Multirate delivery of multiple therapeutic agents from metal-organic frameworks, APL Materials 2 (12) (2014) 124108−124116.

[81] S. Chavan, J.G. Vitillo, E. Groppo, F. Bonino, C. Lamberti, P.D. Dietzel, et al., CO adsorption on CPO-27-Ni coordination polymer: spectroscopic features and interaction energy, Journal of Physical Chemistry C 113 (8) (2009) 3292−3299.

[82] M. Ma, H. Noei, B. Mienert, J. Niesel, E. Bill, M. Muhler, et al., Iron metal−organic frameworks MIL-88B and NH_2-MIL-88B for the loading and delivery of the

gasotransmitter carbon monoxide, Chemistry − A European Journal 19 (21) (2013) 6785−6790.

[83] L. Hamon, C. Serre, T. Devic, T. Loiseau, F. Millange, G. Férey, et al., Comparative study of hydrogen sulfide adsorption in the MIL-53(Al, Cr, Fe), MIL-47(V), MIL-100(Cr), and MIL-101(Cr) metal− organic frameworks at room temperature, Journal of the American Chemical Society 131 (25) (2009) 8775−8777.

[84] L. Hamon, H. Leclerc, A. Ghoufi, L. Oliviero, A. Travert, J.-C. Lavalley, et al., Molecular insight into the adsorption of H_2S in the flexible MIL-53(Cr) and rigid MIL-47(V) MOFs: infrared spectroscopy combined to molecular simulations, Journal of Physical Chemistry C 115 (5) (2011) 2047−2056.

[85] P.K. Allan, P.S. Wheatley, D. Aldous, M.I. Mohideen, C. Tang, J.A. Hriljac, et al., Metal−organic frameworks for the storage and delivery of biologically active hydrogen sulfide, Dalton Transactions 41 (14) (2012) 4060−4066.

[86] J.B. DeCoste, M.H. Weston, P.E. Fuller, T.M. Tovar, G.W. Peterson, M.D. LeVan, et al., Metal−organic frameworks for oxygen storage, Angewandte Chemie 126 (51) (2014) 14316−14319.

[87] C.G. Piscopo, F. Trapani, A. Polyzoidis, M. Schwarzer, A. Pace, S. Loebbecke, Positive effect of the fluorine moiety on the oxygen storage capacity of UiO-66 metal−organic frameworks, New Journal of Chemistry 40 (10) (2016) 8220−8224.

[88] P.Z. Moghadam, T. Islamoglu, S. Goswami, J. Exley, M. Fantham, C.F. Kaminski, et al., Computer-aided discovery of a metal−organic framework with superior oxygen uptake, Nature Communications 9 (2018) 1378−1386.

[89] M. Miller, I. Megson, Recent developments in nitric oxide donor drugs, British Journal of Pharmacology 151 (3) (2007) 305−321.

[90] L.E. Otterbein, The evolution of carbon monoxide into medicine, Respiratory Care 54 (7) (2009) 925−932.

[91] R. Wang, Hydrogen sulfide: the third gasotransmitter in biology and medicine, Antioxidants and Redox Signaling 12 (9) (2010) 1061−1064.

[92] W. Liu, N. Khatibi, A. Sridharan, J.H. Zhang, Application of medical gases in the field of neurobiology, Medical Gas Research 1 (13) (2011) 1−18.

[93] D. Alezi, Y. Belmabkhout, M. Suyetin, P.M. Bhatt, Ł.J. Weselinski, V. Solovyeva, et al., MOF crystal chemistry paving the way to gas storage needs: aluminum-based soc-MOF for CH_4, O_2, and CO_2 storage, Journal of the American Chemical Society 137 (41) (2015) 13308−13318.

[94] T. Engel, P.J. Reid, Thermodynamics, Statistical Thermodynamics, & Kinetics, third ed., Pearson, Prentice Hall, Boston, 2013.

[95] D. Dubbeldam, A. Torres-Knoop, K.S. Walton, Monte Carlo codes, tools and algorithms: on the inner workings of Monte Carlo codes, Molecular Simulation 39 (14−15) (2013) 1253−1292.

[96] B. Smit, Phase diagrams of Lennard-Jones fluids, The Journal of Chemical Physics 96 (11) (1992) 8639−8640.

[97] F.-X. Coudert, A.H. Fuchs, Computational characterization and prediction of metal-organic framework properties, Coordination Chemistry Reviews 307 (2) (2016) 211−236.

[98] A.K. Rappe, C.J. Casewit, K.S. Colwell, W.A. Goddard, W.M. Skiff, UFF, a full periodic table force field for molecular mechanics and molecular dynamics simulations, Journal of the American Chemical Society 114 (25) (1992) 10024−10035.

[99] S.L. Mayo, B.D. Olafson, W.A. Goddard, Dreiding: a generic force field for molecular simulations, Journal of Physical Chemistry 94 (26) (1990) 8897—8909.

[100] H.-C. Guo, F. Shi, Z.-F. Ma, X.-Q. Liu, Molecular simulation for adsorption and separation of CH_4/H_2 in zeolitic imidazolate frameworks, Journal of Physical Chemistry C 114 (28) (2010) 12158—12165.

[101] S. Keskin, J. Liu, J.K. Johnson, D.S. Sholl, Atomically detailed models of gas mixture diffusion through CuBTC membranes, Microporous and Mesoporous Materials 125 (1—2) (2009) 101—106.

[102] D. Berthelot, Sur le melange des gaz, Comptes Rendus Hebdomadaires des Seances de l'Academie des Sciences 126 (1898) 1703—1855.

[103] H. Lorentz, Ueber die anwendung des satzes vom virial in der kinetischen theorie der gase, Annalen der Physik 248 (1) (1881) 127—136.

[104] Q. Xu, C. Zhong, A general approach for estimating framework charges in metal-organic frameworks, Journal of Physical Chemistry C 114 (11) (2010) 5035—5042.

[105] C.E. Wilmer, K.C. Kim, R.Q. Snurr, An extended charge equilibration method, Journal of Physical Chemistry Letters 3 (17) (2012) 2506—2511.

[106] Q. Yang, D. Liu, C. Zhong, J.-R. Li, Development of computational methodologies for metal-organic frameworks and their application in gas separations, Chemical Reviews 113 (10) (2013) 8261—8323.

[107] D. Frenkel, B. Smit, Understanding Molecular Simulation: From Algorithms to Applications, second ed., Academic press, London, 2002.

[108] B. Smit, T.L.M. Maesen, Molecular simulations of zeolites: adsorption, diffusion, and shape selectivity, Chemical Reviews 108 (10) (2008) 4125—4184.

[109] A. Einstein, Zur theorie der brownschen bewegung, Annalen der Physik 324 (2) (1906) 371—381.

[110] R. Babarao, J. Jiang, Unraveling the energetics and dynamics of ibuprofen in mesoporous metal—organic frameworks, Journal of Physical Chemistry C 113 (42) (2009) 18287—18291.

[111] T. Chalati, P. Horcajada, P. Couvreur, C. Serre, M. Ben Yahia, G. Maurin, et al., Porous metal organic framework nanoparticles to address the challenges related to busulfan encapsulation, Nanomedicine 6 (10) (2011) 1683—1695.

[112] J.-Q. Liu, X.-F. Li, C.-Y. Gu, J.C. da Silva, A.L. Barros, S. Alves Jr., et al., A combined experimental and computational study of novel nanocage-based metal—organic frameworks for drug delivery, Dalton Transactions 44 (44) (2015) 19370—19382.

[113] J.-Q. Liu, J. Wu, Z.-B. Jia, H.-L. Chen, Q.-L. Li, H. Sakiyama, et al., Two isoreticular metal—organic frameworks with $CdSO_4$-like topology: selective gas sorption and drug delivery, Dalton Transactions 43 (46) (2014) 17265—17273.

[114] L. Bei, L. Yuanhui, L. Zhi, C. Guangjin, Molecular simulation of drug adsorption and diffusion in bio-MOFs, Acta Chimica Sinica 72 (8) (2014) 942—948.

[115] M.C. Bernini, D. Fairen-Jimenez, M. Pasinetti, A.J. Ramirez-Pastor, R.Q. Snurr, Screening of bio-compatible metal—organic frameworks as potential drug carriers using Monte Carlo simulations, Journal of Materials Chemistry B 2 (7) (2014) 766—774.

[116] R. Bueno-Perez, A. Martin-Calvo, P. Gómez-Álvarez, J.J. Gutiérrez-Sevillano, P.J. Merkling, T.J. Vlugt, et al., Enantioselective adsorption of ibuprofen and lysine in metal—organic frameworks, Chemical Communications 50 (74) (2014) 10849—10852.

[117] E.N. Koukaras, T. Montagnon, P. Trikalitis, D. Bikiaris, A.D. Zdetsis, G.E. Froudakis, Toward efficient drug delivery through suitably prepared metal—organic frameworks: a first-principles study, Journal of Physical Chemistry C 118 (17) (2014) 8885—8890.

[118] D.-Y. Ma, Z. Li, J.-X. Xiao, R. Deng, P.-F. Lin, R.-Q. Chen, et al., Hydrostable and nitryl/methyl-functionalized metal—organic framework for drug delivery and highly selective CO_2 adsorption, Inorganic Chemistry 54 (14) (2015) 6719—6726.

[119] J. Wang, J. Jin, F. Li, B. Li, J. Liu, J. Jin, et al., Combined experimental and theoretical insight into the drug delivery of nanoporous metal—organic frameworks, RSC Advances 5 (104) (2015) 85606—85612.

[120] M. Kotzabasaki, E. Tylianakis, E. Klontzas, G.E. Froudakis, OH-functionalization strategy in metal-organic frameworks for drug delivery, Chemical Physics Letters 685 (2017) 114—118.

[121] M. Kotzabasaki, I. Galdadas, E. Tylianakis, E. Klontzas, Z. Cournia, G.E. Froudakis, Multiscale simulations reveal IRMOF-74-III as a potent drug carrier for gemcitabine delivery, Journal of Materials Chemistry B 5 (18) (2017) 3277—3282.

[122] F. Wang, J. Wang, S. Yang, C. Gu, X. Wu, J. Liu, et al., A combination of experiment and molecular simulation studies on a new metal-organic framework showing pH-triggered drug release, Russian Journal of Coordination Chemistry 43 (2) (2017) 133—137.

[123] W. Chen, C. Wu, Synthesis, functionalization, and applications of metal—organic frameworks in biomedicine, Dalton Transactions 47 (7) (2018) 2114—2133.

[124] S. Beg, M. Rahman, A. Jain, S. Saini, P. Midoux, C. Pichon, et al., Nanoporous metal organic frameworks as hybrid polymer—metal composites for drug delivery and biomedical applications, Drug Discovery Today 22 (4) (2017) 625—637.

Transition metal dichalcogenides for biomedical applications

Rekha Rani Dutta, PhD [1], Rashmita Devi[2], Hemant S. Dutta[2], Satyabrat Gogoi, PhD [2]

[1]*Department of Chemistry, School of Basic Sciences, Assam Kaziranga University, Jorhat, Assam;*
[2]*Analytical Chemistry Group, Material Sciences & Technology Division, Academy of Scientific and Innovative Research, CSIR North-East Institute of Science & Technology, Jorhat, Assam, India*

List of abbreviation

0D	Zero-dimensional
1D	One-dimensional
2D	Two-dimensional
CT	Computed tomography
CVD	Chemical vapor deposition
DFT	Density functional theory
DNA	Deoxyribonucleic acid
DOX	Doxorubicin
EV	Electron volt
FAM	Fluorescein amidite
FET	Field-effect transistor
HRP	Horseradish peroxidase
KSCN	Potassium thiocyanate
LOD	Limit of detection
MR	Magnetic resonance
MTase	Methyltransferase
NIR	Near-infrared
NSs	Nanosheets
PA	Photoacoustic
PDT	Photodynamic therapy
PEG	Polyethylene glycol
PET	Positron emission tomography
PL	Photoluminescence
pM	Picomolar
PSA	Prostate-specific antigen
PTT	Photothermal therapy
RNA	Ribonucleic acid
ROS	Reactive oxygen species
T4 PNK	T4 polynucleotide kinase
TMD	Transition metal dichalcogenide
μM	Micromolar

Two-Dimensional Nanostructures for Biomedical Technology. https://doi.org/10.1016/B978-0-12-817650-4.00007-3

211

7.1 Introduction

Two-dimensional transition metal dichalcogenides (2D TMDs) have emerged as a new kind of nanosheets (NSs) over the past years, acquiring the substantial attraction of audiences from multidisciplinary domains. Inspired by the successful discovery and widespread utilization of 2D graphene NSs, material scientists have explored to find novel 2D nanomaterials based on TMDs. Typical construction of 2D TMDs includes a layer of transition metal atoms sandwiched between two layers of chalcogen atoms and can be presented by the general formula MX_2. Here, M represents transition metals belonging to groups 4—10 in the periodic table (Ti, Zr, Hf, V, Nb, Ta, Mo, W, etc.) and X stands for chalcogen elements (S, Se, or Te). The unique optical, mechanical, and electronic properties of 2D TMDs arising from their atom-scale thickness promote their utility in electronics, photonics, catalysis, and biomedical sector.

The concept of using TMDs in biomedical applications is relatively new. The utilization of 2D NSs, especially graphene-based sheets in various forms (graphene oxide [GO], reduced GO, etc.), has been practiced for quite some time. 2D NSs provide interesting material properties, which include mechanical, structural, and biological attributes, and have been recognized as key factors to achieve favorable interactions at bio-nano interface for various biobased utilities. 2D TMDs also share properties with graphene NSs. Furthermore, they are believed to possess special properties that are difficult to obtain with other conventional 2D materials. Therefore 2D TMDs have been investigated for use in various biomedical applications. In particular, the electronic and optical properties of TMDs are very special. Monolayers of TMD NSs contain multiple adsorption sites owing to the presence of three atomic layers that can interact with different biomotifs, which eventually alters their electronic and optical properties. The reduction of layer thickness of TMDs leads to the emergence of direct electronic bandgap. For a monolayer TMD, such bandgap may vary between 1 and 2 eV. Such type of materials provides its proposition to be used in high-end electronic and biomedical devices. Giving a further edge, the presence of optical bandgap in single- to few-layered TMDs provides extensive opportunities in optical biosensing and photoresponsive therapy. TMD NSs having ultrasmall lateral dimensions possess photoluminescent (PL) properties and can therefore be used in fluorescence-based sensing probes. They are also used in the study of enzymatic activity and cell viability. TMD NSs possess strong absorption in the near-infrared (NIR) region (650—900 nm) and demonstrate an excellent photothermal effect, which has significant applications in photothermal therapy (PTT) and photoacoustic (PA) imaging. TMD offers the scope to dope with paramagnetic particles, which extend their utility in bioimaging techniques such as magnetic resonance imaging (MRI). The metallic edges and defects on the TMDs act as efficient antibacterial material, which may be even stronger than traditional antimicrobial agents such as Ag.

Considering the huge viability, this chapter provides a concise depiction of the synthesis, properties, and biological applications of 2D TMDs.

7.2 Structure

The fascinating properties of TMDs have driven many groundbreaking types of research on the development of newer TMD nanostructures. Compared to the bulk counterpart, 2D TMD sheets exhibit unique chemical, electronic, and optical properties and therefore have been explored for use in various applications including biomedical, catalysis, sensing, etc. These properties of TMD nanostructures depend on various attributes such as crystalline phases, dimensional structure, etc. TMDs are layered structures and consist of a large library of materials that includes the diverse range from insulators (HfS_2, etc.) and semiconductors (MoS_2, WS_2, etc.) to metals (NbS_2, VSe_2, etc.) [1].

The material structure of TMD materials is akin to that of GO, in which a very weak van der Waals force is responsible for holding the layers together, thus permitting the bulk crystal to split into layered surface easily. The layers have a thickness of 6−7 Å approximately [2]. TMDs can exist as several polytypes and polymorphs viz. 1T, 2H, and 3R that refer to octahedral, trigonal prismatic, and rhombohedral geometries, respectively [3]. However, monolayer TMDs have only two kinds of polymorphs: trigonal prismatic (H) that has D_{3h} point group symmetry and octahedral geometry (T) that possesses D_{3d} point group [4] (Fig. 7.1). The coordination mode is decided by the selected combination of metal and chalcogenide elements. The group 4 TMDs are all in the octahedral structure, whereas group 5 TMDs have both octahedral and trigonal prismatic phases, group 6 TMDs exhibit trigonal prismatic geometry, group 7 TMDs are primarily found in a distorted octahedral geometry, and group 10 TMDs are all in an octahedral structure. In a TMD compound, the metal atom donates four electrons and thus exhibits +4 oxidation state, whereas chalcogen shows −2 oxidation state. The lone pair of electrons present in chalcogens prevents the layers from reacting with foreign species by terminating the surface of the layers.

Moreover, the bandgap energy varies between TMD compounds depending on their composition and structure [2]. In the bulk crystal, TMDs show indirect bandgap, but the bandgap gets widened with the reducing number of layers, from bulk to monolayer [5]. Exfoliation of bulk TMDs into a thin layer results in unbounding the adjacent layers from the orbital interaction, which causes the bandgap to widen. Changes in the number of layers switch the bandgap from an indirect bandgap of 1.28 eV to a direct bandgap of 1.80 eV because of the quantum confinement effect [6]. Furthermore, by engineering the desirable thickness of TMD NSs, the bandgap can be tuned, which is useful to control the electronic and optical properties, resulting in the tuning of the realized devices.

After the great success of 2D graphene, in the rich family of 2D nanomaterials, 2D TMDs are able to attract tremendous research interest in the field of biomedical applications, such as drug delivery, biosensing, and imaging. The ultrathin 2D TMDs possess high surface-to-volume ratio, making them the most promising platform to immobilize different biomolecules or drugs for sensing purposes and in drug delivery.

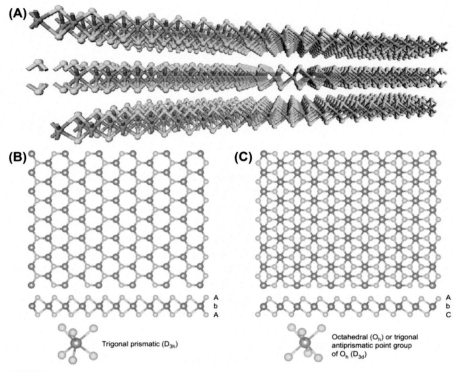

FIGURE 7.1

(A) Layered two-dimensional transition metal dichalcogenide (TMD) structure. (B) Single-layer TMD with trigonal prismatic (D_{3h}) coordination and (C) single-layer TMD with octahedral coordination (D_{3d}) [2].

Copyright 2013 Springer Nature, Reproduced with permission from Chhowalla et al., 2013.

7.3 Synthesis

Numerous methods have been reported to fabricate single- and few-layer 2D TMDs. Methods such as micromechanical exfoliation, chemical and electrochemical intercalation, ultrasonication-assisted liquid exfoliation, physical vapor deposition, chemical vapor deposition (CVD), and colloidal synthesis including solvothermal and hydrothermal methods have been extensively used to prepare few-layer and monolayer TMDs. Typically, these methods can be classified into two categories: top-down and bottom-up approaches (Fig. 7.2) [8]. A brief description of these synthesis techniques is provided in the following sections.

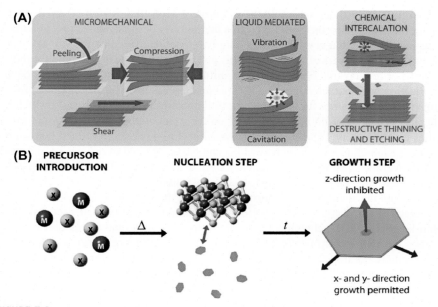

FIGURE 7.2

(A) Top-down approach of synthesis of two-dimensional transition metal dichalcogenide nanosheets (2D TMD NSs) by the exfoliation method. (B) Bottom-top approach of synthesis of 2D TMD NSs by the growing an NS from molecular precursors [7].

Reproduced under Creative Commons Attribution License (CC BY) from Brent et al., 2017.

7.3.1 Top-down approach

The lamellar structures of TMD crystals contain strong interlayer covalent bonds and weak interlayer van der Waals forces [9]. These weak van der Waals forces in the crystals can be disrupted to produce thinner materials by imparting a shear force. This thinning process can be used repeatedly to exfoliate the lamellar flakes for isolating single layers of the crystal NSs. The methods to affect this process can be categorized into different types. A peeling force is applied in micromechanical exfoliation techniques to separate the layers. In liquid medium, exfoliation generally works by applying a vibrational energy to the crystal. In chemical intercalation, forced expansion of the intercalate layers by insertion of small atoms or molecules within the interlayer space of the bulk separates the flakes. However, in the etching technique, the unwanted layers are destructively removed from the bulk to fabricate a monolayered or few-layered NS on a substrate. The top-down exfoliation methods have been described briefly here with an emphasis on the quality of the 2D TMDs produced [7].

7.3.1.1 Micromechanical exfoliation

Mechanical exfoliation has been extensively used to obtain atomically thin 2D TMDs. Initially, the scotch tape method was used to prepare thin-layered materials.

Frindt in 1966 [10] demonstrated the isolation of MoS_2 flakes with 10 nm thickness using the scotch tape method. Later, this method has been advanced to a process of cleaving the lamellar crystal by attaching two tapes to overcome the van der Waals force within the layers of the crystal. This process can be repeated to produce mono-layered or few-layered NSs that can be shifted onto the desired substrate for device fabrication and characterization [11]. This process has been further extended by using viscoelastic stamps for mechanical exfoliation of the lamellar crystals, which has the added benefit of no adhesive traces on the exfoliated sheets that further intensifies its utilization [12]. Moreover, Novoselov et al. [11] demonstrated that simply rubbing a lamellar crystal, such as MoS_2 and $NbSe_2$, placed over a desired substrate can deliver the shear forces required to exfoliate the crystals. Techniques such as the Chalkboard method [7] and sandpaper-assisted rubbing [13] have also been explored to generate the shear forces required to exfoliate the structures.

Electrostatic force—assisted exfoliation method has also been used to obtain thin-layered TMDs, wherein electric field at elevated temperatures and pressure causes the exfoliation. Generally, isolation of a single layer is not possible using this method, but the existing micromechanical methods can further thin the adhered flakes. However, this process produces significantly larger NSs because of the application of reasonably uniform shear force throughout the crystal. This inhibits the tearing of the NSs and the resulting flakes are of greater lateral dimension. The technique has been exploited to produce significantly larger graphene NSs (lateral dimensions of 100s of micrometers were documented) and moderately large dimensions of MoS_2 and $NbSe_2$ NSs (up to 30 μm observed) [14].

Mechanical exfoliation yields pristine, defect-free, and good-quality mono- and multilayer crystal NSs, which can be studied to obtain their intrinsic properties and can be used to fabricate various high-end electronic components. However, issues have been documented in the literature for utilizing mechanically exfoliated NSs in view of irregular thicknesses of the desired flakes. It has also been reported that single-layered areas are usually found sparsely with irregular shapes, which complicates their recognition and integration with devices. This limits the adaptability of micromechanical processes for the production of NSs. Henceforth, alternative processes, such as ultrasonication-assisted liquid exfoliation, have been explored, which yet involves the fascinating micromechanical exfoliation step [15].

7.3.1.2 Ultrasonication-assisted liquid exfoliation
The ultrasonication-assisted liquid exfoliation method utilizes energetic waves to proliferate through the solvent producing low- and high-pressure cycles that help in exfoliation [16—19]. Continuous exposure to these energetic forces thins the crystal continually, which can be separated from the unexfoliated material by centrifugation [15]. There has been persistent growth in liquid-mediated exfoliation method leading to increased flexibility of NS production. Techniques such as high-shear mixing and jet cavitation have been presented as potential alternatives to the conventional methods of exfoliation [7]. In recent developments, researchers have demonstrated efficient exfoliation of various TMDs into ultrathin NSs using a laboratory

high-shear mixer and a common household blender [20,21]. Use of stabilizing agents in water has been able to produce concentrated TMD NS dispersions [22]. Nevertheless, the organic solvents or surfactants in the dispersions used are difficult to remove from the NS surface because of the specific surface energies. This limits the implications of exfoliation in high-performance devices [8]. To circumvent this problem, Zhou et al. [23] proposed a "mixed-solvent method" wherein a mixture of ethanol and water is used as solvent for exfoliating 2D TMDs.

However, 2D TMDs prepared by ultrasonication-assisted liquid exfoliation are generally few-layered NSs with a low concentration (<1 mg mL^{-1}) and small lateral dimension (typically 200−400 nm). O'Neill et al. [24] enhanced the concentration of MoS$_2$ NSs in organic solvents to 40 mg mL^{-1} by increasing the time of sonication. An increase in the duration of ultrasonication usually initiates sonication-induced scission that reduces the lateral dimensions of the flakes. In addition, Guan et al. [25] developed a simplistic method to flake off the lamellar structures by using bovine serum albumin (BSA) as an effective exfoliating agent. The polar and nonpolar groups of BSA help in the exfoliation of the bulk crystal.

7.3.1.3 Chemical intercalation

Molecules or ions can intercalate the TMD bulk crystals because of their lamellar structure and weak van der Waals forces [26]. Alkali metals, especially lithium, have been extensively used to intercalate TMD structures for their implications in fundamental studies and their prospects in energy storage applications [27,28]. Lithium-intercalated TMDs have high water reactivity and form H$_2$, which can be exploited for exfoliation [29−31]. Conventionally, TMDs are chemically intercalated by using organolithium reagents such as n-butyllithium (n-BuLi). The reagents are chemically reacted with TMDs, followed by reaction with water to flake off the lamellar structure. However, the process consumes a lot of time and also sometimes requires a high temperature [2]. Thereby, the ultrasonication-enhanced lithium intercalation strategy has been demonstrated to fabricate TMD NSs [32]. The ultrasonication-enhanced lithium intercalation method provides extraordinary reaction conditions because of the use of ultrasonication, which activates the hexamer of n-BuLi for enhancing the mass transfer of reactants at the solid-liquid interfaces. Apart from lithium, hydrazine salts have also been used to exfoliate TMDs through the chemical intercalation method [33]. Hydrazine salts reduce the MoS$_x$ bulk and cause lattice deformation and sulfur loss, which finally leads to its exfoliation. However, the characterization results show that the ratio of S to Mo was 1.2, demonstrating the alteration of stoichiometric ratio during the reduction process and the PL spectra showed a wider bandgap than its pristine single-layer MoS$_2$ NSs.

7.3.1.4 Electrochemical intercalation

An effective and controllable method to produce TMD NSs is the electrochemical intercalation technique [34]. As compared to the conventional lithium intercalation process, which is very tough to control, electrochemical intercalation offers a controlled environment for the execution of exfoliation with a higher yield. Efforts have been undertaken to improve the quality of the exfoliated NSs. The influence of various parameters during electrochemical intercalation has been thoroughly studied by Zeng and coworkers [35]. Liu et al. [36] demonstrated the synthesis of high-quality MoS_2 NSs by exfoliating their bulk material in Na_2SO_4 solution. Herein, O_2^-, OH^-, and SO_4^{2-} have been inserted into the MoS_2 flakes instead of lithium ions. The MoS_2 layers exfoliate when O_2, SO_2, or both are released in the form of gas from the anode, making them to expand significantly in the presence of ultrasonication. The obtained MoS_2 flakes are 5–50 µm in lateral dimension, making them much bigger than the samples exfoliated using ultrasonication-assisted liquid exfoliation or traditional chemical intercalation techniques [21,31,37].

7.3.2 Bottom-up approach

Bottom-up methods for synthesizing TMD NSs rely on the reaction of precursors containing metals and chalcogens. In such approaches, Mn^+ and nX^{2-} ions interact with each other to form sheetlike nanostructures. Basically, NSs possess large surface areas, high energy levels, and unstable structures. Therefore synthesis of 2D single-layer to few-layer TMDs imposes challenges to choose optimized reaction conditions so as to achieve selective formation of NSs over more stable atomscale structures or simple bulk crystals [38]. Generally, bottom-up approaches offer many tunable reaction parameters to achieve control over the formation of the desired nanostructures. The bottom-up approaches for TMD synthesis can be categorized into two broad classes: gas-phase synthesis and wet chemical techniques. Both methods have their own pros and cons.

7.3.2.1 Gas-phase synthesis/vapor-phase growth process

Gas-phase synthesis of 2D TMDs shows great potential in the recent research owing to its intrinsic compatibility with existing technologies as well as its ability to provide pristine, single- to few-layer nanostructures with high quality and large surface area [39,40]. Consistent growths of monolayer NSs of lateral sizes ranging from atomic scales [41] to several centimeters have been documented [42,43]. In the gas-phase synthesis, a wide variety of parameters need to be optimized, which have a direct influence on the morphology, thickness, and size of 2D TMD nanomaterials. One of the most common gas-phase synthesis methods of 2D TMDs is the CVD technique. CVD relies on high-temperature chemical reactions, in which NSs of TMDs are deposited in vapor form on the substrates. Generally, vapor-phase sulfur is forced to react with a thin film of metal at a high temperature under an inert atmosphere, which leads to the formation of TMDs. Zhan et al. synthesized MoS_2 NSs by using this method. Initially, Mo was coated on a SiO_2/Si substrate, which was then allowed to react with evaporated sulfur in a CVD furnace [44]. Dumcenco et al. [45] demonstrated the applicability of CVD by synthesizing large-area monolayer of MoS_2 NSs on an atomically smooth sapphire surface. The same method is also utilized in the synthesis of monolayer WSe_2. In an alternate approach, group VI oxides such as MoO_3 and WO_3 were also used as metal resources. These metal oxides are suitable to be used as metal precursors, as they possess a relatively low melting point and evaporation temperature [46]. Due to easy volatility, CVD can be accessible at atmospheric and low-pressure conditions. The oxides (MO_3) of TMDs were initially converted into a suboxide, MO_{3-x}, upon reacting with sulfur vapor at elevated temperatures, which undergoes further reaction to produce the 2D MS_2 nanostructure. Balendhran et al. [47] synthesized monolayer MoS_2 NSs by simultaneously evaporating MoO_3 nanoparticles and sulfur powder, which was followed by annealing in a sulfur-rich environment. Here, it is pertinent to mention that CVD-based procedures have been mainly used for the production of MoS_2; however, the growth of other TMD NSs such as $MoSe_2$, WS_2, and WSe_2 NSs has also been reported [48−54]. The most obvious advantage of CVD in the synthesis of 2D TMDs is the purity of the product and the control over the product formation. However, on many occasions, requirements of sophisticated reaction conditions, such as high temperature, inert atmosphere, and reduced pressure, limit its practical applicability to produce 2D TMDs on a large scale.

7.3.2.2 Wet chemical methods

Wet chemical methods are widely used in the synthesis of various nanostructures. These synthesis routes provide fine-tuning of the reaction conditions, such as temperature, reactant concentration, reaction time, and pH, to afford the nanomaterials with desired properties. Compared with other methods, wet chemical methods are both relatively easy to adopt and cost effective.

Solvothermal synthesis

In solvothermal synthesis, TMDs are prepared by subjecting the precursors to an organic medium with high boiling point to initiate the nucleation and crystal growth process. The size and morphologic form of the NSs can be tuned by the use of organic ligands. It also helps to improve the dispensability. The report on solvothermal synthesis of 2D WS_2 NSs came out in 2007, when Seo et al. [55] reported the in situ sulfidation of one-dimensional tungsten oxide nanorods in the presence of carbon disulfide and hexadecylamine. Thereafter, the "one-pot synthesis" technique was also reported for the successful synthesis of MoS_2 and WS_2 NSs by using the wet chemical approach [56]. The uniqueness of this method was the utilization of a single-source precursor in oleylamine, which reacts at a relatively low temperature (360°C). The other advantage of this type of synthesis is the formation of high-quality NSs. The layer thickness is inversely proportional to the reaction time. However, the method works well only for MoS_2 and WS_2, and not for other TMD NSs because of the unavailability of single-source precursors. It is interesting that the reactivity of the precursor may influence the crystal structure of TMDs while synthesized via the solvothermal route. For, example, WS_2 monolayers with both prismatic 2H structure and octahedral 1T structure can be synthesized by reacting WCl_6 and CS_2 in the presence of oleylamine at 320°C [57]. It has been demonstrated that less reactive tungsten precursors favor the formation of 1T structure, whereas the highly reactive ones form the 2H structure. Thus the solvothermal method provides a wide control over the formation of the desired nanostructure by controlling various reaction conditions. Such approaches have been widely used for large-scale production of 2D TMDs. However, in many instances, it is difficult to control the layer thickness. Furthermore, the ligands used in the synthesis process are sometimes difficult to remove, which may create toxicity-related issues in their biomedical applications.

Hydrothermal synthesis

Hydrothermal technique is one of the most widely used wet chemical methods for the synthesis of 2D TMDs because of its simplicity [55,58]. The hydrothermal process involves the treatment of TMD precursors at a high temperature under high pressure within a sealed autoclave [59]. Such treatment results in the growth of single crystals from an aqueous solution under the reaction conditions. The hydrothermal method is particularly suitable to obtain biocompatible TMD NSs for biomedical applications. One of the classic examples of the synthesis of hydrothermal TMDs is the formation of MoS_2 NSs from ammonium molybdate precursor and thiourea at 200°C [60]. A wide variety of TMD precursors can be used. Chen and Fan synthesized MoX_2 (X = S, Se) NSs using the transition metal salts and $Na_2S_2O_3$ in the presence of hydrazine. The reaction was carried out at a much low temperature (130−140°C) [61]. Similarly, Rao and coworkers [62] synthesized MoS_2 NSs by reacting MoO_3 with KSCN. It is pertinent to mention that the formation of 2D TMDs is highly dependent on the conditions of hydrothermal synthesis, such as temperature, time, and pH, which can be explored to prepare various structures of MoS_2 NSs [63]. The hydrothermal method is also suitable to prepare 2D TMDs with lower lateral dimensions, which possess fluorescence property. Despite its simplicity, the hydrothermal method sometimes has the limitation of lack of control over the precise structural parameters, such as lateral dimension and layer thickness.

7.4 Properties

7.4.1 Electronic properties

2D TMD materials having a wide range of properties, from metallic to semiconducting and superconductive, with varying direct and indirect bandgaps have attracted tremendous applications in electronic and optical domains [64,65]. Unfolding their properties is inevitable to determine their applications in various electronic devices. TMDs as bulk are indirect bandgap materials. The electronic transition occurs between the Γ and a point located between K and Γ high symmetry points, namely, Λ, which is the conduction band minimum, as shown in Fig. 7.3. However, the position of the conduction band minimum at the Γ point is dependent on the interlayer interactions [67,68]. Henceforth, the bandgap of TMDs can be tuned by thinning the bulk material to multilayers. It can be seen from Fig. 7.3 that as the material is

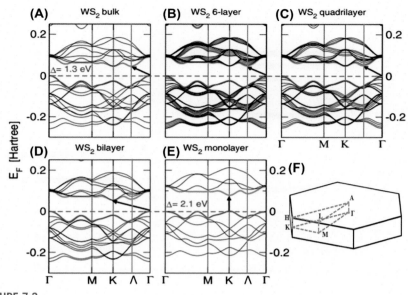

FIGURE 7.3

(A—E) Band structures of bulk WS_2, its monolayer, and its polylayers calculated using the Perdew-Burke-Ernzerhof exchange-correlation functionals. Red dashed horizontal lines show the fermi level. The arrows indicate the fundamental bandgap for a given system. The top of valence band (blue/dark gray) and bottom of conduction band (green/light gray) are highlighted [66]. (F) Brillouin zone and high-symmetry points of the WS_2 reciprocal lattice.

Copyright 2011 American Physical Society, Reproduced with permission from Kuc et al., 2011.

thinned, the valence band valley at the Γ point and the conduction band valley at the Λ point increases. Interestingly, when the material goes from bilayer to monolayer ultimately, the valence band maximum and the conduction band minimum gets positioned at the K point, resulting in an indirect-direct bandgap transition [66]. Mak et al. [69] observed the occurrence of a direct bandgap in monolayer MoS_2 initially, which was experimentally demonstrated and justified using density functional theory calculations by Splendiani et al. [70]. Padilha et al. [72] studied the variations in bandgap of MoS_2 multilayers for the number of layers $n = 1-7$ and showed a monotonic decrease in the bandgaps with increase in the number of layers, which rapidly converges to the bulk values (Fig. 7.4A). They also inspected the band edges on absolute energy scale to provide an insight on the physical origins of bandgap variations (Fig. 7.4D). These analyses can be used to determine band alignments in heterostructures and to identify ohmic and Schottky contact materials [74,75].

The bandgap of 2D TMDs can also be varied by alloying and straining. Komsa and Krasheninnikov [76] investigated band structures of several molybdenum-based dichalcogenide alloys and reported that the bandgaps can be continuously tuned

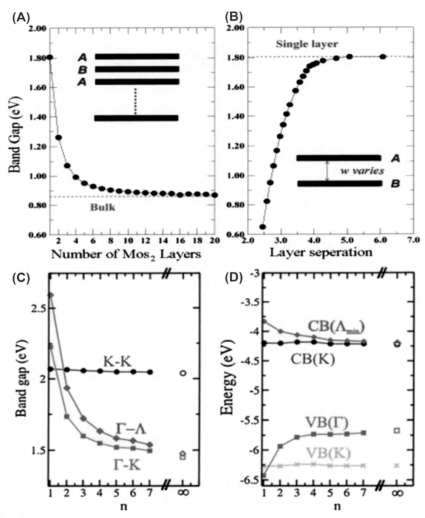

FIGURE 7.4

Variations in bandgap in multilayer MoS_2 sheets with (A) the number of sheets, n, and (B) the separation distance, w (in angstrom), where w indicates the minimum z-axis distance been two consecutive sheets [71]. (C) Evolution of bandgap function of n. The filled black circles (K-K), red squares (Γ-K), and green diamonds (Γ-Λ_{min}) indicate the magnitude of the different bandgaps. Hollow symbols indicate the bulk bandgaps. (D) Position of the band edge with respect to absolute energy for the valence band (VB) at K (*orange crosses*), VB at Γ (*red squares*), conduction band (CB) at K (*black circles*), and CB at Λ_{min} (*green diamonds*) [72].

(A) and (B) Copyright 2007 American Chemical Society, Reproduced with permission from Li et al., 2007; (C) and (D) Copyright 2014 American Physical Society, Reproduced with permission from Padilha et al., 2014.

while retaining the direct bandgap character. Kang et al. [77] extended the investigation to other semiconducting TMD monolayer alloys and explained the electronic process behind the tunable direct bandgap property. The study of strain on the band properties of semiconducting TMDs has been carried out by various researchers [78−80]. Integration of 2D materials leads to mechanical strain due to the lattice parameter or thermal expansion coefficient mismatch and thus is crucial in rational device design. Guzman et al. [81] theoretically investigated the bandgap of several 2H-TMDs as a function of lattice constant stimulated through the application of biaxial strain and showed the possibility of creating heterostructures using the property. It is noteworthy to mention that the bandgaps can be varied almost in the range of 0−2 eV by alloying and straining these TMDs [82].

The combination of the abovementioned properties and countless alloy possibilities of these materials offer a wide range of implications in the area of electronics and other related applications. The performance of field-effect transistors (FETs) in digital electronics can be improved by using TMDs due to their 2D morphologic form and the lack of surface dangling bonds [83,84]. The ability of tuning bandgaps can be explored for developing novel low-power devices based on tunneling. The property of direct bandgap can be used to fabricate efficient optoelectronic devices. Furthermore, its large surface-to-volume ratio enables use in various applications in the area of sensitive chemical and biological sensors.

7.4.2 Optical properties

The change in thickness in TMDs leading to indirect-direct bandgap transition has a significant implication on luminescence. Researchers have demonstrated a pronounced enhancement in the PL intensity as the thickness is decreased to monolayer, which is accompanied by the direct bandgap generation. The PL spectra of mechanically exfoliated MoS_2 are shown in Fig. 7.5. The monolayer MoS_2 has a single peak in the PL spectra that is attributed to the direct-gap luminescence. However, for multilayer samples, emission peaks A and B are attributed to direct-gap hot luminescence, which occurs due to excitonic splitting, and the peak I is obtained as a result of indirect-gap luminescence [69]. Similar effects have also been observed in tungsten-based dichalcogenide materials, in which the PL intensity of monolayers was 100−1000 times stronger than that of bulk materials [73]. However, the absolute emission intensity of WS_2 and WSe_2 monolayers is 20−40 times stronger than that of exfoliated MoS_2 monolayers [43]. The enhancement can be attributed to both the material composition and morphologic factors such as interlayer coupling strength [4].

The TMD thin films have a very high refractive index in the visible frequency region. Monolayers of MoS_2, WS_2, WSe_2, and $MoSe_2$ have an extraordinary refractive index. The refractive index and the extinction coefficient vary with the wavelength of the TMD monolayers. These characteristics of TMD monolayers attract enormous potential in applications such as antireflection coatings for photonics and optoelectronics, substrates for advanced display devices, and encapsulants for high-performance photovoltaic devices [85].

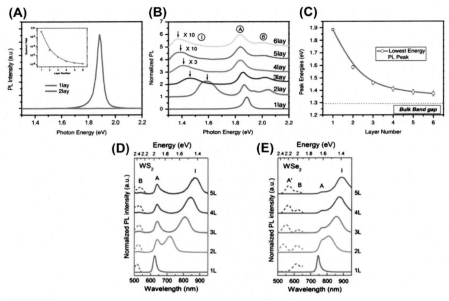

FIGURE 7.5

(A) Photoluminescence (PL) spectra of monolayer and bilayer MoS$_2$. Inset: PL quantum yield of thin layer for $n = 1-6$. (B) Normalized PL spectra by the intensity of peak A of thin-layer MoS$_2$. (C) Bandgap energy of thin-layer MoS$_2$. Dashed line represents the indirect bandgap of the bulk material [69]. (D) Layer-dependent normalized PL spectra of WS$_2$. (E) Layer-dependent normalized PL spectra of WSe$_2$ [73].

(A), (B) and (C) Copyright 2010 American Physical Society, Reproduced with permission from Mak et al., 2010;

(D) and (E) Copyright 2012 American Chemical Society, Reproduced with permission from Zhao et al., 2012.

Henceforth, the unique properties of TMD thin films, such as bandgap alteration and high electron mobility, allow large tuning of photodetectors and phototransistors. Combining TMDs with graphene to create vertical heterostructures opens up opportunities in responsibility and response time, thus allowing a wide range of light detection in optoelectronic devices. Moreover, the PL property of TMDs has been exploited for use in various sensing applications such as electroluminescence-based sensing and fluorescence-based sensing [86].

7.5 Biomedical applications

2D TMDs have emerged as a potential biomaterial owing to their unique structural, physical, and chemical properties. Depending on the choice of transition metal and chalcogen, TMDs can be designed into functional systems having metallic, semiconducting, or insulating characteristics. The stability of chalcogen-based compounds in oxygen and aqueous environments helps them to find applications in liquid-medium-based applications. The low toxicity and nonhazardous nature of 2D TMDs are other important highlights, which enhance their viability in the domain of biomedical science. The most important aspect of 2D TMDs is the inherent biological activity. This type of nanomaterials can interact with bioentities such as enzyme and protein in a favorable manner. Owing to their large surface area, 2D NSs can easily absorb charged biomolecules. Effective interaction with biomolecules can also be brought into by initiating careful intercalation of the layered structure. Such type of bio-nano interactions has been explored for use in different applications, which can be classified into the categories of biosensing, therapeutics, and bioimaging. In this part of the chapter, we are trying to provide a concise description of the biomedical application of 2D TMDs.

7.5.1 Biosensors

A biosensor is an analytical device that can convert biological responses into assessable signals, thereby providing a way to detect minute bio-entities [87,88]. Biosensors are composed of two parts: a sensitive bioreceptor and a transducer or detector part. The bioreceptor exclusively engages with the target analyte by a certain mechanism and recognizes the analyte in the biological environment. While the transducer transforms the recognition events into assessable signals, which are then correlated with the target analyte [89]. High sensitivity, selectivity, ease of functioning, and low cost are the desirable characteristics of a biosensor that can be used in different applications, including food safety, healthcare, and clinical diagnosis [90,91]. In this context, 2D TMDs are very promising. With favorable characteristics such as ultrathin-layered structure, high surface area, and facile functionalization, 2D TMDs effectively interact with various biomolecules. 2D TMD NSs possess an intact basal plane that can adsorb biomolecules either by hydrophobic interactions or through noncovalent interactions such as van der Waals forces [92]. The ultra-high surface-to-volume ratios and the unique 2D structure of these NSs provide considerable possibilities to load functional molecules. Such prominent interactions can easily be assessed in terms of their extremely high electric, electronic, and optical properties, and hence, they can be used in the development of different kinds of biosensors [93—95]. These attributes represent the 2D TMD NSs as suitable candidates in biosensing applications [29]. A large variety of biosensors based on 2D TMD NSs have been studied for different analytes. Most of these biosensors rely on electrochemical, fluorescent, and electronic responses that are generated by the TMD-based transducers during the interaction of bioreceptors and analytes. They can be used to detect nucleic acids, proteins, enzymes, and other biomolecules.

7.5.1.1 Electronic biosensors

2D TMDs are extremely promising to develop electronic biosensors. The most obvious advantage of using TMD NSs for the fabrication of electronic biosensors is the presence of distinct bandgaps as well as their excellent capacitance. 2D TMDs possess varied electronic states with a planar morphologic form, which are exploited to fabricate scalable digital transistors and highly sensitive FET sensors. The state-of-the-art literature showcases utilization of 2D TMD-based FET biosensors in the detection of gases, proteins, and DNA, along with other biomolecules. Generally, MoS_2 and WS_2 have been largely studied in electronic device—based biosensors. Compared with the other TMDs, these two members easily adopt the semiconductor characteristics. In contrary, further modification requires in case of semimetallic and metallic TMDs ($TiSe_2$, WTe_2, VSe_2, NbS_2) for similar adaptation of semiconducting behavior.

The contemporary literature shows the application of TMD electronic biosensors for detecting various gases such as NO_2, NO, NH_3, O_2, and trimethylamine. The detection process relies on the adsorption of reactive gaseous species on the surface of the thin-film TMD-based transistor, thereby altering its resistance through gas-surface interaction and charge-transfer processes. Different responses such as changes in the channel conductance, characteristic transient time, and low-frequency current fluctuations can be measured by such devices corresponding to adsorption characteristics of different gas molecules. He et al. detected toxic NO_2 gas by using thin-film transistor, which was fabricated from MoS_2 NSs [96,97]. Similarly, Late et al. [98] demonstrated FETs by using single-layered and multilayered MoS_2 in the detection of NO_2, NH_3, and humidity (Fig. 7.6A). WS_2-based gas sensors are also known. Bui and coworkers [100] have shown that single layer WS_2 surface is suitable for the detection of NO and O_2 gases. In addition to gases, proteins and other biomolecules can also be detected by using TMD-based electronic sensors. Most importantly, they can be used as label-free sensing devices. Such detection relies on the change in electronic properties of semiconductive 2D TMDs by the binding of charged biomolecules at the biomolecule-device interface. Wang et al. [99] performed a very highly sensitive and selective real-time detection of prostate-specific antigen (PSA), which is a protein cancer marker (Fig. 7.6B). They have designed a field-effect device using antibody-functionalized (specific to PSA) multilayer MoS_2 NSs. Due to the binding of antibody and PSA, a change in the MoS_2 transistor drain current was perceived, which was utilized in PSA detection. DNA molecules can also be detected by using TMD-based electronic transistors. Generally, probe DNAs specific to the target DNAs were immobilized on the thin-film transistor surface. Liu et al. [101] used a nanohybrid system of monolayer MoS_2 and gold nanoparticles immobilized with DNA specific to the target DNA fragments (chromosome 21 or 13). By using this FET-based sensor they measured the target DNA up to a level of 100 aM, which is very promising for the screening of Down syndrome. A similar type of transistor was reported by Lee et al. for the label-free detection of hybridization of DNA by using 2D few-layer MoS_2 as a sensing-channel material.

7.5.1.2 Electrochemical biosensors

Electrochemical biosensors are contemplated as efficient platform for detection of biomolecules owing to their high sensitivity, low cost, and simplicity [102]. 2D nanomaterials have emerged as potent candidates to construct electrochemical

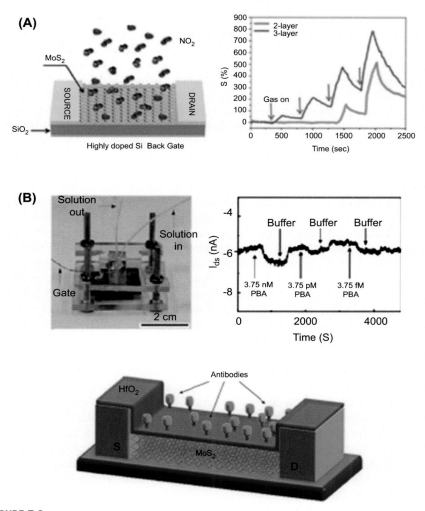

FIGURE 7.6

(A) Thin-layered MoS_2 transistors for gas sensing [98]. (B) Functionalized MoS_2 nanosheet-based field-effect biosensor for label-free sensitive detection of prostate-specific antigen [99].

(A) Copyright 2013 American Chemical Society, Reproduced with permission from Late et al., 2013; (B) Copyright 2014 WILEY-VCH Verlag GmbH & Co. KGaA, Weinheim, Reproduced with permission from Wang et al., 2014.

sensors. Their atomic layer structure with large surface areas provide high electron transfer rate and high conductivity, along with the advantage of facile functionalization [102,103]. Due to their multifaceted structures, unique electric properties, and ease of preparation, 2D TMD NSs have drawn huge research interest. The basic criterion of a transducer of electrochemical sensor is to possess high electric conductivity. Unfortunately, bulk MoS_2 with 2H structure is semiconductive and possesses poor electric conductivity. However, in the exfoliated form, MoS_2 demonstrates good conductivity, which can be explored to design electrochemical biosensors. In this context, the lithium intercalation method is an interesting choice. It is known that ce-MoS_2 NSs produced by the lithium intercalation method can turn 2H structure to metallic 1T phase that shows high conductivity [104–106]. Additionally, the chemical exfoliation method introduces surface defects to the 2D TMD structure in the form of unsaturated metal or chalcogen atoms on the edges. These defect sites make the NSs suitable for bioconjugation, with increasing water dispersibility [107].

The state-of-the-art literature showcases a wide variety of biomolecules including proteins, amino acids, enzymes, etc. that can be detected by TMD-based electrochemical biosensors. Proteins are essential biomolecules present in living organisms. But the irregular appearance of some proteins can cause illness. Therefore responsive and simple assays are essential to determine protein levels for diagnostic purposes [108]. Here, we have mentioned a few works related to the fabrication of electrochemical biosensors based on TMD NSs for protein and amino acid detection. In 2014, Huang and group [109] constructed an electrochemical biosensor by using gold nanoparticles and aptamers functionalized WS_2-graphene composites for the quantification of immunoglobulin E (IgE). The prepared biosensor showed good results, with a limit of detection (LOD) of 0.12 pM. Xia et al. [110] fabricated a MoS_2-NS-based electrochemical biosensor by using silver nanoflakes (Ag NFs) to sense the catalytic oxidation of tryptophan. The sensor showed outstanding catalytic effect toward the oxidation of tryptophan with LOD of 0.05 μM. In 2015, Su and coworkers fabricated an electrochemical immunosensor based on thionine-MoS_2 NSs (thionine-MoS_2) for the detection of carcinoembryonic antigens. The sensor was prepared by decorating gold nanoparticles with MoS_2 NSs and the LOD was found to be 0.52 pg mL^{-1}, with a wide linear range from 1 to 10 ng mL^{-1} under optimal conditions [94]. In addition to macro-biomolecules, small bioentities, such as glucose and H_2O_2, can be detected by using TMD-based electrochemical sensors. Both enzymatic and nonenzymatic producers have been reported. Li et al. used glucose oxidase immobilized Au/MoS_2 NS nanohybrid glassy carbon electrode for glucose detection with a detection limit of 2.8 μM and a linear range of 10–300 μM. Huang et al. used the electrocatalytic activities of Cu nanoparticles in a nonenzymatic electrochemical sensor and constructed a MoS_2-Cu-based electrode, which possesses a linear detection range of 0–4 mM. H_2O_2 is another widely detected molecule by electrochemical means. H_2O_2 is produced in vivo during various metabolic processes and is considered as a major biological exogenous species. In excessive amounts, H_2O_2 can cause damage to the cells and tissues [111]. In a recent development, Wang et al. [112] constructed a H_2O_2 biosensor by embellishing MoS_2 NSs with horseradish peroxidase (HRP) enzyme. Song et al. [113] reported a fabrication method for H_2O_2 detection based on HRP-MoS_2-graphene nanocomposites, with LOD of 0.049 μM (Fig. 7.7A). Chao et al. reported an excellent hemoglobin immobilization technique on gold nanoparticle−based MoS_2 NS

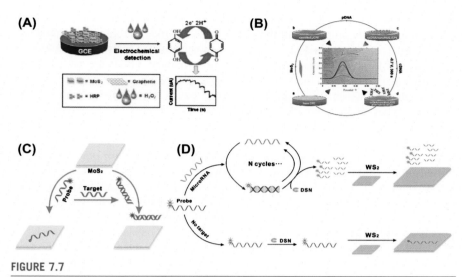

FIGURE 7.7

Two-dimensional transition metal dichalcogenide nanosheets (NSs) for electrochemical and fluorescence biosensing applications: (A) Preparation of horseradish peroxidase–decorated MoS_2/graphene composites for H_2O_2 detection [113], (B) preparation of a label-free DNA biosensor based on MoS_2 NSs [114], (C) DNA detection by fluorescence biosensors based on MoS_2 NSs [115], and (D) microRNA detection based on WS_2 NSs coupled with duplex-specific nuclease (DSN) signal amplification strategy [116]. *cDNA*, complementary DNA; *CPE*, carbon paste electrode; *GEC*, glassy carbon electrode; *pDNA*, plasmid DNA; *ssDNA*, single-stranded DNA.

(A) Copyright 2014, Elsevier (B)V. Reproduced with permission from Song et al., 2014; (B) Copyright 2015 Elsevier (B)V., Reproduced with permission from Wang et al., 2015; (C) Copyright 2013 American Chemical Society, Reproduced with permission from Zhu et al., 2013; (D) Copyright 2014, American Chemical Society, Reproduced with permission from Xi et al., 2014.

film by using a glassy carbon electrode as the sensor probe. The sensor showed good response toward H_2O_2 detection [117].

7.5.1.3 Fluorescence biosensors

Fluorescence is a physical phenomenon that can be explored in the development of novel analytical methods. Extremely high sensitivity, ease of operation, and easy automation are some of the major highlights of fluorescence-based nanosensors [118,119]. Generally, fluorescence-based sensors rely on the *turn-on* or *turn-off* of fluorescence intensity, which can be brought about by the interaction of the probe with the analyte. 2D TMDs, due to their excellent optical properties, have found numerous applications in fluorescence biosensors. TMDs possess extremely high light absorption capability and can be used as a quencher in fluorescence sensors. The quenching ability of chemically exfoliated 2D TMD NSs is even superior to that of graphene [120,121]. On the other hand, low-dimensional TMD NSs exhibit good luminescence property and thus can be used as the fluorophore for the detection of different analytes [13,31,34]. Low-dimensional TMDs also possess the advantage of high water dispersibility. Zero-dimensional TMDs are perhaps more attractive as fluorophores than their 2D analogues, owing to the property of high fluorescence. However, this type of TMDs is beyond the scope of this chapter. Different types of biomolecules including proteins, nucleic acids, enzymes, etc. can be detected by using TMD-based fluorescence sensors.

The first 2D-TMD-NS-based fluorescence biosensor was reported in 2013 [115] (Fig. 7.7C). Zhu and coworkers constructed a *turn-on/turn-off* fluorescence biosensor for single-stranded DNA (ssDNA) by utilizing the variable affinity of ssDNA and double-stranded DNA (dsDNA) toward MoS_2 NSs. The fluorescence probe is composed of a dye-labeled ssDNA, which is specific to the target ssDNA. The researchers have demonstrated that in the single-stranded form, DNA molecule possesses high affinity toward MoS_2 NSs and binds on its surface. As a result, the dye-labeled ssDNA suffered quenching of its fluorescence intensity. But in the presence of target ssDNA, dye-labeled ssDNA forms a duplex and the resulting dsDNA showed very weak affinity to bind to the MoS_2 surface. As a result, fluorescence intensity gets restored, which was corroborated to the amount of target ssDNA present. The work provides an efficient method to discriminate between ssDNA and dsDNA. Not only MoS_2 but also the other members of TMDs, such as WS_2, TiS_2, and TaS_2, can be used in the development of fluorescence biosensors. Similar to the work of Zhu and coworkers, Xi et al. developed a WS_2 based sensing platform for the detection of microRNA (miRNA). They utilized the length-dependent affinity of ssDNA oligonucleotide to bind onto the surface of WS_2 NSs. In brief, long ssDNA oligonucleotides showed higher affinity toward WS_2 NSs than short ssDNA oligonucleotides. When a ssDNA probe was hybridized with the target miRNA, duplex-specific nuclease could cleave the DNA strand in the DNA-RNA complex and release the target miRNA. The released miRNA is further hybridized with other ssDNA, thus continuing the cycle as cleaving then releasing and again hybridizing. It is also reported that the fluorescence signal of short ssDNA can be conserved, but in case of long ssDNA, it will be quenched. Huang and group [122] reported a MoS_2-based fluorometric sensing method for DNA detection with an LOD value of 15 pM.

TMD-based fluorescence biosensors have also been used in the detection of enzymes. Ge et al. [123] constructed a fluorescence biosensor for T4 polynucleotide kinase (T4 PNK) using WS_2 NSs. The working principle of the biosensor depends on specific fluorescence quenching of ssDNA over the surface of WS_2 NSs and polynucleotide kinase—catalyzed phosphorylation-specific exonuclease reaction. The fluorescence probe consisted of dye-labeled dsDNA/WS_2 NSs, which showed high fluorescence intensity due to the weak dsDNA-WS_2 interaction. But in the presence of T4 PNK, dsDNA was phosphorylated and degraded by λ exonuclease to form ssDNA. As mentioned earlier, ssDNA was strongly adsorbed onto the WS_2 surface causing quenching of the fluorescence. The sensor showed outstanding performance with an LOD of 0.01 U mL^{-1} and linearity in the range of 0.01−10 U mL^{-1}.

Deng and coworkers reported an easy and sensitive method for fabricating DNA methyltransferase (MTase) fluorescence biosensor by utilizing MoS_2 NSs as the quencher. The fluorometric DNA MTase biosensor showed linear response in the range of 0.2−20 U mL^{-1}, with an LOD of 0.14 U mL^{-1}. In brief, the substrate DNA was first designed with a fluorescein amidite (FAM)-labeled dsDNA segment having the specific recognizing sequence of DNA adenosine methyltransferase (called Dam) and a ssDNA segment that can graft the substrate DNA on the surface of MoS_2 NSs. The substrate DNA was adsorbed on the 2D nanosurface through the interaction between the ssDNA segment and the MoS_2 basal plane. This caused the quenching of fluorescence of dsDNA. The substrate DNA was then methylated by Dam and finally cleaved by the methylation-sensitive restriction endonuclease DpnI, releasing FAM-labeled ssDNA fragments. As a result, fluorescence was recovered again. The restored fluorescence was directly related to the methylation level and thus quantified the Dam MTase activity [124].

In the recent years, combined aptamers/2D TMD NSs have been proven as extremely sensitive fluorescence biosensors for detection of various bioanalytes [125]. Kong and coworkers developed such fluorescence biosensors to detect PSA. The fluorescence probe is based on MoS_2 NSs functionalized with FAM-labeled aptamer. This work showed that the binding interaction of the FAM-labeled aptamer with the target PSA is stronger than that with MoS_2 NSs. As a result, the FAM-labeled aptamer gets easily detached from the surface of NSs in the presence of PSA, with a recovery in fluorescence. The LOD of this MoS_2-NS-based PSA biosensor was 0.2 ng mL^{-1}, making it a very sensitive detection technique [121]. Xiang et al. [126] developed a MoS_2-NS-based fluorescence biosensor to detect streptavidin. The work explained the advantages of exonuclease III-assisted DNA recycling procedure and the terminal protection strategy of the small molecule linked. The prepared biosensor showed a good response toward protein detection with an LOD of 0.67 ng mL^{-1}. In addition, Yin and group successfully fabricated an ultrasensitive biosensor for cytochrome c detection. The sensor probe was prepared by combining an aptamer-based bioassay with VS_2 NSs and it was very sensitive with an LOD down to 0.5 nM [127].

7.5.2 Cancer therapy

2D TMDs can be used in different therapeutic applications. Cancer has always been a major threat to human health, although significant efforts have been made to fight it for a very long time [128]. Conventional treatments such as chemotherapy, surgery, and radiation therapies often result in poor efficiency, harsh side effects, and so on [129,130]. To circumvent these problems, advanced and new therapeutic modalities have been developed. Gene therapy, PTT, photodynamic therapy (PDT), etc. [130,131] possess a huge potential for treating a wide range of diseases including cancer, heart diseases, diabetes, etc. In recent developments, there are so many efforts to improve cancer therapy through nanotechnology and nanoscience. A variety of nanomaterials are recently emerging as promising tools for cancer diagnostics owing to their small size, ultrahigh surface-to-volume ratios, and unique chemical and physical properties. So far, many types of inorganic and organic nanomaterials, including graphene, carbon nanotubes, quantum dots, magnetic iron oxides, and upconversion nanoparticles, have widely been investigated for biomedical applications [132–135]. Lately, 2D TMDs have drawn considerable interest as ideal nanoplatforms for multimodal therapy, PTT, and drug delivery. Their ultrathin 2D structure, high aspect ratio, and comparatively high biocompatibility, along with their unique chemical and physical properties, have made this type of material viable in advanced therapeutic applications.

7.5.2.1 Photothermal therapy

PTT is a newly developed and encouraging therapeutic strategy that utilizes the NIR laser photoabsorbing agents to generate heat for thermal excision of cancer cells and finally cell death upon NIR laser irradiation [130]. The utilization of wavelength of the NIR laser source has several advantages. NIR possesses high tissue penetration depth and is very weakly absorbed by biological tissues, thus eliminating any adverse effect to the normal biological environment [136,137]. Following are a few examples of recent developments made by using TMDs in PTT for the treatment of cancer.

The working principle of ce-MoS_2 NSs as competent NIR photothermal agents was first described by Chou and his coworkers in 2013. Their work explained that ce-MoS_2 NSs could be a better photothermal agent than gold nanorods and graphene derivatives [120]. At 800 nm, the mass extinction coefficient value of ce-MoS_2 NSs was found to be 29.2 $Lg^{-1}cm^{-1}$ which was much greater than that of GO, nanorods, and reduced GO [138]. The development of ultrathin 1T-WS_2 NSs through ammonium intercalation utilizing the hydrothermal approach was described in 2015 by Liu and coworkers. The prepared 1T-WS_2 measured 150 nm in size and showed good hydrophilicity and intense optical absorption in the NIR region. In vivo experiments were performed by intrathecal injection of 1T-WS_2 NSs in NOD/SCID mice under the irradiation of 808 nm laser at a power density of 0.6 W cm^{-2}. The results were encouraging, and the temperature of HeLa tumors in NOD/SCID mice raised up to 50 °C within a period of 4 min. The temperature at this stage showed a hold for the next 6 min. All tumors treated with PTT disappeared after 4 days without reappearance [139]. The state-of-the-art literature showcases very interesting approaches on the utilization of 2D TMDs for PTT applications. One such approach is the effect of size reduction of TMDs in PTT. Wang et al. developed a bottom-up approach for the preparation of PEGylated MoS_2 NSs with 50–300 nm size. The work illustrated the photothermal effect of MoS_2 NSs with a variation in the size of MoS_2 NSs as well as in PEGylation. Results showed that the photothermal effect increased with increasing size reduction. In vivo experiments were also performed and results showed that after using the intravenous injection of MoS_2-PEG in the presence of NIR laser irradiation at 808 nm, the size of the tumor (4T1) in Balb/c mice was significantly reduced. In addition, the MoS_2-PEG injection–treated mice stayed healthy after 40 days of treatment. The results are conclusive showing high photothermal efficiency of dimensionally small MoS$_2$ NSs and their nontoxic behavior for in vitro application [140].

7.5.2.2 Gene therapy

Gene therapy is a very modern therapeutic technique that relies on the introduction of genetic materials such as DNA or RNA into cells by using a suitable carrier to treat the abnormal genes or to initiate the production of a beneficial protein. Gene-therapy-based modification of human DNA was first attempted in 1980, but success was achieved in 1989, which was approved by the National Institutes of Health. Despite its huge viability, gene therapy has certain limitations. One such hindrance is the lack of competent, sensitive, and safe gene delivery carriers [131,141]. In the past decades, consistent efforts have been devoted to develop effective gene carrier vesicles. In general, gene therapy proceeds through three basic phases. The first phase involves delivery of extracellular therapeutic genetic materials into cells through endocytosis. In the second phase, genes are liberated from the nanocarriers after being delivered from endosomes into the cytosol. The therapeutic genetic materials then bind to the targeted mRNA, thus finally ensuing the silence of a specific protein to attain the aim of disease treatment. For a successful gene therapy, the gene releasing and loading efficiency of nanocarriers is the prime factor. Broadly, the gene carriers can be classified into two categories: viral and nonviral vectors [142]. The genomes of viruses can initiate various mechanisms and thereby penetrate into the target cell's cytoplasm during the gene delivery process. Unfortunately, these viral nanocarriers are pathogenic to the organism and often carry the risk of oncogenic effects, adverse immune response, and mutagenesis of host cell genes [142,143]. Therefore in the recent times, nonviral gene delivery systems have been targeted [144]. Nonviral vectors possess the benefits of low pathogenicity and biosafety, as well as the advantage of facile preparation and reusability. 2D nanomaterials have gained sustainable interest as gene delivery carriers because of their high surface area and ease of surface functionalization. Along with conventional 2D nanomaterials such as graphene, TMDs have also drawn a huge interest in this regard. Among the different TMDs, MoS$_2$ perhaps has gained the majority of attention. In 2014, Kou and his colleagues proposed a novel method to fabricate positively charged polyethylenimine (PEI)-PEG-MoS$_2$ nanocomposite as a gene carrier. They have successfully loaded the carrier with negatively charged small interfering RNA (siRNA) through electrostatic interaction and utilized it for gene therapy. Kim et al., reported single-layered MoS$_2$-PEI-PEG nanocomposite as a stimulus-responsive carrier. The carrier can form a stable polyplex with DNA molecules through electrostatic iterations and enter the cell by endocytosis. Application of NIR radiation brings the endosomal escape of the nanocomposite by restrained heat-mediated rupture of the endosomal membrane. In a reducing environment, the disulfide bond between MoS$_2$ and PEI/PEG begins to detach from the MoS$_2$ surface resulting in the release of the gene to the host environment. The presence of intracellular reducing agents such as glutathione could further enhance the gene release process by helping in the detachment of polymer matrices. However, widespread use of TMDs in their full potential is yet to be achieved. The critical challenge is the high crystalline structure, which renders their degradation in the physiologic environment, as well as toxicity concern by the release of metal cations to the host environment [145].

7.5.2.3 Combination therapy or polytherapy

In combination therapy, more than one medication or modality is used simultaneously in the treatment of a single disease. In many instances, combination therapy is much more effective than conventional therapy. Combination therapy can help address issues related to drug resistance in a patient because the development of simultaneous resistance by a pathogen or tumor is less viable during the application of parallel multiple drugs against the same pathogen or tumor. In combination therapy, 2D TMDs possess a huge potential owing to their multifunctional properties. Different types of combinations of therapies have been explored in the recent years. For example, the combination mechanism of PTT with gene therapy was discovered accidentally in 2016. In this work, hyperthermia was induced in the synthesis of heat shock protein 70. A relevant strategy was commenced by gene therapy to downregulate the expression of heat shock protein 70. Multifunctional WS_2-polyetherimide was developed as a magnificent bioimaging-therapeutic nanoprobe for gene-photothermal synergistic therapy of tumors at a moderate condition [146]. Poly(amidoamine) dendrimer and MoS_2 composite was targeted by Kong et al. [147] for combinational PTT and gene silencing in tumor therapy. Similarly, Xu et al. performed a combination of phototherapy and bioimaging for cancer cells simultaneously. They found that MoS_2 NSs are quite responsive for 808-nm NIR radiation, which can be effectively used for PTT and PDT. In PDT, cancer cells are killed by toxic reactive oxygen species, which are generated by a photosensitizer. Furthermore, simultaneous imaging is quite possible, which includes upconversion luminescence and computed tomography (CT)/MRI [148]. PTT and chemotherapy together form a very important combinational approach. In such an approach, the PTT agent is also loaded with the drug. The local heat produced under irradiation of the PTT agent can directly kill the cancer cells. At the same time, the dissipated heat energy enhanced the release of drug from the surface of the PTT agent. Under such circumstances, the cancer cell easily uptakes the drug as local heat also increases the membrane permeability. Yin and coworkers developed chitosan-functionalized MoS_2 NSs in PTT and simultaneous chemotherapy. The nanocomposite was loaded with doxorubicin, and at 808-nm laser irradiation, it demonstrated thermal ablation of cancer cells and triggered drug release [149]. Fig. 7.8A–C depicts different combinational approaches reported based on TMD based 2D NSs.

7.5.3 Bioimaging application

So far various noninvasive imaging strategies like photoacoustic imaging (PA), positron emission tomography (PET), computed tomography (CT) and magnetic resonance imaging (MRI) have widely been developed for bioimaging applications

[151,152]. In the recent times, developments of a variety of contrast agents for the recovery of the resolution in bioimaging for tissues and vessels have recognized the domain. It further helps in the progress of functional molecular bioimaging with significantly enhanced sensitivity and selectivity that will help in the identification of important biomolecules [151].

Nanomaterial-supported contrast agents have shown great potential in both bioimaging and therapy. The 2D TMDs can be employed not only as a photothermal platform for combinations of PTT with other therapeutic modalities, but also in biomedical imaging because of their inherent physical and optical properties [153]. The favorable attributes, such as strong NIR absorption capability, high light-to-heat transformation capability, and subsequent generation of ultrasound signal, make 2D TMDs suitable for PA imaging. Nowadays, PA has become an advanced technology offering the benefits of in-depth cross-sectional listening of entire tumors. These advantages are not available in ordinary microscopic techniques. The sensitivity is also higher than that of CT imaging [153].

To date, MoS_2 and WS_2 have widely been used as contrast agents in CT and superior results are obtained when compared with the other contrast agents. 2D TMDs containing high-atomic-number elements, W and Bi, have high X-ray absorption coefficient and are reported as suitable contrast agents for CT imaging [153].

Liu et al. reported prominent PA signal developments after 24 h in tumors that were subjected to PEGylated MoS_2 intravenous injection. It was observed that PEGylated MoS_2 mostly accumulated in tumors [139]. Chen et al. reported the PA effect of MoS_2 NSs. The MoS_2 NSs are extremely dependent on the number of layers for their PA effect. Single-layered MoS_2 shows outstanding biocompatible behavior and can produce stronger PA signals than few-layer and multilayer MoS_2. These unique properties make them very sensitive for the detection of brain tumor cells. The main cause of this process is that as the layer number decreases, the absorption of NIR light increases and also changes its elastic properties [154].

Each of the above-mentioned imaging probe has its own merits and demerits, resulting in limited use of single-mode imaging technique for practical purposes. A combination of different imaging techniques with one contrast agent helps develop a new multimodal imaging method with potentially promising policy [152]. The combination of PA and X-ray CT imaging could result in a one-system imaging modality. By using the adaptable properties of 2D TMDs, along with high X-ray and NIR absorbance, complete utilization of the deep penetration power of CT imaging and sensitivity of PA imaging could be achieved [153]. Yin et al. observed that MoS_2 NSs could be employed to combine dual-modal PA and CT imaging of tumor tissues in vivo. Cheng et al. reported on CT and PA imaging characteristics of WS_2-PEG NSs. In this work, in vivo experiments were performed by using mice bearing 4T1 tumors. The tumors were then treated with an intravenous injection of WS_2-PEG NSs and observed under a PA and CT imaging system. Their results could provide intense PA signals and the noticeably improved contrast showed high accumulation of WS_2-PEG NSs in the tumors [149]. Owing to the potential viability and easy modification skill in surfaces, some other functional

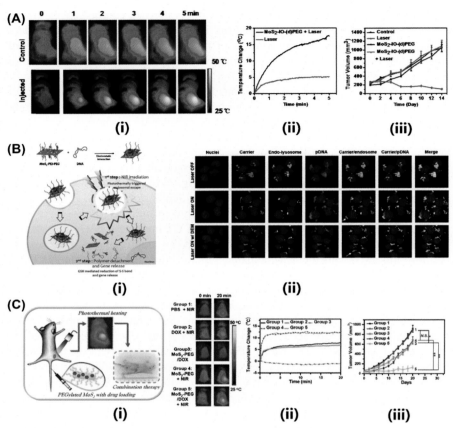

FIGURE 7.8

(A) Fe$_2$O$_3$@MoS$_2$ nanosheets with double polyethylene glycol (PEG)ylation for photothermal therapy (PTT): (i) infrared (IR) thermal images of 4T1 tumor, (ii) average temperature change at the tumor site laser irradiation, and (iii) tumor volume growth [150]. (B) gene therapy: (i) MoS$_2$-polyethylenimine (PEI)-PEG nanocomposite for gene therapy, (ii) confocal laser scanning microscopic images of HCT116 cells treated with MoS$_2$-PEI-PEG/DNA. White arrows and yellow arrow represent endosomal escape of MoS$_2$-PEI-PEG nanocomposite and intracellular gene release from MoS$_2$-PEI-PEG carrier, respectively [144]. (C) Combination therapy: (i) MoS$_2$-PEG/doxorubicin (DOX) for in vivo PTT and chemotherapy, (ii) IR thermal images of 4T1 tumor—bearing mice, and (iii) temperature change and tumor volume growth curves [36]. *GSH*, reduced glutathione; *IO*, iron oxide; *NIR*, near-infrared; *pDNA*, plasmid DNA; *PBS*, Phosphate buffered saline.

(A) Copyright 2015 American Chemical Society, Reproduced with permission from Liu et al., 2015; (B) Copyright 2015 WILEY-VCH Verlag GmbH & Co. KGaA, Weinheim, Reproduced with permission from Kim et al., 2015; (C) Copyright 2014 WILEY-VCH Verlag GmbH & Co. KGaA, Weinheim, Reproduced with permission from Liu et al., 2014.

moieties with different imaging techniques could be ornamented on 2D TMDs to improve their functionalities [153].

Yu et al. reported a facile two-step hydrothermal approach for the preparation of superparamagnetic MoS_2-Fe_3O_4 nanocomposites functionalized with PEG. The MoS_2-Fe_3O_4 nanocomposites could be employed as photothermal and contrast agent in combined-mode PA imaging and MRI [155]. Yang and his group [156] observed that the Fe_3O_4 nanoparticles can be attached noncovalently to the surface of WS_2 to form suspensions of WS_2/Fe_3O_4, which can be used for multimodal MRI and PA imaging of tumor tissues. These two studies have revealed that Fe_3O_4-decorated MoS_2 and WS_2 could magnetically target photothermal excision of cancer signified in MRI and PA imaging. Additionally, Cheng et al. fabricated a WS_2-based Gd-doped NS that could be able to combine PTT and radiotherapy of cancer under a trimodal (CT, PA imaging, and MRI) image control [157].

The development of a $FeSe_2$/Bi_2Se_3 nanocomposite–based multifunctional nanoprobe modified by PEG through hydrophobic interactions was reported in 2016 by Cheng and his group. The prepared nanocomposite showed high NIR and X-ray absorbance, along with powerful magnetic property and radioisotope labeling. The in vivo studies were done by using ^{64}Cu-$FeSe_2$/Bi_2Se_3-treated tumor imaging with three modal techniques including PA imaging, MRI, CT, and PET [71]. Fig. 7.9 depicts dual-mode imaging by WS_2@PEI-siRNA for in vivo CT and PA imaging in BEL-7402 tumor–bearing mice and MoS_2-iron oxide-PEG nanocomposite for multimodal in vivo PA and MRI of 4T1 tumor in mice. These examples represent the versatility of diagnostics-therapeutics combination based on 2D TMDs as a potential candidate in bioimaging-based diagnosis.

7.5.4 Future direction and conclusions

Undoubtedly, 2D TMDs have drawn serious attention over the past few years. They are the newly emerging nanomaterials possessing huge potentiality for biomedical utilities. The unique properties of TMDs, especially in terms of electronic and optical attributes, have made them the alternative for conventional 2D nanostructures such as graphene. The viability of TMD nanomaterials in the biomedical sector has been highly recognized. Characteristics such as tunable bandgaps, monolayer to few-layer morphology, fluorescence quenching ability, and high photothermal conversion efficiency enable TMD NSs to find applications in biosensing, PTT, and gene therapy, as well as in bioimaging. However, from a commercial aspect, TMD-based nanomaterials are relatively new in the healthcare sector. The exploration of TMDs has just begun for their biomedical applicability. There are many questions that need to be answered before TMD NSs could go into the translational stage. In this context, one of the prime issues is biosafety. Many reports suggest that 2D TMD NSs are nontoxic and biocompatible. However, in most of these reports, conclusions were drawn from basic in vitro experiments. Therefore the detailed long-term in vivo effect of TMD NSs needs to be investigated thoroughly. In vivo biodegradability of TMDs may pose other serious hindrances, as they are of

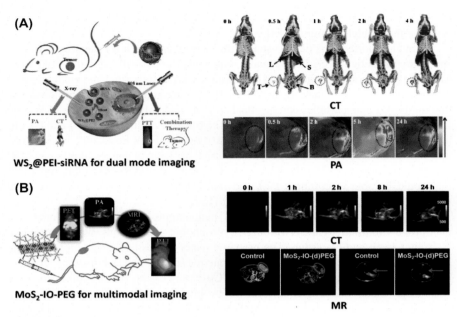

FIGURE 7.9

(A) Dual-mode imaging by WS$_2$@PEI-siRNA (polyethylenimine-small interfering RNA): in vivo computed tomographic (CT) and photoacoustic (PA) imaging in BEL-7402 tumor—bearing mice [146]. (B) MoS$_2$-IO-PEG (iron oxide-polyethylene glycol) for multimodal imaging: in vivo PA imaging and magnetic resonance imaging (MRI) of 4T1 tumor in mice [150]. *PET*, positron emission tomography; *PTT*, photothermal therapy.

(A) Copyright 2016 WILEY-VCH Verlag GmbH & Co. KGaA, Weinheim, Reproduced with permission from Zhang et al., 2016; (B) Copyright2015 American Chemical Society, Reproduced with permission from Liu et al., 2015.

nonorganic origin. Despite all these concerns, we can presume that the potentiality of 2D TMD NSs is huge and the current trend of research is going on in the right path unrevealing the new dimensions of these type of materials. In the near future, TMD NSs can be made a revolutionary biomaterial by understanding the in vivo biodistribution, metabolic pathways, and excretion within a biological system.

Acknowledgment

SG acknowledges SERB, DST, India for financial support (grant no. PDF/2016/003142) and HSD acknowledges DBT, India for financial support through grant no. BT/PR16223/NER/95/494/2016.

References

[1] X. Li, J. Shan, W. Zhang, S. Su, L. Yuwen, L. Wang, Recent advances in synthesis and biomedical applications of two-dimensional transition metal dichalcogenide nanosheets, Small 13 (5) (2017) 1602660−1602687.

[2] M. Chhowalla, H.S. Shin, G. Eda, L.J. Li, K.P. Loh, H. Zhang, The chemistry of two-dimensional layered transition metal dichalcogenide nanosheets, Nature Chemistry 5 (4) (2013) 263−275.

[3] S. Kretschmer, H.P. Komsa, P. Bøggild, A.V. Krasheninnikov, Structural transformations in two-dimensional transition-metal dichalcogenide MoS_2 under an electron beam: insights from first-principles calculations, Journal of Physical Chemistry Letters 8 (13) (2017) 3061−3067.

[4] A.V. Kolobov, J. Tominaga, Two-Dimensional Transition-Metal Dichalcogenides, vol. 239, Springer, 2016.

[5] M. Javaid, S.P. Russo, K. Kalantar-Zadeh, A.D. Greentree, D.W. Drumm, Band structure and giant Stark effect in two-dimensional transition-metal dichalcogenides, Electronic Structure 1 (1) (2018) 015005−015024.

[6] L. Debbichi, O. Eriksson, S. Lebègue, Electronic structure of two-dimensional transition metal dichalcogenide bilayers from ab initio theory, Physical Review B 89 (20) (2014) 205311−205315.

[7] J.R. Brent, N. Savjani, P. O'Brien, Synthetic approaches to two-dimensional transition metal dichalcogenide nanosheets, Progress in Materials Science 89 (2017) 411−478.

[8] H. Li, Y. Li, A. Aljarb, Y. Shi, L.J. Li, Epitaxial growth of two-dimensional layered transition-metal dichalcogenides: growth mechanism, controllability, and scalability, Chemical Reviews 118 (13) (2017) 6134−6150.

[9] T. Stephenson, Z. Li, B. Olsen, D. Mitlin, Lithium ion battery applications of molybdenum disulfide (MoS_2) nanocomposites, Energy & Environmental Science 7 (1) (2014) 209−231.

[10] R.F. Frindt, Single crystals of MoS_2 several molecular layers thick, Journal of Applied Physics 37 (4) (1966) 1928−1929.

[11] K.S. Novoselov, D. Jiang, F. Schedin, T.J. Booth, V.V. Khotkevich, S.V. Morozov, A.K. Geim, Two-dimensional atomic crystals, Proceedings of the National Academy of Sciences 102 (30) (2005) 10451−10453.

[12] A. Castellanos-Gomez, E. Navarro-Moratalla, G. Mokry, J. Quereda, E. Pinilla-Cienfuegos, N. Agraït, G. Rubio-Bollinger, Fast and reliable identification of atomically thin layers of $TaSe_2$ crystals, Nano Research 6 (3) (2013) 191−199.

[13] Y. Yu, S. Jiang, G. Zhang, W. Zhou, X. Miao, Y. Zeng, L. Zhang, Universal ultrafast sandpaper assisting rubbing method for room temperature fabrication of two-dimensional nanosheets directly on flexible polymer substrate, Applied Physics Letters 101 (7) (2012) 073113.

[14] K. Gacem, M. Boukhicha, Z. Chen, A. Shukla, High quality 2D crystals made by anodic bonding: a general technique for layered materials, Nanotechnology 23 (50) (2012) 505709.

[15] J.H. Bang, K.S. Suslick, Applications of ultrasound to the synthesis of nanostructured materials, Advanced Materials 22 (10) (2010) 1039−1059.

[16] V. Nicolosi, M. Chhowalla, M.G. Kanatzidis, M.S. Strano, J.N. Coleman, Liquid exfoliation of layered materials, Science 340 (6139) (2013) 1226419.

[17] Z. Shen, J. Li, M. Yi, X. Zhang, S. Ma, Preparation of graphene by jet cavitation, Nanotechnology 22 (36) (2011) 365306.

[18] M. Dular, B. Stoffel, B. Širok, Development of a cavitation erosion model, Wear 261 (5—6) (2006) 642—655.

[19] D. Lohse, Sonoluminescence: cavitation hots up, Nature 434 (7029) (2005) 33.

[20] K.R. Paton, E. Varrla, C. Backes, R.J. Smith, U. Khan, A. O'Neill, T. Higgins, Scalable production of large quantities of defect-free few-layer graphene by shear exfoliation in liquids, Nature Materials 13 (6) (2014) 624.

[21] J.N. Coleman, M. Lotya, A. O'Neill, S.D. Bergin, P.J. King, U. Khan, I.V. Shvets, Two-dimensional nanosheets produced by liquid exfoliation of layered materials, Science 331 (6017) (2011) 568—571.

[22] D. Gao, M. Si, J. Li, J. Zhang, Z. Zhang, Z. Yang, D. Xue, Ferromagnetism in free-standing MoS_2 nanosheets, Nanoscale Research Letters 8 (1) (2013) 129.

[23] K.G. Zhou, N.N. Mao, H.X. Wang, Y. Peng, H.L. Zhang, A mixed-solvent strategy for efficient exfoliation of inorganic graphene analogues, Angewandte Chemie International Edition 50 (46) (2011) 10839—10842.

[24] A. O'Neill, U. Khan, J.N. Coleman, Preparation of high concentration dispersions of exfoliated MoS_2 with increased flake size, Chemistry of Materials 24 (12) (2012) 2414—2421.

[25] G. Guan, S. Zhang, S. Liu, Y. Cai, M. Low, C.P. Teng, Y. Zheng, Protein induces layer-by-layer exfoliation of transition metal dichalcogenides, Journal of the American Chemical Society 137 (19) (2015) 6152—6155.

[26] E. Benavente, M.A. Santa Ana, F. Mendizábal, G. González, Intercalation chemistry of molybdenum disulfide, Coordination Chemistry Reviews 224 (1—2) (2002) 87—109.

[27] J.K. Huang, J. Pu, C.L. Hsu, M.H. Chiu, Z.Y. Juang, Y.H. Chang, L.J. Li, Large-area synthesis of highly crystalline WSe2 monolayers and device applications, ACS Nano 8 (1) (2013) 923—930.

[28] M.B. Dines, Lithium intercalation via n-butyllithium of the layered transition metal dichalcogenides, Materials Research Bulletin 10 (4) (1975) 287—291.

[29] Q.H. Wang, K. Kalantar-Zadeh, A. Kis, J.N. Coleman, M.S. Strano, Electronics and optoelectronics of two-dimensional transition metal dichalcogenides, Nature Nanotechnology 7 (11) (2012) 699.

[30] P. Joensen, R.F. Frindt, S.R. Morrison, Single-layer MoS_2, Materials Research Bulletin 21 (4) (1986) 457—461.

[31] G. Eda, H. Yamaguchi, D. Voiry, T. Fujita, M. Chen, M. Chhowalla, Photoluminescence from chemically exfoliated MoS_2, Nano Letters 11 (12) (2011) 5111—5116.

[32] L. Yuwen, H. Yu, X. Yang, J. Zhou, Q. Zhang, Y. Zhang, L. Wang, Rapid preparation of single-layer transition metal dichalcogenide nanosheets via ultrasonication enhanced lithium intercalation, Chemical Communications 52 (3) (2016) 529—532.

[33] T. Daeneke, R.M. Clark, B.J. Carey, J.Z. Ou, B. Weber, M.S. Fuhrer, K. Kalantar-zadeh, Reductive exfoliation of substoichiometric MoS_2 bilayers using hydrazine salts, Nanoscale 8 (33) (2016) 15252—15261.

[34] Z. Zeng, Z. Yin, X. Huang, H. Li, Q. He, G. Lu, H. Zhang, Single-layer semiconducting nanosheets: high-yield preparation and device fabrication, Angewandte Chemie 123 (47) (2011) 11289—11293.

[35] H. Zeng, J. Dai, W. Yao, D. Xiao, X. Cui, Valley polarization in MoS_2 monolayers by optical pumping, Nature Nanotechnology 7 (8) (2012) 490.

[36] T. Liu, C. Wang, X. Gu, H. Gong, L. Cheng, X. Shi, Z. Liu, Drug delivery with PEGylated MoS_2 nano-sheets for combined photothermal and chemotherapy of cancer, Advanced Materials 26 (21) (2014) 3433–3440.

[37] R.J. Smith, P.J. King, M. Lotya, C. Wirtz, U. Khan, S. De, J. Chen, Large-scale exfoliation of inorganic layered compounds in aqueous surfactant solutions, Advanced Materials 23 (34) (2011) 3944–3948.

[38] A. Pimpinelli, J. Villain, Physics of Crystal Growth, Cambridge University Press, Cambridge, 1998.

[39] D.J. Lewis, A.A. Tedstone, X.L. Zhong, E.A. Lewis, A. Rooney, N. Savjani, J.M. Raftery, Thin films of molybdenum disulfide doped with chromium by aerosol-assisted chemical vapor deposition (AACVD), Chemistry of Materials 27 (4) (2015) 1367–1374.

[40] G. Wahl, P.B. Davies, R.F. Bunshah, B.A. Joyce, C.D. Bain, G. Wegner, P.K. Bachmann, Thin films, Ullmann's Encyclopedia of Industrial Chemistry (2000) 1–75.

[41] H.G. Füchtbauer, A.K. Tuxen, P.G. Moses, H. Topsøe, F. Besenbacher, J.V. Lauritsen, Morphology and atomic-scale structure of single-layer WS_2 nanoclusters, Physical Chemistry Chemical Physics 15 (38) (2013) 15971–15980.

[42] A.L. Elías, N. Perea-López, A. Castro-Beltrán, A. Berkdemir, R. Lv, S. Feng, H.R. Gutiérrez, Controlled synthesis and transfer of large-area WS_2 sheets: from single layer to few layers, ACS Nano 7 (6) (2013) 5235–5242.

[43] H.R. Gutiérrez, N. Perea-López, A.L. Elías, A. Berkdemir, B. Wang, R. Lv, M. Terrones, Extraordinary room-temperature photoluminescence in triangular WS_2 monolayers, Nano Letters 13 (8) (2012) 3447–3454.

[44] Y. Zhan, Z. Liu, S. Najmaei, P.M. Ajayan, J. Lou, Large-area vapor-phase growth and characterization of MoS_2 atomic layers on a SiO_2 substrate, Small 8 (7) (2012) 966–971.

[45] D. Dumcenco, D. Ovchinnikov, K. Marinov, P. Lazic, M. Gibertini, N. Marzari, S. Bertolazzi, Large-area epitaxial monolayer MoS_2, ACS Nano 9 (4) (2015) 4611–4620.

[46] Y.H. Lee, X.Q. Zhang, W. Zhang, M.T. Chang, C.T. Lin, K.D. Chang, T.W. Lin, Synthesis of large-area MoS_2 atomic layers with chemical vapor deposition, Advanced Materials 24 (17) (2012) 2320–2325.

[47] S. Balendhran, J.Z. Ou, M. Bhaskaran, S. Sriram, S. Ippolito, Z. Vasic, K. Kalantar-Zadeh, Atomically thin layers of MoS_2 via a two step thermal evaporation–exfoliation method, Nanoscale 4 (2) (2012) 461–466.

[48] Y. Li, A. Chernikov, X. Zhang, A. Rigosi, H.M. Hill, A.M. van der Zande, T.F. Heinz, Measurement of the optical dielectric function of monolayer transition-metal dichalcogenides: MoS_2, $MoSe_2$, WS_2, and WSe_2, Physical Review B 90 (20) (2014) 205422.

[49] T. Wang, R. Zhu, J. Zhuo, Z. Zhu, Y. Shao, M. Li, Direct detection of DNA below ppb level based on thionin-functionalized layered MoS_2 electrochemical sensors, Analytical Chemistry 86 (24) (2014) 12064–12069.

[50] X. Wang, Y. Gong, G. Shi, W.L. Chow, K. Keyshar, G. Ye, B.K. Tay, Chemical vapor deposition growth of crystalline monolayer $MoSe_2$, ACS Nano 8 (5) (2014) 5125–5131.

[51] X. Lu, M.I.B. Utama, J. Lin, X. Gong, J. Zhang, Y. Zhao, W. Zhou, Large-area synthesis of monolayer and few-layer $MoSe_2$ films on SiO_2 substrates, Nano Letters 14 (5) (2014) 2419–2425.

[52] Y.H. Chang, W. Zhang, Y. Zhu, Y. Han, J. Pu, J.K. Chang, T. Takenobu, Monolayer MoSe$_2$ grown by chemical vapor deposition for fast photodetection, ACS Nano 8 (8) (2014) 8582−8590.

[53] Y. Zhang, Y. Zhang, Q. Ji, J. Ju, H. Yuan, J. Shi, X. Song, Controlled growth of high-quality monolayer WS$_2$ layers on sapphire and imaging its grain boundary, ACS Nano 7 (10) (2013) 8963−8971.

[54] X. Huang, Z. Zeng, H. Zhang, Metal dichalcogenide nanosheets: preparation, properties and applications, Chemical Society Reviews 42 (5) (2013) 1934−1946.

[55] J.W. Seo, Y.W. Jun, S.W. Park, H. Nah, T. Moon, B. Park, J. Cheon, Two-dimensional nanosheet crystals, Angewandte Chemie International Edition 46 (46) (2007) 8828−8831.

[56] C. Altavilla, M. Sarno, P. Ciambelli, A novel wet chemistry approach for the synthesis of hybrid 2D free-floating single or multilayer nanosheets of MS$_2$@ oleylamine (M] Mo, W), Chemistry of Materials 23 (17) (2011) 3879−3885.

[57] B. Mahler, V. Hoepfner, K. Liao, G.A. Ozin, Colloidal synthesis of 1T-WS$_2$ and 2H-WS$_2$ nanosheets: applications for photocatalytic hydrogen evolution, Journal of the American Chemical Society 136 (40) (2014) 14121−14127.

[58] C.N.R. Rao, H.S.S. Ramakrishna Matte, U. Maitra, Graphene analogues of inorganic layered materials, Angewandte Chemie International Edition 52 (50) (2013) 13162−13185.

[59] X. Zhang, Z. Lai, C. Tan, H. Zhang, Solution-processed two-dimensional MoS$_2$ nanosheets: preparation, hybridization, and applications, Angewandte Chemie International Edition 55 (31) (2016) 8816−8838.

[60] J. Xie, H. Zhang, S. Li, R. Wang, X. Sun, M. Zhou, Y. Xie, Defect-rich MoS$_2$ ultrathin nanosheets with additional active edge sites for enhanced electrocatalytic hydrogen evolution, Advanced Materials 25 (40) (2013) 5807−5813.

[61] X. Chen, R. Fan, Low-temperature hydrothermal synthesis of transition metal dichalcogenides, Chemistry of Materials 13 (3) (2001) 802−805.

[62] H.S.S. Ramakrishna Matte, A. Gomathi, A.K. Manna, D.J. Late, R. Datta, S.K. Pati, C.N.R. Rao, MoS$_2$ and WS$_2$ analogues of graphene, Angewandte Chemie International Edition 49 (24) (2010) 4059−4062.

[63] M. Li, D. Wang, J. Li, Z. Pan, H. Ma, Y. Jiang, Z. Tian, Facile hydrothermal synthesis of MoS$_2$ nano-sheets with controllable structures and enhanced catalytic performance for anthracene hydrogenation, RSC Advances 6 (75) (2016) 71534−71542.

[64] K.F. Mak, K. He, J. Shan, T.F. Heinz, Control of valley polarization in monolayer MoS$_2$ by optical helicity, Nature Nanotechnology 7 (8) (2012) 494.

[65] Z. Zeng, T. Sun, J. Zhu, X. Huang, Z. Yin, G. Lu, H. Zhang, An effective method for the fabrication of Few-layer-Thick inorganic nanosheets, Angewandte Chemie International Edition 51 (36) (2012) 9052−9056.

[66] A. Kuc, N. Zibouche, T. Heine, Influence of quantum confinement on the electronic structure of the transition metal sulfide TS$_2$, Physical Review B 83 (24) (2011) 245213.

[67] S.W. Han, H. Kwon, S.K. Kim, S. Ryu, W.S. Yun, D.H. Kim, S.C. Hong, Band-gap transition induced by interlayer van der Waals interaction in MoS$_2$, Physical Review B 84 (4) (2011) 045409.

[68] S.W. Han, G.B. Cha, E. Frantzeskakis, I. Razado-Colambo, J. Avila, Y.S. Park, W.S. Yun, Band-gap expansion in the surface-localized electronic structure of MoS$_2$, Physical Review B 86 (11) (2012) 115105.

[69]   K.F. Mak, C. Lee, J. Hone, J. Shan, T.F. Heinz, Atomically thin MoS_2: a new direct-gap semiconductor, Physical Review Letters 105 (13) (2010) 136805.

[70]   A. Splendiani, L. Sun, Y. Zhang, T. Li, J. Kim, C.Y. Chim, F. Wang, Emerging photo-luminescence in monolayer MoS_2, Nano Letters 10 (4) (2010) 1271−1275.

[71]   T. Li, G. Galli, Electronic properties of MoS_2 nanoparticles, Journal of Physical Chemistry C 111 (44) (2007) 16192−16196.

[72]   J.E. Padilha, H. Peelaers, A. Janotti, C.G. Van de Walle, Nature and evolution of the band-edge states in MoS_2: from monolayer to bulk, Physical Review B 90 (20) (2014) 205420.

[73]   W. Zhao, Z. Ghorannevis, L. Chu, M. Toh, C. Kloc, P.H. Tan, G. Eda, Evolution of electronic structure in atomically thin sheets of WS_2 and WSe_2, ACS Nano 7 (1) (2012) 791−797.

[74]   A.K. Geim, I.V. Grigorieva, Van der Waals heterostructures, Nature 499 (7459) (2013) 419.

[75]   I. Popov, G. Seifert, D. Tománek, Designing electrical contacts to MoS_2 monolayers: a computational study, Physical Review Letters 108 (15) (2012) 156802.

[76]   H.P. Komsa, A.V. Krasheninnikov, Two-dimensional transition metal dichalcogenide alloys: stability and electronic properties, Journal of Physical Chemistry Letters 3 (23) (2012) 3652−3656.

[77]   J. Kang, S. Tongay, J. Li, J. Wu, Monolayer semiconducting transition metal dichalco-genide alloys: stability and band bowing, Journal of Applied Physics 113 (14) (2013) 143703.

[78]   J. Kang, S. Tongay, J. Zhou, J. Li, J. Wu, Band offsets and heterostructures of two-dimensional semiconductors, Applied Physics Letters 102 (1) (2013) 012111.

[79]   P. Johari, V.B. Shenoy, Tuning the electronic properties of semiconducting transition metal dichalcogenides by applying mechanical strains, ACS Nano 6 (6) (2012) 5449−5456.

[80]   Y. Liang, S. Huang, R. Soklaski, L. Yang, Quasiparticle band-edge energy and band offsets of monolayer of molybdenum and tungsten chalcogenides, Applied Physics Letters 103 (4) (2013) 042106.

[81]   D.M. Guzman, A. Strachan, Role of strain on electronic and mechanical response of semiconducting transition-metal dichalcogenide monolayers: an ab-initio study, Journal of Applied Physics 115 (24) (2014) 243701.

[82]   E.M. Vogel, J.A. Robinson, Two-dimensional layered transition-metal dichalcogenides for versatile properties and applications, MRS Bulletin 40 (7) (2015) 558−563.

[83]   T. Roy, M. Tosun, J.S. Kang, A.B. Sachid, S.B. Desai, M. Hettick, A. Javey, Field-effect transistors built from all two-dimensional material components, ACS Nano 8 (6) (2014) 6259−6264.

[84]   K. Majumdar, C. Hobbs, P.D. Kirsch, Benchmarking transition metal dichalcogenide MOSFET in the ultimate physical scaling limit, IEEE Electron Device Letters 35 (3) (2014) 402−404.

[85]   H.L. Liu, C.C. Shen, S.H. Su, C.L. Hsu, M.Y. Li, L.J. Li, Optical properties of mono-layer transition metal dichalcogenides probed by spectroscopic ellipsometry, Applied Physics Letters 105 (20) (2014) 201905.

[86]   D. Zappa, Molybdenum dichalcogenides for environmental chemical sensing, Materials 10 (12) (2017) 1418.

[87]   T. Anthony, A. Turner, G. Wilson, I. Kaube, Biosensors: Fundamentals and Applications, Oxford University Press, UK, 1987, p. 770.

[88] F.G. Banica, Chemical Sensors and Biosensors: Fundamentals and Applications, John Wiley & Sons, 2012.

[89] M. Pumera, Graphene in biosensing, Materials Today 14 (7–8) (2011) 308–315.

[90] P. Hong, W. Li, J. Li, Applications of aptasensors in clinical diagnostics, Sensors 12 (2) (2012) 1181–1193.

[91] A. Amine, H. Mohammadi, I. Bourais, G. Palleschi, Enzyme inhibition-based biosensors for food safety and environmental monitoring, Biosensors and Bioelectronics 21 (8) (2006) 1405–1423.

[92] K.C. Knirsch, N.C. Berner, H.C. Nerl, C.S. Cucinotta, Z. Gholamvand, N. McEvoy, S. Sanvito, Basal-plane functionalization of chemically exfoliated molybdenum disulfide by diazonium salts, ACS Nano 9 (6) (2015) 6018–6030.

[93] M. Pumera, A.H. Loo, Layered transition-metal dichalcogenides (MoS_2 and WS_2) for sensing and biosensing, TRAC Trends in Analytical Chemistry 61 (2014) 49–53.

[94] S. Su, M. Zou, H. Zhao, C. Yuan, Y. Xu, C. Zhang, L. Wang, Shape-controlled gold nanoparticles supported on MoS_2 nanosheets: synergistic effect of thionine and MoS_2 and their application for electrochemical label-free immunosensing, Nanoscale 7 (45) (2015) 19129–19135.

[95] Y. Huang, J. Guo, Y. Kang, Y. Ai, C.M. Li, Two dimensional atomically thin MoS_2 nanosheets and their sensing applications, Nanoscale 7 (46) (2015) 19358–19376.

[96] D.W. Lee, J. Lee, I.Y. Sohn, B.Y. Kim, Y.M. Son, H. Bark, N.E. Lee, Field-effect transistor with a chemically synthesized MoS_2 sensing channel for label-free and highly sensitive electrical detection of DNA hybridization, Nano Research 8 (7) (2015) 2340–2350.

[97] Q. He, Z. Zeng, Z. Yin, H. Li, S. Wu, X. Huang, H. Zhang, Fabrication of flexible MoS_2 thin-film transistor arrays for practical gas-sensing applications, Small 8 (19) (2012) 2994–2999.

[98] D.J. Late, Y.K. Huang, B. Liu, J. Acharya, S.N. Shirodkar, J. Luo, C.N.R. Rao, Sensing behavior of atomically thin-layered MoS_2 transistors, ACS Nano 7 (6) (2013) 4879–4891.

[99] L. Wang, Y. Wang, J.I. Wong, T. Palacios, J. Kong, H.Y. Yang, Functionalized MoS_2 nanosheet-based field-effect biosensor for label-free sensitive detection of cancer marker proteins in solution, Small 10 (6) (2014) 1101–1105.

[100] V.Q. Bui, T.T. Pham, D.A. Le, C.M. Thi, H.M. Le, A first-principles investigation of various gas (CO, H_2O, NO, and O_2) absorptions on a WS_2 monolayer: stability and electronic properties, Journal of Physics: Condensed Matter 27 (30) (2015) 305005.

[101] J. Liu, X. Chen, Q. Wang, M. Xiao, D. Zhong, W. Sun, Z. Zhang, Ultrasensitive monolayer MoS_2 field-effect transistor based DNA sensors for screening of Down Syndrome, Nano Letters (2019).

[102] S. Su, J. Chao, D. Pan, L. Wang, C. Fan, Electrochemical sensors using two-dimensional layered nanomaterials, Electroanalysis 27 (5) (2015) 1062–1072.

[103] Y. Liu, X. Dong, P. Chen, Biological and chemical sensors based on graphene materials, Chemical Society Reviews 41 (6) (2012) 2283–2307.

[104] M.A. Py, R.R. Haering, Structural destabilization induced by lithium intercalation in MoS_2 and related compounds, Canadian Journal of Physics 61 (1) (1983) 76–84.

[105] Y. Yuan, R. Li, Z. Liu, Establishing water-soluble layered WS_2 nanosheet as a platform for biosensing, Analytical Chemistry 86 (7) (2014) 3610–3615.

[106] X. Fan, P. Xu, D. Zhou, Y. Sun, Y.C. Li, M.A.T. Nguyen, T.E. Mallouk, Fast and efficient preparation of exfoliated 2H MoS_2 nanosheets by sonication-assisted lithium

intercalation and infrared laser-induced 1T to 2H phase reversion, Nano Letters 15 (9) (2015) 5956−5960.

[107] S. Presolski, M. Pumera, Covalent functionalization of MoS$_2$, Materials Today 19 (3) (2016) 140−145.

[108] Y. Zhang, Y. Guo, Y. Xianyu, W. Chen, Y. Zhao, X. Jiang, Nanomaterials for ultrasensitive protein detection, Advanced Materials 25 (28) (2013) 3802−3819.

[109] K.J. Huang, Y.J. Liu, H.B. Wang, T. Gan, Y.M. Liu, L.L. Wang, Signal amplification for electrochemical DNA biosensor based on two-dimensional graphene analogue tungsten sulfide−graphene composites and gold nanoparticles, Sensors and Actuators B: Chemical 191 (2014) 828−836.

[110] X. Xia, Z. Zheng, Y. Zhang, X. Zhao, C. Wang, Synthesis of Ag-MoS$_2$/chitosan nanocomposite and its application for catalytic oxidation of tryptophan, Sensors and Actuators B: Chemical 192 (2014) 42−50.

[111] C.C. Winterbourn, The biological chemistry of hydrogen peroxide, in: Methods in Enzymology, vol. 528, Academic Press, 2013, pp. 3−25.

[112] G.X. Wang, W.J. Bao, J. Wang, Q.Q. Lu, X.H. Xia, Immobilization and catalytic activity of horseradish peroxidase on molybdenum disulfide nanosheets modified electrode, Electrochemistry Communications 35 (2013) 146−148.

[113] H. Song, Y. Ni, S. Kokot, Investigations of an electrochemical platform based on the layered MoS$_2$−graphene and horseradish peroxidase nanocomposite for direct electrochemistry and electrocatalysis, Biosensors and Bioelectronics 56 (2014) 137−143.

[114] X. Wang, F. Nan, J. Zhao, T. Yang, T. Ge, K. Jiao, A label-free ultrasensitive electrochemical DNA sensor based on thin-layer MoS$_2$ nanosheets with high electrochemical activity, Biosensors and Bioelectronics 64 (2015) 386−391.

[115] C. Zhu, Z. Zeng, H. Li, F. Li, C. Fan, H. Zhang, Single-layer MoS$_2$-based nanoprobes for homogeneous detection of biomolecules, Journal of the American Chemical Society 135 (16) (2013) 5998−6001.

[116] Q. Xi, D.M. Zhou, Y.Y. Kan, J. Ge, Z.K. Wu, R.Q. Yu, J.H. Jiang, Highly sensitive and selective strategy for microRNA detection based on WS$_2$ nanosheet mediated fluorescence quenching and duplex-specific nuclease signal amplification, Analytical Chemistry 86 (3) (2014) 1361−1365.

[117] J. Chao, M. Zou, C. Zhang, H. Sun, D. Pan, H. Pei, L. Wang, A MoS$_2$−based system for efficient immobilization of hemoglobin and biosensing applications, Nanotechnology 26 (27) (2015) 274005.

[118] K.R. Rogers, Recent advances in biosensor techniques for environmental monitoring, Analytica Chimica Acta 568 (1−2) (2006) 222−231.

[119] K.D. Wegner, Z. Jin, S. Linden, T.L. Jennings, N. Hildebrandt, Quantum-dot-based forster resonance energy transfer immunoassay for sensitive clinical diagnostics of low-volume serum samples, ACS Nano 7 (8) (2013) 7411−7419.

[120] S.S. Chou, B. Kaehr, J. Kim, B.M. Foley, M. De, P.E. Hopkins, V.P. Dravid, Chemically exfoliated MoS$_2$ as near-infrared photothermal agents, Angewandte Chemie 125 (15) (2013) 4254−4258.

[121] R.M. Kong, L. Ding, Z. Wang, J. You, F. Qu, A novel aptamer-functionalized MoS$_2$ nanosheet fluorescent biosensor for sensitive detection of prostate specific antigen, Analytical and Bioanalytical Chemistry 407 (2) (2015) 369−377.

[122] J. Huang, L. Ye, X. Gao, H. Li, J. Xu, Z. Li, Molybdenum disulfide-based amplified fluorescence DNA detection using hybridization chain reactions, Journal of Materials Chemistry B 3 (11) (2015) 2395−2401.

[123] J. Ge, L.J. Tang, Q. Xi, X.P. Li, R.Q. Yu, J.H. Jiang, X. Chu, A WS_2 nanosheet based sensing platform for highly sensitive detection of T4 polynucleotide kinase and its inhibitors, Nanoscale 6 (12) (2014) 6866−6872.

[124] H. Deng, X. Yang, Z. Gao, MoS_2 nanosheets as an effective fluorescence quencher for DNA methyltransferase activity detection, Analyst 140 (9) (2015) 3210−3215.

[125] J. Liu, Z. Cao, Y. Lu, Functional nucleic acid sensors, Chemical Reviews 109 (5) (2009) 1948−1998.

[126] X. Xiang, J. Shi, F. Huang, M. Zheng, Q. Deng, J. Xu, MoS_2 nanosheet-based fluorescent biosensor for protein detection via terminal protection of small-molecule-linked DNA and exonuclease III-aided DNA recycling amplification, Biosensors and Bioelectronics 74 (2015) 227−232.

[127] X. Yin, J. Cai, H. Feng, Z. Wu, J. Zou, Q. Cai, A novel VS_2 nanosheet-based biosensor for rapid fluorescence detection of cytochrome c, New Journal of Chemistry 39 (3) (2015) 1892−1898.

[128] M. Ferrari, Cancer nanotechnology: opportunities and challenges, Nature Reviews Cancer 5 (3) (2005) 161.

[129] D. Peer, J.M. Karp, S. Hong, O.C. Farokhzad, R. Margalit, R. Langer, Nanocarriers as an emerging platform for cancer therapy, Nature Nanotechnology 2 (12) (2007) 751.

[130] L. Cheng, C. Wang, L. Feng, K. Yang, Z. Liu, Functional nanomaterials for phototherapies of cancer, Chemical Reviews 114 (21) (2014) 10869−10939.

[131] L. Naldini, Gene therapy returns to centre stage, Nature 526 (7573) (2015) 351.

[132] D. Chen, C.A. Dougherty, K. Zhu, H. Hong, Theranostic applications of carbon nanomaterials in cancer: focus on imaging and cargo delivery, Journal of Controlled Release 210 (2015) 230−245.

[133] J.M. Klostranec, W.C. Chan, Quantum dots in biological and biomedical research: recent progress and present challenges, Advanced Materials 18 (15) (2006) 1953−1964.

[134] Z. Gu, L. Yan, G. Tian, S. Li, Z. Chai, Y. Zhao, Recent advances in design and fabrication of upconversion nanoparticles and their safe theranostic applications, Advanced Materials 25 (28) (2013) 3758−3779.

[135] L.H. Reddy, J.L. Arias, J. Nicolas, P. Couvreur, Magnetic nanoparticles: design and characterization, toxicity and biocompatibility, pharmaceutical and biomedical applications, Chemical Reviews 112 (11) (2012) 5818−5878.

[136] R. Weissleder, A clearer vision for in vivo imaging, Nature Biotechnology 19 (2001) 316−317.

[137] S.M. Lee, H. Park, K.H. Yoo, Synergistic cancer therapeutic effects of locally delivered drug and heat using multifunctional nanoparticles, Advanced Materials 22 (36) (2010) 4049−4053.

[138] J.T. Robinson, S.M. Tabakman, Y. Liang, H. Wang, H. Sanchez Casalongue, D. Vinh, H. Dai, Ultrasmall reduced graphene oxide with high near-infrared absorbance for photothermal therapy, Journal of the American Chemical Society 133 (17) (2011) 6825−6831.

[139] T. Liu, C. Wang, W. Cui, H. Gong, C. Liang, X. Shi, Z. Liu, Combined photothermal and photodynamic therapy delivered by PEGylated MoS_2 nanosheets, Nanoscale 6 (19) (2014) 11219−11225.

[140] S. Wang, K. Li, Y. Chen, H. Chen, M. Ma, J. Feng, J. Shi, Biocompatible PEGylated MoS_2 nanosheets: controllable bottom-up synthesis and highly efficient photothermal regression of tumor, Biomaterials 39 (2015) 206−217.

[141] K. Yang, L. Feng, X. Shi, Z. Liu, Nano-graphene in biomedicine: theranostic applications, Chemical Society Reviews 42 (2) (2013) 530−547.

[142] Z. Kou, X. Wang, R. Yuan, H. Chen, Q. Zhi, L. Gao, L. Guo, A promising gene delivery system developed from PEGylated MoS_2 nanosheets for gene therapy, Nanoscale research letters 9 (1) (2014) 587.

[143] Y. Chao, G. Wang, C. Liang, X. Yi, X. Zhong, J. Liu, Z. Liu, Rhenium-188 labeled tungsten disulfide nanoflakes for self-sensitized, near-infrared enhanced radioisotope therapy, Small 12 (29) (2016) 3967−3975.

[144] J. Kim, H. Kim, W.J. Kim, Single-layered MoS2−PEI−PEG nanocomposite-mediated gene delivery controlled by photo and redox stimuli, Small 12 (9) (2016) 1184−1192.

[145] F. Yin, B. Gu, Y. Lin, N. Panwar, S.C. Tjin, J. Qu, K.T. Yong, Functionalized 2D nanomaterials for gene delivery applications, Coordination Chemistry Reviews 347 (2017) 77−97.

[146] C. Zhang, Y. Yong, L. Song, X. Dong, X. Zhang, X. Liu, Z. Hu, Multifunctional WS_2@ poly (ethylene imine) nanoplatforms for imaging guided gene-photothermal synergistic therapy of cancer, Advanced Healthcare Materials 5 (21) (2016) 2776−2787.

[147] L. Kong, L. Xing, B. Zhou, L. Du, X. Shi, Dendrimer-modified MoS_2 nanoflakes as a platform for combinational gene silencing and photothermal therapy of tumors, ACS Applied Materials & Interfaces 9 (19) (2017) 15995−16005.

[148] J. Xu, A. Gulzar, Y. Liu, H. Bi, S. Gai, B. Liu, P. Yang, Integration of IR-808 sensitized upconversion nanostructure and MoS_2 nanosheet for 808 nm NIR light triggered phototherapy and bioimaging, Small 13 (36) (2017) 1701841.

[149] W. Yin, L. Yan, J. Yu, G. Tian, L. Zhou, X. Zheng, Y. Zhao, High-throughput synthesis of single-layer MoS_2 nanosheets as a near-infrared photothermal-triggered drug delivery for effective cancer therapy, ACS Nano 8 (7) (2014) 6922−6933.

[150] T. Liu, S. Shi, C. Liang, S. Shen, L. Cheng, C. Wang, Z. Liu, Iron oxide decorated MoS_2 nanosheets with double PEGylation for chelator-free radiolabeling and multimodal imaging guided photothermal therapy, ACS Nano 9 (1) (2015) 950−960.

[151] E.C. Cho, C. Glaus, J. Chen, M.J. Welch, Y. Xia, Inorganic nanoparticle-based contrast agents for molecular imaging, Trends in Molecular Medicine 16 (12) (2010) 561−573.

[152] P. Sharma, S. Brown, G. Walter, S. Santra, B. Moudgil, Nanoparticles for bioimaging, Advances in Colloid and Interface Science 123 (2006) 471−485.

[153] L. Gong, L. Yan, R. Zhou, J. Xie, W. Wu, Z. Gu, Two-dimensional transition metal dichalcogenide nanomaterials for combination cancer therapy, Journal of Materials Chemistry B 5 (10) (2017) 1873−1895.

[154] J. Chen, C. Liu, D. Hu, F. Wang, H. Wu, X. Gong, H. Zheng, Single-layer MoS_2 nanosheets with amplified photoacoustic effect for highly sensitive photoacoustic imaging of orthotopic brain tumors, Advanced Functional Materials 26 (47) (2016) 8715−8725.

[155] J. Yu, W. Yin, X. Zheng, G. Tian, X. Zhang, T. Bao, Y. Zhao, Smart MoS_2/Fe_3O_4 nanotheranostic for magnetically targeted photothermal therapy guided by magnetic resonance/photoacoustic imaging, Theranostics 5 (9) (2015) 931.

[156] G. Yang, H. Gong, T. Liu, X. Sun, L. Cheng, Z. Liu, Two-dimensional magnetic WS_2@ Fe_3O_4 nanocomposite with mesoporous silica coating for drug delivery and imaging-guided therapy of cancer, Biomaterials 60 (2015) 62−71.

[157] L. Cheng, C. Yuan, S. Shen, X. Yi, H. Gong, K. Yang, Z. Liu, Bottom-up synthesis of metal-ion-doped WS2 nanoflakes for cancer theranostics, ACS nano 9 (11) (2015) 11090−11101.

Polymer nanocomposites based on two-dimensional nanomaterials

Rajarshi Bayan, BSc, MSc[1], Niranjan Karak, MTech, PhD[1,2]

[1]*Advanced Polymer & Nanomaterial Laboratory, Department of Chemical Sciences, Tezpur University, Tezpur, Assam, India;* [2]*Professor. Department of Chemical Sciences, Tezpur University, Tezpur, Assam, India*

8.1 Introduction

Polymer nanocomposite is one of the popular advances in the field of polymer science. The unique combination of a polymer and a nanoscale material results in not only structural and morphologic variations but also changes in physicochemical properties of the material [1]. These structural variations and changes in physicochemical properties can be tuned significantly by the choice of appropriate polymers and nanomaterials. It is found that incorporation of even a minute amount of nanomaterial (less than 5 wt%) to a polymer matrix can boost the material properties dramatically [2]. As a result, these nanomaterials are being employed as "nano"fillers, and in hindsight, bridging the gap between polymer chemistry and nanotechnology.

The renewed interest in polymer nanocomposites is due to several reasons. First, nanofillers often display properties that are different from their bulk counterparts of the same material. For example, single-walled carbon nanotubes exhibit strength, rigidity, and strain to failure that significantly surpass those of traditional carbon fibers [3]. Second, these nanoscale fillers are of smaller magnitude of defects than their bulk counterparts. They can prevent early failure, resulting in nanocomposites with enhanced toughness and ductility [4]. Third, these nanofillers offer high surface aspect ratio, leading to nanocomposites with a large volume of interfacial matrix material with contrasting properties from that of bulk polymer. This interfacial nanopolymer matrix can radically alter the mechanical, thermal, and electric properties of the overall nanocomposite. Thus, in short, all these features provide an opportunity for creating polymer nanocomposites with unique properties.

8.2 Two-dimensional nanomaterials

The term "nanomaterial" generally refers to materials with external dimensions or an internal structure, measured in nanoscale, exhibiting additional or different

Two-Dimensional Nanostructures for Biomedical Technology. https://doi.org/10.1016/B978-0-12-817650-4.00008-5

unique properties. According to the International Organization for Standardization (ISO), nanomaterial is defined as a material with any external dimension in the nanoscale or having internal structure or surface structure in the nanoscale [5]. These nanomaterials having one, two, or three extra dimensions in the nanoscale region are engineered, manufactured, or incidental nanoparticles, nanofibers, nanorods, nanosheets, nanoribbons, nanotubes, nanocubes, core-shell nanoparticles, etc. [6]. Various definitions can be found in prior art, albeit retaining the core of the nanodimension.

Nanomaterials are of three different classes based on their dimensions. Among these, two-dimensional (2D) nanomaterials carve a unique niche of their own in polymer nanocomposites. However, the other dimensional nanomaterials, such as zero- and one-dimensional nanomaterials, are also fabricated with polymers.

2D nanomaterials include nanosheets, nanofilms, and nanoribbons (ultrafine-grained overlayers or buried layers) (Fig. 8.1). Free particles with large aspect ratio, having any one dimension in the nanoscale range, are also considered as 2D nanomaterials. 2D nanomaterials can also be amorphous or crystalline and composed of metallic, ceramic, or polymeric matrices. In these nanostructures, electrons are confined within one dimension, indicating the inability of electrons to move freely within the associated dimension [7]. In addition, these nanomaterials possess atomic thickness that grants them high mechanical flexibility and optical transparency. These nanostructures are promising for applications, including sensors, in electronics/optoelectronics, and in biomedicine [7,8].

In the recent years, nanomaterials are carving out a niche for themselves and have garnered copious amount of attention from the scientific community all over

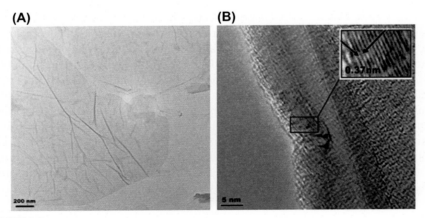

FIGURE 8.1

Transmission electron microscopic image of graphene nanosheet at a resolution of (A) 200 nm and (B) 5 nm (Inset: High-resolution image (blue box) showing lattice fringes in graphene nanosheet).

Reproduced with permission from S. Thakur, N. Karak, Green reduction of graphene oxide by aqueous phytoextracts, Carbon 50(14) (2012) 5331–5339.

because of their myriad features. Nanomaterials can exhibit unique optical, mechanical, magnetic, and conductive properties in comparison to their bulk equivalents of the same chemical nature [8]. As of now, nanomaterials are slowly becoming commercialized and used in many innovative technologic applications and products, including a wide range of consumer products.

8.3 Classification

In order to understand the diversity of nanomaterials, they are classified mainly on the basis of their elemental origin, morphology, and dimensions. Depending on the elemental composition, nanomaterials can be classified into three subclasses: (1) organic, (2) inorganic, and (c) composite/hybrid (containing both organic and inorganic constituents) [8,9]. Again, nanomaterials can be distinguished in terms of morphology, depending on parameters such as sphericity, flatness, and aspect ratio. Small aspect ratio morphologies usually come in the form of sphere, oval, cube, helix, or rod, whereas high aspect ratio morphologies are found in the shape of zigzag, helices, and belts [9]. Again, based on dimensions, nanomaterials are classified into three subclasses: (1) zero-dimensional, (2) one-dimensional, and (c) 2D, as mentioned earlier [9]. These different types of nanomaterials possess their unique physical, chemical, optical, electric, and biological characteristics, which can be harnessed for desired properties and target-specific applications.

In this chapter, we are limiting our overview to only 2D nanomaterials and their polymer nanocomposites. Some of the commonly employed nanomaterials in polymer nanocomposites are briefly described in the following.

8.3.1 Carbon-based nanomaterials

Carbon-based nanomaterials are basically nanomaterials in which carbon is the sole or main constituting element. Owing to their abundance and facile preparation, they are the most significant nanomaterials in recent times, with tunable physical, mechanical, chemical, electric, optical, thermal, and biological properties. Graphitic nanostructures such as graphene oxide (GO), reduced graphene oxide (rGO), and graphene constitute the most frequently used 2D nanomaterials for polymer nanocomposite fabrication (Fig. 8.2).

8.3.1.1 Graphene

Graphene is a crystalline allotrope of carbon, consisting of one-atom-thick single 2D layer of sp^2 hybridized carbon atoms, arranged in a hexagonal lattice structure. It can be considered as an indefinitely large aromatic molecule and serves as the basic structural motif of many other allotropes of carbon, such as graphite, diamond, charcoal, carbon nanotubes, and fullerenes [10]. Graphene possesses unique properties such as high elastic modulus, large theoretic specific surface area, excellent strength,

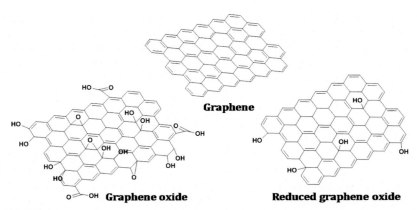

FIGURE 8.2

Carbon-based two-dimensional nanomaterials.

and high thermal and electric conductivities [11]. These qualities endow graphene to be an apt choice for the fabrication of polymer nanocomposites.

8.3.1.2 Graphene oxide

GO is an oxidized variant of graphene, which is made by the powerful oxidation of graphite. It is a unique material that can be viewed as a single monomolecular layer of graphite, laced with various oxygen-containing functionalities such as epoxide, carbonyl, carboxyl, and hydroxyl groups [12]. The mechanical, chemical, and electronic properties of GO are strongly influenced by the presence of various oxygeneous groups, contrasting itself from graphene in many aspects [13]. However, the presence of these oxygeneous groups accounts for better compatibility and dispersion of GO in polymer nanocomposites, especially for polar polymers.

8.3.1.3 Reduced graphene oxide

rGO is another important variant of graphene, consisting of few-atom-thick 2D sp^2 hybridized carbon layers with fewer oxygeneous functionalities [14]. When GO is reduced in a suitable process, the rGO formed resembles graphene but contains residual oxygen and other heteroatoms, as well as structural defects [12]. The qualities of rGO are almost similar to those of graphene in most of the cases, barring electric conductivity, as rGO inevitably contains lattice defects that degrade its electric properties [15]. The many chemical and structural defects of rGO may create problem for some applications, but are considered advantages for some others.

8.3.2 Inorganic nanomaterials

Inorganic nanomaterials are some of the naturally found nanomaterials and come in various sizes and morphologies. They include different metals, metal oxides, metal chalcogenides, inorganic minerals, and nanoclays. 2D nanomaterials of inorganic

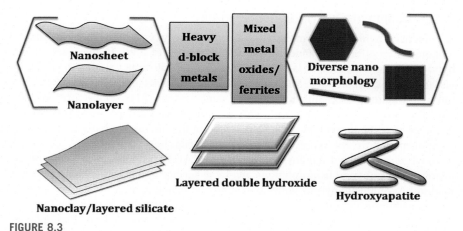

FIGURE 8.3

Inorganic two-dimensional nanomaterials with different morphologies.

origin include mostly nanoclay, layered double hydroxides (LDHs), hydroxyapatite (HAp), and metal/mixed metal oxides (Fig. 8.3). Reduction of heavy metals including silver (Ag), gold (Au), platinum (Pt), palladium (Pd), etc. in the presence of suitable capping agents and using special methods provides a facile way of manufacturing metal-based 2D nanomaterials with unique morphology, such as nanosheets and nanolayers [16]. Again, 2D nano metal oxides of iron (Fe_3O_4), copper (CuO/CuO_2), nickel (NiO), zinc (ZnO), etc. and metal ferrites (MFe_3O_4, M = Fe, Cu, Ni, Co, etc.) are prepared by various techniques such as hydrolysis, solvolysis, wet chemical, and sol-gel methods using organometallic precursors. Inorganic minerals such as nanoclay, layered silicate, and LDHs are some of the most abundant minerals on the earth surface and come in various shapes and sizes. Nanoclays and layered silicates are hydrous aluminum phyllosilicate thin platelets or sheets having a layered structure. They are characterized by their unique intercalated structures that can engineer physical and material properties for polymers with low nanomaterial loading [17,18]. On the other hand, LDHs are mineral and synthetic materials with positively charged brucite-type layers of mixed metal hydroxides. They are also called anionic clays and are promising layered materials due to a host of interesting properties such as intercalated anions with interlayer spaces, swelling properties, the ability to intercalate different types of anions (inorganic, organic, biomolecular), delivery of intercalated anions in a sustained manner, and high biocompatibility. These LDHs exhibit insulating properties that enhance thermal and flame-retardant properties of polymers [19]. In similar lines, HAp is another naturally occurring inorganic nanomaterial having structural and chemical similarity with the mineral phase of bone and teeth. Owing to its bioactive, osteoconductive, nontoxic, and nonimmunogenic properties, HAp can be engineered for biomedical applications [20].

8.3.3 Composite/hybrid nanomaterials

A composite/hybrid is a combination of more than two different materials mixed together in an effort to channelize the best properties of both. In the recent times, hybrid/composite nanomaterials are emerging as a front-runner among the different variants of nanomaterials. Such type of nanomaterials coexists by interacting through a definite mechanism within the same system [21]. Such hybrid systems may also contain both organic and inorganic components and possess exclusive advantages over the individual ones. Depending on the interaction between the nanocomponents, different types of morphologies can be achieved, e.g., decorated nanohybrid, embedded nanohybrid, etc. (Fig. 8.4). Especially, in the case of 2D nanomaterials, favorable morphology and large surface aspect ratio enable easy formation of composite/hybrid with other kinds of nanomaterials. The advantage of such nanohybrid system over conventional 2D nanomaterials lies in the fact that different properties can be imparted within a single nanostructure. Examples of such nanohybrid materials are Ag/graphene nanohybrid with antibacterial, photocatalytic, and sensing properties [22]; Fe_3O_4/graphene nanohybrid with absorptive, electric, and magnetic properties [23,24]; clay/TiO_2 nanohybrid with photocatalytic activity [25]; ZnO/graphitic carbon nitride nanohybrid with electrochemical sensing and photocatalytic applications [26]; HAp/graphene nanohybrid with osteogenetic activity [21]; and many more.

Some of the most encountered 2D nanomaterials in the fabrication of polymer nanocomposites are presented in Table 8.1.

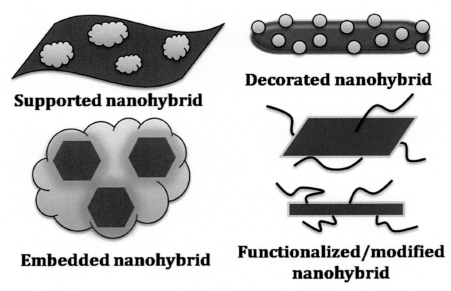

Supported nanohybrid

Decorated nanohybrid

Embedded nanohybrid

Functionalized/modified nanohybrid

FIGURE 8.4

Two-dimensional nanohybrids with different morphologies.

Table 8.1 2D nanomaterials with their properties, used for fabrication of polymer nanocomposites.

2D morphology	Origin	2D nanomaterial	Property
Sheet/ribbon/platelet	Organic	Graphene, GO, rGO, graphitic carbon nitride	Structural stability, electric conductivity, thermal conductivity, optical property
Sheet/platelet/layer/needle/prism	Inorganic	Nanoclay, LDHs, HAp, graphitic boron nitride	Structural stability, thermal insulation, hydrophilicity, flame retardancy, biological property
Sheet/ribbon/needle/platelet	Hybrid	Fe_3O_4-GO/rGO/graphene, Ag-GO/rGO/graphene, ZnO-rGO/graphene, Cu/CuO/Cu_2O-rGO/graphene, HAp/graphene, Ag/Cu-graphitic carbon nitride, Ag-nanoclay, Ag-HAp, etc.	Structural stability, electric conductivity, thermal conductivity, optical property, magnetic property, biological property, catalytic property

2D, two-dimensional; GO, graphene oxide; HAp, hydroxyapatite; LDHs, layered double hydroxides; rGO, reduced graphene oxide.

8.4 Preparative methods

Preparative methods of nanomaterials compose of two major approaches, namely, top-down and bottom-up approaches (Fig. 8.5) [27]. In a top-down approach, nanomaterial is formed by breaking down a larger structure (bulk material) into a nanosized structure. This approach generally requires harsh and extreme conditions. Various methods come under the purview of top-down process, including ball milling, mechanochemical, laser ablation, arc discharge, electrochemical method, etc. On the other hand, the bottom-up approach of synthesis relies on the use of molecular-level precursors, which aggregate through physicochemical processes such as polymerization, condensation, and pyrolysis to form nanostructured

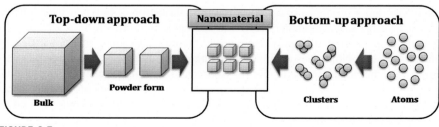

FIGURE 8.5

A schematic representation of "top-down" and "bottom-up" approaches.

materials. Different methods employed under this approach include hydrothermal, solvothermal, microwave-assisted pyrolysis, sonochemical synthesis, coprecipitation method, sol-gel method, etc. This approach requires less extreme conditions and can be tuned to exact desired results. Among the two approaches, the bottom-up approach is the most common compared with the top-down approach, as it is inexpensive, is expeditious, and offers large-scale production. Moreover, the bottom-up approach offers better control over the morphology and size distribution of the nanomaterial [27].

8.5 Polymer nanocomposites

Polymer nanocomposite is generally defined as a composite material comprising of a polymer/copolymer with nanomaterials acting as nanofillers (with at least one dimension in nanoscale) and dispersed within the polymeric matrix (Fig. 8.6) [28]. In contrast to traditional polymer composites with high loadings (30–60 wt %) of micrometer-sized filler particles, polymer nanocomposites are being developed with very low loadings (less than 5 wt%) of well-dispersed nanofillers, which is one of their defining features. These nanocomposites are found to exhibit extraordinarily interesting properties, courtesy of the distinctive nature of the polymer and the nanomaterial. In fact, the advent of new nanofillers in the past 15 years is providing a golden opportunity for the development of high-performance multifunctional nanocomposites. For example, transparent conducting polymer/nanotube composites are under development as solar cell electrodes [29]; nanoparticle-filled amorphous polymers are being used as scratch-resistant, transparent coatings in cell phone and compact disc technology [30]; and plasmonic graphene/polymer nanocomposites are being introduced as excellent candidates for applications in solar steam generation and seawater desalination [31].

8.6 Fabrication methods

Polymer nanocomposites can be fabricated by either physical or chemical process. The uniform and homogeneous dispersion of nanomaterials in the polymer matrix is one of the major problems experienced during fabrication of polymer nanocomposites. This is because nanomaterials tend to aggregate and form clusters, which hinder their dispersion in the polymer matrix, thereby degrading the properties of nanocomposites. Numerous attempts are being made to disperse nanomaterials uniformly and homogeneously in the polymer matrix by using chemical reactions, complicated polymerization reactions, or surface modification of nanomaterials [32–35]. Some of the most commonly encountered fabrication techniques of polymer nanocomposites are briefly discussed in the following.

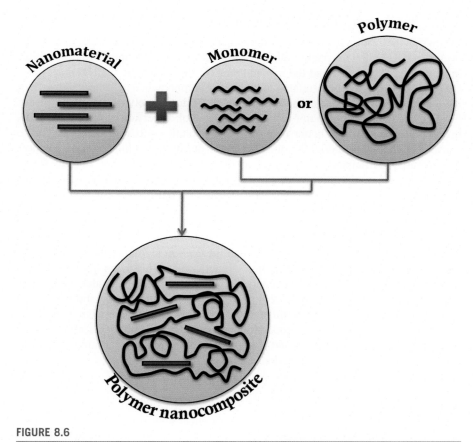

FIGURE 8.6

Schematic representation of the fabrication of polymer nanocomposite.

8.6.1 Solution mixing technique

The solution mixing technique is based on a solvent system in which the polymer or prepolymer is soluble and the nanomaterial is capable of swelling. In this technique the nanomaterial is swollen and dispersed in a suitable solvent or mixture of solvents by using mechanical force and ultrasonication. The properly dispersed nanomaterial is then mixed with a solution of polymer by means of mechanical force and additional ultrasonication. The nanocomposite is obtained either by evaporating the solvent or by precipitating in an immiscible solvent. In this technique the level of dispersion of the nanomaterial in the polymer matrix is dependent on the interactions among polymer, nanomaterial, and solvent molecules [33,34].

8.6.2 In situ polymerization technique

In situ polymerization is a chemical encapsulation technique in which polymerization occurs in the continuous phase, rather than on both sides of the interface between the monomer or prepolymer and the nanomaterial. In this technique the nanomaterial is swollen and dispersed in a suitable monomer or prepolymer. The properly dispersed nanomaterial is then added to the continuous phase during polymerization process to form the nanocomposite. In this technique the level of dispersion of the nanomaterial in the polymer matrix is better than that in the solution mixing technique. This is because as the nanomaterial is added before or during the polymerization reaction, it can easily participate in the reaction or cross-linking process, forging strong polymer-nanomaterial interactions [33,34]. Furthermore, low viscosity of monomer or prepolymer helps in proper dispersion of nanomaterial in the system.

8.6.3 Melt mixing technique

The melt mixing technique is a straightforward method involving the mixing of the nanomaterial and the polymer at molten temperature. In this technique the mixture of polymer and the nanomaterial is annealed, either statically or under shear force. The nanocomposite is ultimately obtained by using suitable mixing and processing equipment such as twin screw mixers, compression molding, injection molding, extrusion molding, and fiber production process. This technique is attuned with the current industrial processes and is also useful for polymers that are not suitable for in situ polymerization or solution mixing technique [32,35].

8.6.4 Sol-gel process

Sol-gel is a wet chemical process, also called chemical solution deposition. In this process, the solid nanomaterial is dispersed in the monomer to form a colloidal suspension (sol) that acts as the precursor for an interconnecting network (or gel) of discrete particles or continuous polymer. The polymer serves as the nucleating agent and promotes the growth of layered crystals. The polymer penetrates between the layered crystals during their nucleation and growth, leading to the formation of nanocomposite [35].

Apart from these, there are numerous ex situ and in situ techniques, such as grafting, chemical vapor deposition, latex fabrication, ultrasonic treatment, template synthesis, plasma treatment, that are employed for the fabrication of polymer nanocomposites [32–35].

8.7 Characterization

Characterization of polymer nanocomposites is an important aspect of the polymeric materials that helps in predicting and understanding their properties. Unlike fine chemicals and materials, polymer nanocomposites contain both polymer matrix and nanomaterial combined in an intricate manner. While polymers typically conform to a macromolecular assembly of coiled and entangled chains in the matrix, nanomaterials arrange and orient themselves in the polymer matrix by means of various physicochemical interactions with the polymer matrix [34]. As a result, the structure of the nanocomposite becomes a complex assemblage of macro- and nanostructures. Hence, characterization of such nanocomposites demands sophisticated analytical techniques and their understanding [33,34]. These sophisticated analytical techniques are briefly mentioned in the following sections.

8.7.1 Spectroscopic techniques

The spectroscopic techniques used for the characterization of polymer nanocomposites include infrared (IR) spectroscopy, ultraviolet-visible (UV-vis) spectroscopy, and nuclear magnetic resonance (NMR) spectroscopy.

IR spectroscopy is used to determine the chemical functional groups present in the structures of the polymer nanocomposites. IR spectroscopy is recorded as a plot of the absorbance or transmittance intensity (in terms of percentage) as a function of energy (in terms of wave number, cm^{-1}). However, as polymer nanocomposites have complex structures, some sophisticated variations such as Fourier transform infrared (FT-IR) spectroscopy and attenuated total reflection FT-IR spectroscopy are employed (Fig. 8.7A). UV-vis spectroscopy is used to detect the electronic excitations by chromophoric groups present in the polymer nanocomposite on interaction with UV light or visible light. UV-vis spectroscopy is usually recorded as a plot of wavelength (in nanometers) versus molar absorptivity (ε, L mol^{-1} cm^{-1}) or absorbance (A) of a unit concentration of polymer nanocomposite (Fig. 8.7B). NMR spectroscopy is a very crucial technique for elucidating the structure of the polymer matrix of the nanocomposite. NMR spectroscopy is usually employed to determine the presence, position, and orientation of ^{1}H nuclei and ^{13}C nuclei present in the polymer, even though other nuclei such as ^{19}F, ^{29}Si, and ^{31}P can also be detected, if present. NMR spectroscopy is normally observed as a plot of intensity (in percentage) versus chemical shift (in parts per million), with different types of ^{1}H and ^{13}C nuclei identified consequently (Fig. 8.7C).

8.7.2 Microscopic techniques

Microscopic techniques including scanning electron microscopy (SEM), transmission electron microscopy (TEM), atomic force microscopy (AFM), etc. are some of the visual techniques used for the characterization of polymer nanocomposites.

SEM is an important microscopic technique for viewing the surface morphology of polymer nanocomposites. This technique involves imaging of the surface of polymer nanocomposite with the help of a focused beam of electrons and gives information about the surface topography and elemental composition of the nanocomposite (Fig. 8.8A). TEM is a highly sophisticated microscopic technique for observing the bulk morphology of polymer nanocomposites. This technique involves imaging of the bulk matrix of the nanocomposite with the help of a focused beam of electrons

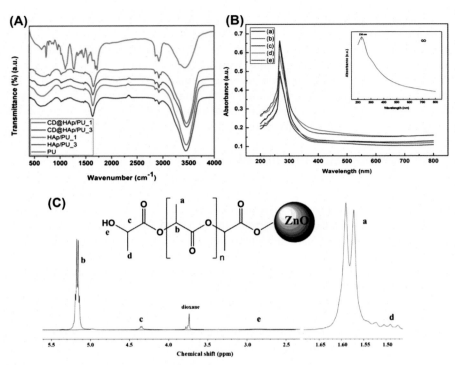

FIGURE 8.7

(A) Fourier transform infrared spectra of pure waterborne polyurethane (PU) and carbon dot (CD)-decorated hydroxyapatite (HAp)/waterborne PU nanocomposites. (B) Ultraviolet-visible (UV-vis) spectra of graphene oxide/poly(3,4-ethylenedioxythiophene)-*block*-poly(ethylene glycol) (PEG)/poly(vinylidene fluoride) nanocomposite (inset: UV-vis spectra of graphene oxide). (C) Nuclear magnetic resonance spectrum of poly-(L-lactic acid)/ZnO nanocomposite.

(A) Reproduced from S. Gogoi, M. Kumar, B.B. Mandal, N. Karak, A renewable resource based carbon dot decorated hydroxyapatite nanohybrid and its fabrication with waterborne hyperbranched polyurethane for bone tissue engineering, RSC Advances 6 (2016) 26066–26076 with permission from The Royal Society of Chemistry. (B) Reproduced from K. Deshmukh, G.M. Joshi, Novel nanocomposites of graphene oxide reinforced poly (3, 4-ethylenedioxythiophene)-block-poly (ethylene glycol) and poly(vinylidene fluoride) for embedded capacitor applications, RSC Advances 4(71) (2014) 37954–37963 with permission from The Royal Society of Chemistry. (C) Reproduced with permission from H. Kaur, A. Rathore, S. Raju, A study on ZnO nanoparticles catalyzed ring opening polymerization of L-lactide, Journal of Polymer Research 21(9) (2014) 537.

and acquires information about its morphology, crystalline structure, and elemental composition (Fig. 8.8B). AFM is another high-resolution scanning probe microscopic technique used for observing the surface topology of nanocomposites. AFM obtains information pertaining to the surface topology and related local properties of the nanocomposite, such as height/thickness, friction, and magnetism (Fig. 8.8C).

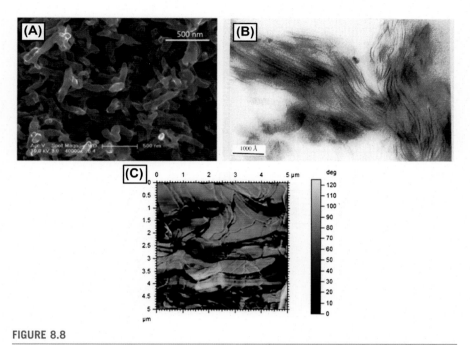

FIGURE 8.8

(A) Scanning electron microscopic image of graphene-polyaniline nanofiber composite. (B) Transmission electron microscopic image of epoxy/clay nanocomposite. (C) Atomic force microscopic image of graphene oxide—filled ethylene methyl acrylate hybrid nanocomposite.

(A) Reproduced from Q.Wu, Y.X. Xu, Z.Y. Yao A.R. Liu, G.Q. Shi, Supercapacitors based on flexible graphene/ polyaniline nanofiber composite films, ACS Nano 4 (2010) 1963—1970. (B) Reproduced from T. Lan, T.J. Pinnavaia, Clay-reinforced epoxy nanocomposites, Chemistry of Materials 6 (12) (1994) 2216—2219. (C) Reproduced from P. Bhawal, S. Ganguly, T.K. Chaki, N.C. Das, Synthesis and characterization of graphene oxide filled ethylene methyl acrylate hybrid nanocomposites, RSC Advances 6(25) 20781—20790 with permission from The Royal Society of Chemistry.

8.7.3 Other techniques

There are several other techniques that reveal the physical, chemical, structural, and elemental characteristics of polymer nanocomposites.

Elemental analysis using CHN (carbon, hydrogen, and nitrogen), heteroatoms (halogens, sulfur, phosphorus, etc.), atomic absorption spectrometry (metals, metalloids, halogen, sulfur, phosphorus, etc.), energy-dispersive X-ray spectrometry, X-ray photoelectron spectroscopy, etc. delivers information about the elemental composition of polymer nanocomposites (Fig. 8.9B). X-ray diffraction (XRD) analysis of polymer nanocomposites predicts their crystallinity or amorphousness. This technique involves scanning of powdered or thin sheets of polymer nanocomposites by an X-ray beam of specific wavelength (CuKα, wavelength of 1.54 nm) over a

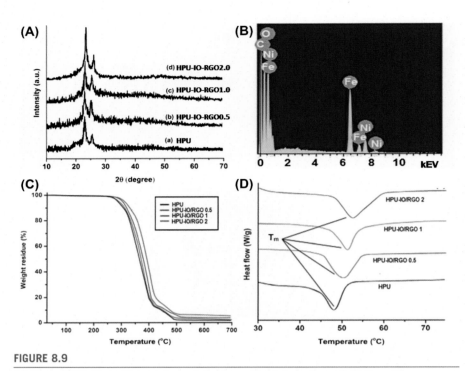

FIGURE 8.9

(A) X-ray diffraction patterns of pure hyperbranched polyurethane (HPU) and its nanocomposite. (B) Energy-dispersive X-ray spectrum of NiFe$_2$O$_4$@reduced graphene oxide (rGO) nanohybrid. (C) Thermogravimetric thermograms of HPU and its nanocomposites. (D) Differential scanning calorimetric curve showing the melting temperature of the soft segment of HPU and its nanocomposites. *IO*, iron oxide.

(A), (C), and (D) Reproduced from S. Thakur, N. Karak, A tough, smart elastomeric bio-based hyperbranched polyurethane nanocomposite, New Journal of Chemistry 39(3) (2015) 2146–2154 with permission from the Centre National de la Recherche Scientifique (CNRS) and The Royal Society of Chemistry. (B) Reproduced from S. Gogoi, N. Karak, Biobased waterborne hyperbranched polyurethane/NiFe2O4@rGO nanocomposite with multi-stimuli responsive shape memory attributes, RSC Advances 6(97) (2016) 94815-94825with permission from The Royal Society of Chemistry.

range of incident angle (θ). XRD is generally derived as a plot of peak intensity versus scattering angle (2θ), with sharp diffraction peaks indicating crystallinity (Fig. 8.9A). Gel permeation chromatography (GPC) is the most popular technique to determine the molecular weight and distribution, i.e., number average, weight average molecular weight, etc. of polymers. As polymers are macromolecules, chromatographic techniques are preferred over other techniques. GPC plots the refractive index or UV absorption intensity as a function of elution volume, from which molecular weights can be determined. Thermal characteristics of polymer nanocomposites are evaluated by using thermogravimetric analysis (TGA) and differential scanning calorimetry (DSC). TGA is used to study the thermal stability of polymer nanocomposites, in terms of weight loss of polymeric materials with respect to temperature. Thermogravimetric thermogram is recorded as the function of change of weight (weight loss or weight residue percentage) of the sample versus temperature or time. Thermogravimetric thermogram also provides information such as the amount of moisture or any other volatiles, plasticizers, fillers, etc. present in the nanocomposite (Fig. 8.9C). DSC is used to study the heat changes in a polymeric nanocomposite with temperature. The DSC thermogram is recorded as a plot of change in heat energy or enthalpy against temperature. DSC also provides certain significant information about polymeric materials, such as glass-transition temperature, melting temperature, percentage crystallinity, specific heat, and amount of endothermic/exothermic energy (Fig. 8.9D).

8.8 Properties

The main objective of fabricating polymer nanocomposites translates to gain a deep knowledge and understanding of their properties, and ultimately their utility for suitable applications [33,34]. Polymeric nanocomposites based on 2D nanomaterials possess versatile properties, especially owing to the inherent nature of the nanomaterial and its interaction with the polymer matrix. These nanomaterials not only structurally secure/reinforce the polymer matrix but also augment their sustainability and longevity. As discussed in Section 8.3, 2D nanomaterials imbibe their special features in the nanocomposite, which augments the overall properties of the polymer nanocomposite. These properties are classified into physical, mechanical, chemical, thermal, optical, biological, electric, and magnetic parameters and are briefly discussed in the following.

8.8.1 Physical properties

The physical properties are the inherent characteristic of polymeric nanocomposites, including solubility, viscosity, crystallinity, etc. Solubility or dissolution of polymer nanocomposites in a suitable solvent is slow and difficult because of their high molecular weight and three-dimensional coiled and entangled structures. As a result, during solubilization of polymer nanocomposites in any solvent, the chains of polymer matrix are unfolded and solvated by the penetration of solvent molecules in between polymer chains. This state of dissolution is called swelling. This swelling process is further augmented by the presence of 2D nanomaterials such as GO and clay. These 2D nanomaterials help in better penetration of solvent molecules in the polymer chain and swelling of the nanocomposite [33,34]. Melt and solution viscosity of polymer nanocomposites are different owing to several factors such as high molecular weight, three-dimensional coiled and entangled structure, crosslinking, and secondary interactive forces. Presence of 2D nanomaterials helps in enhancing the melt and solution viscosity of the polymer nanocomposites by physicochemical interactions with the polymer chains, thereby altering their orientation and mobility [34]. Crystallinity in polymer nanocomposites occurs due to the presence of long range order in the molecular arrangement of polymer chain segments. Polymers with simple and highly regular structural units usually tend to crystallize. This crystallinity is further influenced by the presence of nanomaterials in the polymer matrix. In the polymer matrix, the 2D nanomaterial serves as nucleating centers that may induce crystallization. 2D nanomaterials of carbon, metal, and inorganic origin act as good reinforcing fillers that increase structural compactness, cross-linking density, secondary interactions, etc., resulting in enhanced crystallinity of nanocomposites [35].

8.8.2 Mechanical properties

Mechanical properties are the most important property of the polymer nanocomposites, as most of their future applications are dependent on them. The mechanical properties of polymer nanocomposites include parameters such as tensile strength, tensile modulus, elongation-at-break, scratch hardness, shore A hardness, pencil hardness, and impact resistance [1]. Each of these parameters demands special attention, as they are useful for different applications. However, it is sometimes observed that pure polymers exhibit poor mechanical properties compared with other materials such as metals and ceramics. As a result, pristine polymers cannot meet the demands of suitable applications. In a bid to improve the mechanical properties, polymer nanocomposites offer a genuine and effective answer for fabricating polymeric materials for these suitable applications. Consequently, polymer nanocomposites are fabricated using 2D nanomaterials of organic and inorganic origin that not only compensate for the poor mechanical properties of polymers but also impart dimensional stability to the material [1,28,35]. The mechanical properties of the nanocomposites are mostly dependent on their molecular weight, arrangement of polymer chains, dispersion of nanomaterials, and various physicochemical interactions thereof in the polymer matrix [34,35].

8.8.3 Chemical properties

Polymer nanocomposites may show certain reactivity toward chemicals such as acid, alkali, salt, and organic solvent. This reactivity is mainly due to their chemical composition and the presence of free reactive functionalities in the polymer structure [34]. The chemical reactivity of polymer nanocomposites is crucial for their modification to obtain new properties. Polymer nanocomposites are chemically modified by various entities to suit different applications. However, in terms of future prospects, such reactivity is undesirable for their long-term application, as it changes the state and properties of the nanocomposite. In this context the presence of 2D nanomaterials provides structural stability to the polymer matrix. This can be attributed to the proper reinforcement of the polymer matrix by various physicochemical interactions that strengthen the polymer matrix and make it chemically resistant to acid, alkali, salt, etc. [34,35]. As a result, polymer nanocomposites exhibit better resistance to chemical reactivity under different media, in comparison to pristine polymers.

8.8.4 Thermal properties

Thermal stability is a significant drawback of pure polymers, in view of their inherent covalent nature and high thermolabile linkages. Pure polymers are primarily composed of covalent bonds between the molecules, which are comparatively weaker than other type of bonds. However, other key factors such as chemical linkages, molecular weight, cross-linking, and crystallinity also play significant roles toward the thermal performance of pure polymers [1,2]. In order to improve the thermal stability of polymers, 2D nanomaterials can be used for the fabrication of polymer nanocomposites. 2D nanomaterials of carbon nanostructures, inorganic minerals, and metal nanoparticles are suitable choices for designing such polymer nanocomposites, as they offer mechanical robustness and high thermal stability. Moreover, these 2D nanomaterials reinforce the polymer matrix by means of various physicochemical interactions and amplify the thermal stability by occupying the free volume in the matrix and restricting the thermal motion of polymer chains [34,35]. Thermal study of nanocomposites reveals important properties such as degradation temperature, degradation pattern, flammability, flame retardancy, glass-transition temperature, heat capacity, and specific heat, which are crucial for their processing and service life.

8.8.5 Optical properties

Optical properties of polymer nanocomposites are observed upon their interaction with light. Depending on the nature of the polymer matrix and the nanomaterial, polymer nanocomposites may interact with electromagnetic radiation in the UV or visible range. Polymer matrices, by means of chromophoric polar functional groups in the molecular chains, can absorb UV or visible light, while 2D nanomaterials such as carbon nanostructures and metal nanoparticles also absorb UV or visible light [35]. As a result, the nanocomposite may show color under visible light, as well as certain luminescence under UV light.

8.8.6 Biological properties

Polymer nanocomposites can demonstrate biological response such as biodegradation, antibacterial activity, or even antimicrobial activity under various biological environments. Most of the synthetic polymers do not show the desired biological responses unless they contain any biocomponents [36]. Biopolymers such as starch, cellulose, chitosan (CS), and vegetable oil-based polymers attract microorganisms such as bacteria and fungi and are biodegradable, owing to their labile chemical linkages (ester, amide, etc.). In general, the biodegradability of such polymeric materials is dependent on their degree of hydrophilicity. As biopolymers contain hydrophilic functional groups, this makes them easily susceptible to microorganisms [34]. Polymer nanocomposites containing hydrophilic 2D nanomaterials, such as GO and nanoclay, are the apt choice for designing such polymer bio-nanocomposites, as they provide hydrophilicity as well mechanical strength. On the other hand, polymer nanocomposites containing 2D metals/metalloid nanomaterials such as Cu, Ag, Zn, and Sn show antibacterial and antimicrobial properties [37–41]. Although the actual mechanism of their action is still unclear, it is believed that these metal nanoparticles play a crucial role by altering the bacterial metabolism, thereby killing them or making them dormant in the process [33,34].

8.8.7 Magnetic properties

Polymer nanocomposites may show magnetic behavior, depending on the nature of the polymer matrix or nanomaterial. This magnetic behavior appears when the magnetic dipoles of the material are activated by an external magnetic field. As most of the polymers are notably unresponsive toward magnetic behavior, nanomaterials with magnetic property are designed and incorporated into the polymer matrix. Especially, 2D nanomaterials containing Fe, Co, and Ni are the ideal choice for building magnetic nanostructures [42–44]. Such polymer nanocomposites are exploited for their magnetic properties in smart and biomedical applications [34].

8.9 Applications

Polymer nanocomposites based on 2D nanomaterials are in the forefront on numerous innovations and applications. In this context, polymer nanocomposites can be specially tuned to suit an interdisciplinary domain encompassing biology, polymer science, and nanotechnology. The combined attributes of polymer and 2D nanomaterials can be utilized to introduce biological activities in polymer nanocomposites, which can be applied in biomedical applications (Fig. 8.10). These 2D-nanomaterial-based nanocomposites are found to be useful in biomedical technologies such as tissue engineering, wound sutures, controlled drug delivery, and artificial muscles/medical implants [34,35]. Henceforth, prioritizing the biological impact and the need for life-science-based technologies in the benefit of human health, these applications of polymer nanocomposites are discussed in accordance with the recent state of art literature.

8.9.1 Tissue engineering materials

Polymer nanocomposites are being considered as excellent materials for tissue engineering, as they possess biocompatible surfaces and favorable mechanical properties. For example, bone fixation/repair is based on implants that mimic the natural bone material. Normally, screws and rods are used as the support for internal

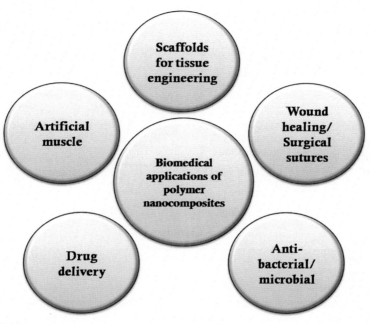

FIGURE 8.10

Biomedical applications of polymer nanocomposites based on two-dimensional nanomaterials.

bone fixation that brings the bone surfaces in close proximity to promote healing. This support must persist for weeks to months without any damage. The screws and rods must be noncorrosive, nontoxic, and easy to remove if necessary. Moreover, the mechanical strength of the implant must be close to that of the bone for efficient load transfer [45]. Thus a polymer nanocomposite implant must meet certain design and functional criteria, including biocompatibility, biodegradability, mechanical properties, and, in some cases, aesthetic demands (Fig. 8.11).

In this scenario, polymer nanocomposites based on biodegradable polymers containing HAp, nanometals, and carbon nanostructures are found to be effective. HAp is widely used as a biocompatible ceramic 2D material for contact with bone tissue owing to its resemblance to mineral bone [46]. It is the major mineral constituent (69% wt.) of human hard tissues and possesses excellent biocompatibility with bones, teeth, skin, and muscles, promoting bone regeneration and hardening [45,46]. As a result, polymer-HAp nanocomposites are being used clinically as biocompatible and osteoconductive substitutes for bone repair and implantation. To date, primarily polysaccharide and polypeptidic matrices have been used with HAp nanoparticles for nanocomposite formation. Farokhi et al. reported a nanocomposite of nano-HAp with natural silk fibroin for bone tissue engineering. Silk fibroin/nano-HAp nanocomposite combined the extraordinary material features of silk fibroin with those

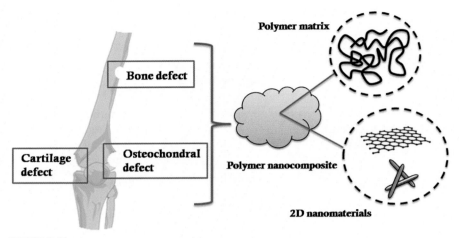

FIGURE 8.11

Two-dimensional (2D)-nanomaterial-based polymer nanocomposites for bone tissue engineering.

of nano-HAp for bone construct scaffolds [47]. Sharma and coworkers [48] reported the fabrication of a novel bio-nanocomposite scaffold using a combination of natural polymers, such as CS, gelatin, alginate, and nano-HAp, with high mechanical stability. In a similar vein, Nazeer and coworkers [49] presented the fabrication of intercalated CS/nano-HAp nanocomposite using the sol-gel process, with the nanocomposite as the promising material for bone tissue regeneration. The intercalation of nano-HAp by CS provided the formidable mechanical sturdy through enhanced physicochemical interactions. In another instance, the nanocomposite of gelatin/CS/polycaprolactone with nano-HAp was fabricated using co-electrospinning process by Arabahmadi and group members. The process demonstrated the potential of electrospun-nanofiber scaffold for bone tissue engineering [50]. Pavia et al. [51] reported the production of poly(L-lactic acid)/nano-HAp porous scaffolds by thermally induced phase separation for bone tissue engineering applications. In another example, the author's group was successful in fabricating a biobased waterborne polyurethane with nano-HAp/carbon nanodots and peptide-functionalized nano-HAp/carbon nanodots and in utilizing them as bone-regenerating material [52,53]. Reports by Yu et al. [54] also demonstrated that controllable three-dimensional porous shape memory polyurethane/nano-HAp nanocomposites could be used as bone regeneration scaffolds. Again, carbon 2D nanomaterials, as discussed in Section 8.3.1, are very good reinforcing materials for polymers. Polymer nanocomposites with carbon-based nanomaterials are investigated for use in tissue engineering, as they provide the needed structural reinforcement for such biomedical scaffolds. Biodegradable and biocompatible poly(-propylene fumarate) (PPF) nanocomposites with various 2D nanostructures, such as single- and multiwalled GO and GO nanoplatelets, showed high mechanical stabilities for bone tissue engineering [55]. Again, the mechanical properties and in vitro cytotoxicity of porous PPF nanocomposites reinforced with 2D GO nanoribbons and nanoplatelets revealed favorable cytocompatibility and increased protein adsorption, thus opening avenues for in vivo safety and efficacy studies for bone tissue engineering applications [56]. In another instance, a self-assembled graphene/HAp/CS hydrogel nanocomposite with high mechanical strength, high fixing capacity from HAp, and high porosity displayed good promise for use in bone tissue engineering [57]. In addition, inorganic minerals such as 2D nanosilicates are also found to provide significant enhancement of material properties that are required for generating bone tissue scaffolds. Xavier and coworkers [58] developed collagen-based hydrogel—containing 2D nanosilicates that demonstrated increased network stiffness and porosity, injectability, and enhanced mineralized matrix formation in a growth-factor-free microenvironment, as well as was conducive to the regeneration of bone in nonunion defects. In another report by Gaharwar and group, nanoclay-enriched electrospun poly(ε-caprolactone) scaffolds were developed that showed musculoskeletal tissue formation. The role of nanoclay was pertinent for both enhancing the mechanical strength of the electrospun fibers and promoting in vitro biomineralization [59]. In similar lines, supercritical-CO_2-processed nanocomposites of organically modified montmorillonite clay/poly-D-lactide were found to show structural and mechanical properties analogous to those of corticocancellous bone. The processed nanoclay/poly-D-lactide constructs were found to elicit in vivo osteogenesis and antiinflammatory response, making them ideal for bone grafting [60]. Nanoclay-incorporated PEG hydrogel nanocomposites with enhanced mechanical properties were found to facilitate in vivo and in vitro osteogenetic activity. The presence of nanoclay in the hydrogel nanocomposite was found to improve the adsorption and spreading ability of cells on the hydrogel, leading to efficient new bone formation [61]. The other most recent examples include nanocomposites of GO/HAp nanohybrid with spermine-based high-strength thermoplastic polyurethane-urea as porous bone tissue scaffolds with in vivo and in vitro cytocompatibility [62] and CS/GO nanocomposite as cross-linked cartilage tissue scaffolds [63].

8.9.2 Surgical sutures/wound healing materials

Surgical sutures are one of the most widely used medical devices nowadays. These sutures should possess not only good mechanical properties and biocompatibility but also the ability to keep microorganisms away from the surgical site [64]. In this milieu, a polymer nanocomposite must be adapted to include these features. 2D metal nanostructures show very good antibacterial or even antimicrobial properties and can also be easily combined with carbon nanostructures. Hence, 2D carbon nanomaterials, nanoclays, and metal/hybrid nanoparticles are considered to be an apt choice for fabricating such surgical sutures, as they can offer structural rigidity to keep the wounded area safely intact, along with antimicrobial activity (Fig. 8.12).

Ma and coworkers reported a nanocomposite of mechanically exfoliated graphene in natural honey and its fabrication with poly(vinyl alcohol). The nanocomposite afforded not only excellent mechanical properties but also low cytotoxicity and

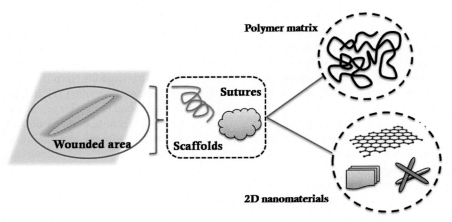

FIGURE 8.12

Two-dimensional (2D)-nanomaterial-based polymer nanocomposites for surgical sutures and wound healing.

antibacterial properties and served as the ideal material for surgical suture [65]. Biswas et al. demonstrated a nanocomposite of synthetic polyurethane/nanoclay for use as self-tightening sutures with bioabsorbable property. The nanocomposites showed excellent efficiency in wound healing through their self-tightening ability at near body temperature (37°C), without leaving behind any scar after the period of healing [66]. In another instance, a castor-oil-based polymer matrix was fabricated with CS-modified ZnO nanoparticles by Pascual and Vicente to obtain wound healing sutures. The nanocomposite showed strong bioactivity against gram-positive bacteria and enabled faster wound healing [67]. Noori and coworkers [68] developed a smart nanocomposite hydrogel based on poly(vinyl alcohol)/CS/honey/clay to be used as novel antibacterial wound dressing materials. Lu et al. showed Ag/ZnO nanohybrid-loaded CS composite as a sponge-shaped wound dressing material with enhanced antibacterial activity and low cytotoxicity. The nanocomposite displayed enhanced blood clotting and enhanced wound healing properties by promoting reepithelialization and collagen deposition [69]. In another example, Khalid et al. [70] combined the wound healing characteristics of bacterial cellulose and the antimicrobial properties of zinc oxide nanoparticles to fashion a bacterial cellulose-zinc oxide nanocomposite for in vivo burn wound healing and fine tissue regeneration.

8.9.3 Drug delivery

Drug delivery systems are engineered technologies for the targeted delivery and/or controlled release of therapeutic agents. Efficient and controlled drug delivery plays a crucial role in disease treatment and remains an important challenge in medicine [71].

Recent advances in the field of polymer nanocomposites offer the possibility of developing controlled-release systems for drug delivery. Drug delivery systems using polymer nanocomposites provide growth factors and deliver drugs directly to the target site to encourage reparation and regeneration and prevent infection (Fig. 8.13). For such systems, the surface of the nanocomposite or the nanofiller is modified for drug delivery by plasma treatment or wet chemical method, surface graft polymerization, or co-electrospinning for drug loading [72].

As discussed in Section 8.3.1Section 8.3.1, 2D carbon nanomaterials are suitable candidates for drug loading and release because of their favorable morphology and susceptibility to functionalization. For example, the PEGylated/GO nanocomposite prepared by surface modification of GO via single-electron transfer living radical polymerization in aqueous solution, using PEG methyl ether methacrylate as the monomer and 11-bromoundecanoic acid as the initiator, was found to exhibit high water dispersibility, excellent biocompatibility, and high efficient drug loading capability [73]. Again, inorganic 2D nanomaterials such as nanoclay, silicates, and HAp are also apt materials for targeted drug delivery. Owing to their intercalated nanostructures, drug molecules can be easily loaded in their interlayer spaces. For example, the biopolymer CS/montmorillonite clay nanocomposite loaded with silver sulfadiazine was prepared via the intercalation solution technique. The drug was found to be effectively loaded in the three-dimensional nanocomposite structure with CS chains adsorbed in monolayers into the clay mineral interlayer spaces [74]; in another work, a CS-grafted-poly(acrylic acid-*co*-acrylamide)/HAp nanocomposite scaffold demonstrated cytocompatibility without any cytotoxicity, cell viability, and proliferation. The nanocomposite displayed efficient loading of celecoxib as a model drug because of its large specific surface area, with a sustained in vitro release of up to 14 days. The results suggested that the cytocompatible and nontoxic nanocomposite scaffolds can act as efficient implants and drug carriers [75]. Among the metal and hybrid 2D nanomaterials, magnetic iron oxide nanoparticles, Zn

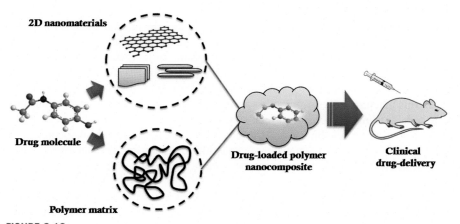

FIGURE 8.13

Two-dimensional (2D)-nanomaterial-based polymer nanocomposites for targeted drug delivery.

nanoparticles, and Cu nanoparticles are known to be efficient drug delivery systems. The basic premise involves therapeutic agents being attached to, or encapsulated within, the magnetic nanoparticle. These particles may either have magnetic cores with a polymer or metal coating that can be functionalized or consist of porous polymers that contain magnetic nanoparticles precipitated within the pores. By functionalizing the polymer or metal coating, it is possible to attach drug molecules for targeted applications [76]. As an example, poly(lactic-*co*-glycolic acid) nanoparticles were converted into polymer/iron oxide nanocomposites by attaching colloidal iron oxide nanoparticles onto the surface, via a simple surface modification method using dopamine polymerization. The nanocomposite was found to be effective transported across physical barriers and into cells and captured under flow conditions under magnetic field gradients. In vivo magnetic resonance imaging, ex vivo fluorescence imaging, and tissue histology confirmed that the uptake of the drug-loaded nanocomposite was higher in tumors exposed to magnetic field gradients [77]. In another instance, CS- PEG-poly(vinyl pyrrolidone) polymer nanocomposites coated with superparamagnetic iron oxide (Fe_3O_4) nanoparticles were loaded with curcumin as the model drug. The encapsulation efficiency, loading capacity, and in vitro drug release behavior of the curcumin-loaded nanocomposite revealed good efficiency with reduced side effects. Additionally, it was observed that the drug release was dependent on pH medium in addition to the nature of matrix [78]. Two works on similar lines by Pathania et al. described CS-*g*-polyacrylamide/CuS (CPA/CS) nanocomposite and CS-*g*-polyacrylamide/Zn (CPA-Zn) nanocomposite loaded with the drug ofloxacin, which showed efficient loading capacity and release profiles. In addition, the nanocomposites also showed antibacterial activity against *Escherichia coli* [79,80].

8.9.4 Artificial muscles

Artificial muscles are materials or devices that mimic natural muscles and can reversibly contract, expand, or rotate within one component in response to an external stimulus (such as voltage, current, pressure, or temperature) [81]. As these materials mimic natural muscles and fibers, they must possess the adequate mechanical rigidity and flexibility for load-bearing capacity (Fig. 8.14).

In this juncture, polymer nanocomposites based on 2D carbon nanomaterials provide suitable reinforcements required for building such artificial muscles. For example, poly(citric acid-octanediol-polyethylene glycol) (PCE)-graphene nanocomposites were developed that demonstrated myogenic differentiation and skeletal muscle tissue regeneration. PCE provided the biomimetic elastomeric behavior, while rGO offered enhanced mechanical strength and conductivity. As a result, highly elastomeric behavior, significantly enhanced modulus (400%−800%), and strength (200% −300%) of the nanocomposites with controlled biodegradability and electrochemical conductivity were achieved. The nanocomposite was found to significantly promote formation of muscle fibers and blood vessels in vivo in a skeletal muscle lesion model of rat [82]. In the recent years, artificial muscle−like biomaterials have gained

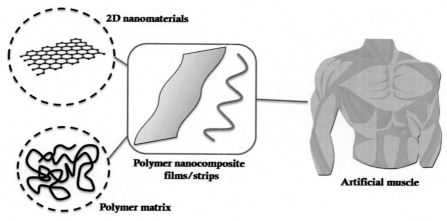

FIGURE 8.14

Two-dimensional (2D)-nanomaterial-based polymer nanocomposites as artificial muscles.

tremendous interests owing to their broad applications in regenerative medicine, as wearable devices, in bioelectronics, and in artificial intelligence.

Besides biomedical applications, 2D-nanomaterial-based polymer nanocomposites have found applications in other futuristic and advanced sectors including smart materials such as shape memory materials [83,84], self-healing materials [85,86], and self-cleaning materials [87,88]; active functional coatings and paints [89–91]; structural materials and construction materials [92–94]; and automobile and space vehicle materials [95–97].

8.10 Conclusion

Polymer nanocomposites based on 2D nanomaterials present an attractive alternative for a variety of products, most importantly in biomedical applications. In the recent times, the unique attributes of various types of polymer nanocomposites fabricated with 2D nanomaterials are exploited for direct and indirect innovations in the biomedical field. Hence, the imminent future of 2D-nanomaterial-based polymer nanocomposite bodes well for neotechnologic advancements.

References

[1] Y.W. Mai, Z.Z. Yu (Eds.), Polymer Nanocomposites, Woodhead Publishing, Cambridge, 2006.

[2] R. Vaithylingam, M.N.M. Ansari, R.A. Shanks, Recent advances in polyurethane-based nanocomposites: a review, Polymer – Plastics Technology and Engineering 56 (14) (2017) 1528–1541.

[3] W.D. Zhang, L. Shen, I.Y. Phang, T. Liu, Carbon nanotubes reinforced nylon-6 composite prepared by simple melt-compounding, Macromolecules 37 (2) (2004) 256–259.

[4] T. Liu, I.Y. Phang, L. Shen, S.Y. Chow, W.D. Zhang, Morphology and mechanical properties of multiwalled carbon nanotubes reinforced nylon-6 composites, Macromolecules 37 (19) (2004) 7214–7222.

[5] ISO/TS 80004-1:2010, Nanotechnologies—Vocabulary—Part 1 Core Terms, 2017.

[6] ISO/TS 27867, Nanotechnologies—terminology and Definitions for Nano-Objects—Nanoparticle, Nanofibre and Nanoplate, 2008.

[7] P.N. Sudha, K. Sangeetha, K. Vijayalakshmi, A. Barhoum, Nanomaterials history, classification, unique properties, production and market, in: A.S.H. Makhlouf, A. Barhoum (Eds.), Emerging Applications of Nanoparticles and Architecture Nanostructures, Elsevier, Amsterdam, 2018, pp. 341–384.

[8] C.N.R. Rao, A. Müller, A.K. Cheetham (Eds.), The Chemistry of Nanomaterials: Synthesis, Properties and Applications, John Wiley & Sons, Weinheim, 2006.

[9] K.E. Geckeler, H. Nishide (Eds.), Advanced Nanomaterials, John Wiley & Sons, Weinheim, 2009.

[10] A.K. Geim, Graphene: status and prospects, Science 324 (5934) (2009) 1530–1534.

[11] C. Lee, X. Wei, J.W. Kysar, J. Hone, Measurement of the elastic properties and intrinsic strength of monolayer graphene, Science 321 (5887) (2008) 385–388.

[12] S. Ray (Ed.), Applications of Graphene and Graphene-Oxide Based Nanomaterials, William Andrew, Waltham, 2015.

[13] S. Guo, S. Dong, Graphene nanosheet: synthesis, molecular engineering, thin film, hybrids, and energy and analytical applications, Chemical Society Reviews 40 (5) (2011) 2644–2672.

[14] S.V. Tkachev, E.Y. Buslaeva, A.V. Naumkin, S.L. Kotova, I.V. Laure, S.P. Gubin, Reduced graphene oxide, Inorganic Materials 48 (8) (2012) 796–802.

[15] J.H. Warner, F. Schaffel, M. Rummeli, A. Bachmatiuk (Eds.), Graphene: Fundamentals and Emergent Applications, Newnes, Waltham, 2012.

[16] C.N.R. Rao, S.R.C. Vivekchand, K. Biswas, A. Govindaraj, Synthesis of inorganic nanomaterials, Dalton Transactions (34) (2007) 3728–3749.

[17] C.N.R. Rao, W.U. Vasudeo (Eds.), 2D Inorganic Materials beyond Graphene, World Scientific, London, 2017.

[18] S.S. Ray, M. Okamoto, Polymer/layered silicate nanocomposites: a review from preparation to processing, Progress in Polymer Science 28 (11) (2003) 1539–1641.

[19] G. Mishra, B. Dash, S. Pandey, Layered double hydroxides: a brief review from fundamentals to application as evolving biomaterials, Applied Clay Science 153 (2018) 172–186.

[20] A. Asefnejad, A. Behnamghader, M.T. Khorasani, B. Farsadzadeh, Polyurethane/fluorhydroxyapatite nanocomposite scaffolds for bone tissue engineering. Part I: morphological, physical, and mechanical characterization, International Journal of Nanomedicine 6 (2011) 93.

[21] C. Sanchez, P. Belleville, M. Popall, L. Nicole, Applications of advanced hybrid organic–inorganic nanomaterials: from laboratory to market, Chemical Society Reviews 40 (2) (2011) 696–753.

[22] K. He, Z. Zeng, A. Chen, G. Zeng, R. Xiao, P. Xu, Z. Huang, J. Shi, L. Hu, G. Chen, Advancement of Ag–graphene based nanocomposites: an overview of synthesis and its applications, Small 14 (32) (2018) 1800871.

[23] Y. Wu, G. Jiang, Z.P. Cano, G. Liu, W. Liu, K. Feng, G. Lui, Z. Zhang, Z. Chen, Highly efficient removal of suspended solid pollutants from wastewater by magnetic Fe_3O_4-graphene oxides nanocomposite, Chemistry Select 3 (41) (2018) 11643−11648.

[24] Y.T. Ng, W. Kong, Magnetite-functionalized graphene nanohybrids: preparation and characterization of electrical and magnetic property, Materials Today: Proceedings 5 (1) (2018) 3202−3210.

[25] V. Belessi, D. Lambropoulou, I. Konstantinou, A. Katsoulidis, P. Pomonis, D. Petridis, T. Albanis, Structure and photocatalytic performance of TiO_2/clay nanocomposites for the degradation of dimethachlor, Applied Catalysis B: Environmental 73 (3−4) (2007) 292−299.

[26] A. Mohammad, K. Ahmad, A. Qureshi, M. Tauqeer, S.M. Mobin, Zinc oxide-graphitic carbon nitride nanohybrid as an efficient electrochemical sensor and photocatalyst, Sensors and Actuators B: Chemical 277 (2018) 467−476.

[27] A. Biswas, I.S. Bayer, A.S. Biris, T. Wang, E. Dervishi, F. Faupel, Advances in top−down and bottom−up surface nanofabrication: techniques, applications & future prospects, Advances in Colloid and Interface Science 170 (1−2) (2012) 2−27.

[28] L.S. Schadler, L.C. Brinson, W.G. Sawyer, Polymer nanocomposites: a small part of the story, Journal of Occupational Medicine 59 (3) (2007) 53−60.

[29] M.W. Rowell, M.A. Topinka, M.D. McGehee, H.J. Prall, G. Dennler, N.S. Sariciftci, L. Hu, G. Gruner, Organic solar cells with carbon nanotube network electrodes, Applied Physics Letters 88 (23) (2006) 233506.

[30] E. Barna, B. Bommer, J. Kürsteiner, A. Vital, O.V. Trzebiatowski, W. Koch, B. Schmid, T. Graule, Innovative, scratch proof nanocomposites for clear coatings, Composites Part A: Applied Science and Manufacturing 36 (4) (2005) 473−480.

[31] F.S. Awad, H.D. Kiriarachchi, K.M. AbouZeid, U. Özgür, M.S. El-Shall, Plasmonic graphene polyurethane nanocomposites for efficient solar water desalination, ACS Applied Energy Materials 1 (3) (2018) 976−985.

[32] M. Tanahashi, Development of fabrication methods of filler/polymer nanocomposites: with focus on simple melt-compounding-based approach without surface modification of nanofillers, Materials 3 (3) (2010) 1593−1619.

[33] J. Koo (Ed.), Polymer Nanocomposites: Processing, Characterization, and Applications, McGraw Hill Professional, New York, 2006.

[34] N. Karak (Ed.), Experimental Methods on Polymers, Nanomaterials and Their Nanocomposites, Nova Science Publishers, New York, 2016.

[35] W.S. Khan, N. Hamadneh, W.A. Khan, Polymer nanocomposites−synthesis techniques, classification and properties, in: P.D. Sia (Ed.), Science and Applications of Tailored Nanostructures, One Central Press, Cheshire, 2016, pp. 50−67.

[36] M.I. Khan, M.R. Islam, True Sustainability in Technological Development and Natural Resource Management, Nova Publishers, New York, 2007.

[37] A. Munoz-Bonilla, M. Fernández-García, Polymeric materials with antimicrobial activity, Progress in Polymer Science 37 (2) (2012) 281−339.

[38] K. Shameli, M.B. Ahmad, W.M.Z.W. Yunus, N.A. Ibrahim, R.A. Rahman, M. Jokar, M. Darroudi, Silver/poly (lactic acid) nanocomposites: preparation, characterization, and antibacterial activity, International Journal of Nanomedicine 5 (2010) 573.

[39] L. Tamayo, M. Azócar, M. Kogan, A. Riveros, M. Páez, Copper-polymer nanocomposites: an excellent and cost-effective biocide for use on antibacterial surfaces, Materials Science and Engineering: C 69 (2016) 1391−1409.

[40] B.S. Rathore, G. Sharma, D. Pathania, V.K. Gupta, Synthesis, characterization and antibacterial activity of cellulose acetate—tin (IV) phosphate nanocomposite, Carbohydrate Polymers 103 (2014) 221—227.

[41] B. Król, K. Pielichowska, M. Sochacka-Pietal, P. Król, Physicochemical and antibacterial properties of polyurethane coatings modified by ZnO, Polymers for Advanced Technologies 29 (3) (2018) 1056—1067.

[42] J.L. Wilson, P. Poddar, N.A. Frey, H. Srikanth, K. Mohomed, J.P. Harmon, S. Kotha, J. Wachsmuth, Synthesis and magnetic properties of polymer nanocomposites with embedded iron nanoparticles, Journal of Applied Physics 95 (3) (2004) 1439—1443.

[43] X. Chen, S. Wei, C. Gunesoglu, J. Zhu, C.S. Southworth, L. Sun, A.B. Karki, D.P. Young, Z. Guo, Electrospun magnetic fibrillar polystyrene nanocomposites reinforced with nickel nanoparticles, Macromolecular Chemistry and Physics 211 (16) (2010) 1775—1783.

[44] H. Wang, N. Ma, Z. Yan, L. Deng, J. He, Y. Hou, Y. Jiang, G. Yu, Cobalt/polypyrrole nanocomposites with controllable electromagnetic properties, Nanoscale 7 (16) (2015) 7189—7196.

[45] I. Armentano, M. Dottori, E. Fortunati, S. Mattioli, J.M. Kenny, Biodegradable polymer matrix nanocomposites for tissue engineering: a review, Polymer Degradation and Stability 95 (11) (2010) 2126—2146.

[46] L.E. Freed, G. Vunjak-Novakovic, R.J. Biron, D.B. Eagles, D.C. Lesnoy, S.K. Barlow, R. Langer, Biodegradable polymer scaffolds for tissue engineering, Biotechnology 12 (7) (1994) 689.

[47] M. Farokhi, F. Mottaghitalab, S. Samani, M.A. Shokrgozar, S.C. Kundu, R.L. Reis, Y. Fatahi, D.L. Kaplan, Silk fibroin/hydroxyapatite composites for bone tissue engineering, Biotechnology Advances 36 (1) (2018) 68—91.

[48] C. Sharma, A.K. Dinda, P.D. Potdar, C.F. Chou, N.C. Mishra, Fabrication and characterization of novel nano-biocomposite scaffold of chitosan—gelatin—alginate—hydroxyapatite for bone tissue engineering, Materials Science and Engineering: C 64 (2016) 416—427.

[49] M.A. Nazeer, E. Yilgör, I. Yilgör, Intercalated chitosan/hydroxyapatite nanocomposites: promising materials for bone tissue engineering applications, Carbohydrate Polymers 175 (2017) 38—46.

[50] S. Arabahmadi, M. Pezeshki Modaress, S. Irani, M. Zandi, Electrospun biocompatible gelatin-chitosan/polycaprolactone/hydroxyapatite nanocomposite scaffold for bone tissue engineering, International Journal of Nano Dimension 10 (2) (2019) 169—179.

[51] F.C. Pavia, G. Conoscenti, S. Greco, V. La Carrubba, G. Ghersi, V. Brucato, Preparation, characterization and in vitro test of composites poly-lactic acid/hydroxyapatite scaffolds for bone tissue engineering, International Journal of Biological Macromolecules 119 (2018) 945—953.

[52] S. Gogoi, M. Kumar, B.B. Mandal, N. Karak, A renewable resource based carbon dot decorated hydroxyapatite nanohybrid and its fabrication with waterborne hyperbranched polyurethane for bone tissue engineering, RSC Advances 6 (31) (2016) 26066—26076.

[53] S. Gogoi, S. Maji, D. Mishra, K.S.P. Devi, T.K. Maiti, N. Karak, Nano-bio engineered carbon dot-peptide functionalized water dispersible hyperbranched polyurethane for bone tissue regeneration, Macromolecular Bioscience 17 (3) (2017) 1600271.

[54] J. Yu, H. Xia, Q.Q. Ni, A three-dimensional porous hydroxyapatite nanocomposite scaffold with shape memory effect for bone tissue engineering, Journal of Materials Science 53 (7) (2018) 4734–4744.

[55] B. Farshid, G. Lalwani, M. Shir Mohammadi, J. Srinivas Sankaran, S. Patel, S. Judex, J. Simonsen, B. Sitharaman, Two-dimensional graphene oxide reinforced porous biodegradable polymeric nanocomposites for bone tissue engineering, Journal of Biomedical Materials Research Part A (2019), https://doi.org/10.1002/jbm.a.36606.

[56] G. Lalwani, A.M. Henslee, B. Farshid, L. Lin, F.K. Kasper, Y.X. Qin, A.G. Mikos, B. Sitharaman, Two-dimensional nanostructure-reinforced biodegradable polymeric nanocomposites for bone tissue engineering, Biomacromolecules 14 (3) (2013) 900–909.

[57] P. Duan, J. Shen, G. Zou, X. Xia, B. Jin, J. Yu, Synthesis spherical porous hydroxyapatite/graphene oxide composites by ultrasonic-assisted method for biomedical applications, Biomedical Materials 13 (4) (2018) 045001.

[58] J.R. Xavier, T. Thakur, P. Desai, M.K. Jaiswal, N. Sears, E. Cosgriff-Hernandez, R. Kaunas, A.K. Gaharwar, Bioactive nanoengineered hydrogels for bone tissue engineering: a growth-factor-free approach, ACS Nano 9 (3) (2015) 3109–3118.

[59] A.K. Gaharwar, S. Mukundan, E. Karaca, A. Dolatshahi-Pirouz, A. Patel, K. Rangarajan, S.M. Mihaila, G. Iviglia, H. Zhang, A. Khademhosseini, Nanoclay-enriched poly (ε-caprolactone) electrospun scaffolds for osteogenic differentiation of human mesenchymal stem cells, Tissue Engineering Part A 20 (15–16) (2014) 2088–2101.

[60] K.C. Baker, T. Maerz, H. Saad, P. Shaheen, R.M. Kannan, In vivo bone formation by and inflammatory response to resorbable polymer-nanoclay constructs, Nanomedicine: Nanotechnology, Biology and Medicine 11 (8) (2015) 1871–1881.

[61] X. Zhai, C. Hou, H. Pan, W.W. Lu, W. Liu, C. Ruan, Nanoclay incorporated polyethylene-glycol nanocomposite hydrogels for stimulating in vitro and in vivo osteogenesis, Journal of Biomedical Nanotechnology 14 (4) (2018) 662–674.

[62] S.K. Ghorai, S. Maji, B. Subramanian, T.K. Maiti, S. Chattopadhyay, Coining attributes of ultra-low concentration graphene oxide and spermine: an approach for high strength, anti-microbial and osteoconductive nanohybrid scaffold for bone tissue regeneration, Carbon 141 (2019) 370–389.

[63] M.A. Shamekhi, H. Mirzadeh, H. Mahdavi, A. Rabiee, D. Mohebbi-Kalhori, M.B. Eslaminejad, Graphene oxide containing chitosan scaffolds for cartilage tissue engineering, International Journal of Biological Macromolecules 127 (2019) 396–405.

[64] B. Joseph, A. George, S. Gopi, N. Kalarikkal, S. Thomas, Polymer sutures for simultaneous wound healing and drug delivery—A review, International Journal of Pharmaceutics 524 (1–2) (2017) 454–466.

[65] Y. Ma, D. Bai, X. Hu, N. Ren, W. Gao, S. Chen, H. Chen, Y. Lu, J. Li, Y. Bai, Robust and antibacterial polymer/mechanically exfoliated graphene nanocomposite fibers for biomedical applications, ACS Applied Materials and Interfaces 10 (3) (2018) 3002–3010.

[66] A. Biswas, A.P. Singh, D. Rana, V.K. Aswal, P. Maiti, Biodegradable toughened nanohybrid shape memory polymer for smart biomedical applications, Nanoscale 10 (21) (2018) 9917–9934.

[67] A.M. Díez-Pascual, A.L. Díez-Vicente, Wound healing bionanocomposites based on castor oil polymeric films reinforced with chitosan-modified ZnO nanoparticles, Biomacromolecules 16 (9) (2015) 2631–2644.

[68] S. Noori, M. Kokabi, Z.M. Hassan, Poly (vinyl alcohol)/chitosan/honey/clay responsive nanocomposite hydrogel wound dressing, Journal of Applied Polymer Science 135 (21) (2018) 46311.

[69] Z. Lu, J. Gao, Q. He, J. Wu, D. Liang, H. Yang, R. Chen, Enhanced antibacterial and wound healing activities of microporous chitosan-Ag/ZnO composite dressing, Carbohydrate Polymers 156 (2017) 460−469.

[70] A. Khalid, R. Khan, M. Ul-Islam, T. Khan, F. Wahid, Bacterial cellulose-zinc oxide nanocomposites as a novel dressing system for burn wounds, Carbohydrate Polymers 164 (2017) 214−221.

[71] T.M. Allen, P.R. Cullis, Drug delivery systems: entering the mainstream, Science 303 (5665) (2004) 1818−1822.

[72] B.L. Banik, J.L. Brown, Polymeric biomaterials in nanomedicine, in: S.G. Kumbar, C.T. Laurencin, M. Deng (Eds.), Natural and Synthetic Biomedical Polymers, Elsevier, Burlighton, 2014, pp. 387−395.

[73] P. Gao, M. Liu, J. Tian, F. Deng, K. Wang, D. Xu, L. Liu, X. Zhang, Y. Wei, Improving the drug delivery characteristics of graphene oxide based polymer nanocomposites through the "one-pot" synthetic approach of single-electron-transfer living radical polymerization, Applied Surface Science 378 (2016) 22−29.

[74] F. Jabeen, Polymer−ceramic nanocomposites for controlled drug delivery, in: A.M. Inamuddi, A. Mohammad Asiri (Eds.), Applications of Nanocomposite Materials in Drug Delivery, Woodhead Publishing, Duxford, 2018, pp. 805−822.

[75] S. Saber-Samandari, S. Saber-Samandari, Biocompatible nanocomposite scaffolds based on copolymer-grafted chitosan for bone tissue engineering with drug delivery capability, Materials Science and Engineering: C 75 (2017) 721−732.

[76] S.C. McBain, H.H. Yiu, J. Dobson, Magnetic nanoparticles for gene and drug delivery, International Journal of Nanomedicine 3 (2) (2008) 169.

[77] J. Park, N.R. Kadasala, S.A. Abouelmagd, M.A. Castanares, D.S. Collins, A. Wei, Y. Yeo, Polymer−iron oxide composite nanoparticles for EPR-independent drug delivery, Biomaterials 101 (2016) 285−295.

[78] G. Prabha, V. Raj, Preparation and characterization of polymer nanocomposites coated magnetic nanoparticles for drug delivery applications, Journal of Magnetism and Magnetic Materials 408 (2016) 26−34.

[79] D. Pathania, D. Gupta, S. Agarwal, M. Asif, V.K. Gupta, Fabrication of chitosan-g-poly (acrylamide)/CuS nanocomposite for controlled drug delivery and antibacterial activity, Materials Science and Engineering: C 64 (2016) 428−435.

[80] D. Pathania, D. Gupta, N.C. Kothiyal, G.E. Eldesoky, M. Naushad, Preparation of a novel chitosan-g-poly (acrylamide)/Zn nanocomposite hydrogel and its applications for controlled drug delivery of ofloxacin, International Journal of Biological Macromolecules 84 (2016) 340−348.

[81] S.M. Mirvakili, I.W. Hunter, Artificial muscles: mechanisms, applications, and challenges, Advanced Materials 30 (6) (2018) 1704407.

[82] Y. Du, J. Ge, Y. Li, P.X. Ma, B. Lei, Biomimetic elastomeric, conductive and biodegradable polycitrate-based nanocomposites for guiding myogenic differentiation and skeletal muscle regeneration, Biomaterials 157 (2018) 40−50.

[83] F. Cao, S.C. Jana, Nanoclay-tethered shape memory polyurethane nanocomposites, Polymer 48 (13) (2007) 3790−3800.

[84] S. Thakur, N. Karak, Multi-stimuli responsive smart elastomeric hyperbranched poly-urethane/reduced graphene oxide nanocomposites, Journal of Materials Chemistry A 2 (36) (2014) 14867−14875.

[85] E. Abdullayev, V. Abbasov, A. Tursunbayeva, V. Portnov, H. Ibrahimov, G. Mukhtarova, Y. Lvov, Self-healing coatings based on halloysite clay polymer composites for protection of copper alloys, ACS Applied Materials and Interfaces 5 (10) (2013) 4464−4471.

[86] S. Thakur, S. Barua, N. Karak, Self-healable castor oil based tough smart hyper-branched polyurethane nanocomposite with antimicrobial attributes, RSC Advances 5 (3) (2015) 2167−2176.

[87] A. Steele, I. Bayer, E. Loth, Adhesion strength and superhydrophobicity of polyure-thane/organoclay nanocomposite coatings, Journal of Applied Polymer Science 125 (S1) (2012) E445−E452.

[88] I. Hejazi, G.M.M. Sadeghi, J. Seyfi, S.H. Jafari, H.A. Khonakdar, Self-cleaning behavior in polyurethane/silica coatings via formation of a hierarchical packed morphology of nanoparticles, Applied Surface Science 368 (2016) 216−223.

[89] A. Kausar, Composite coatings of polyamide/graphene: microstructure, mechanical, thermal, and barrier properties, Composite Interfaces 25 (2) (2018) 109−125.

[90] H. Hu, S. Zhao, G. Sun, Y. Zhong, B. You, Evaluation of scratch resistance of function-alized graphene oxide/polysiloxane nanocomposite coatings, Progress in Organic Coat-ings 117 (2018) 118−129.

[91] M.S. Selim, S.A. El-Safty, M.A. Shenashen, M.A. El-Sockary, O.M.A. Elenien, A.M. EL-Saeed, Robust alkyd/exfoliated graphene oxide nanocomposite as a surface coating, Progress in Organic Coatings 126 (2019) 106−118.

[92] A.K. Naskar, J.K. Keum, R.G. Boeman, Polymer matrix nanocomposites for automotive structural components, Nature Nanotechnology 11 (12) (2016) 1026.

[93] Y. Ruan, W. Zhang, J. Wang, D. Wang, X. Yu, B. Han, Nanocarbon material-filled cementitious composites for construction applications, in: Nanocarbon and its Composites, Woodhead Publishing, 2019, pp. 781−803.

[94] L.M. Chiacchiarelli, Sustainable, nanostructured, and bio-based polyurethanes for energy-efficient sandwich structures applied to the construction industry, in: Biomass Biopolymer-Based Materials, and Bioenergy, Woodhead Publishing, 2019, pp. 135−160.

[95] J. Njuguna, K. Pielichowski, Polymer nanocomposites for aerospace applications: properties, Advanced Engineering Materials 5 (11) (2003) 769−778.

[96] P. Ding, S.F. Tang, H. Yang, L.Y. Shi, PP/LDH nanocomposites via melt-intercalation: synergistic flame retardant effects, properties and applications in automobile industries, in: L. Ma, C. Wang, W. Yang (Eds.), Advanced Materials Research, vol. 87, Trans Tech Publications, Switzerland, 2010, pp. 427−432.

[97] A. Kausar, I. Rafique, B. Muhammad, Aerospace application of polymer nanocompo-site with carbon nanotube, graphite, graphene oxide, and nanoclay, Polymer − Plastics Technology and Engineering 56 (13) (2017) 1438−1456.

Future prospects and commercial viability of two-dimensional nanostructures for biomedical technology

Sujata Pramanik, PhD, Dhriti Sundar Das, MBBS, MD
All India Institute of Medical Sciences, Bhubaneswar, Odisha, India

9.1 Introduction

As an emerging class of the dynamically developed material family, two-dimensional (2D) nanostructures hold the key to the futuristic demands of biomedical technology [1,2]. The escalating demand of clinical biotechnology and fast development of nanobiotechnology have substantially promoted the generation of a multitude of 2D nanostructures [1,2]. These nanostructures possess unparalleled advantages and superior performances and have surpassed zero-dimensional, one-dimensional, or three-dimensional nanostructures in their triumphal march toward biomedical technology [1]. The physicochemical and biological properties of nano-materials primarily depend on their atomic arrangements and morphology [3]. The indubitable charm of these 2D nanostructures can be ascribed to their exceptional properties including high degree of anisotropy; unique compositional and planar topography; ultrathin thickness; excellent thermal, optical, mechanical, electric, and physicochemical attributes; and chemical functionality [4]. The 2D topology is also instrumental in imparting unusual electronic and optical properties to varied applications [4]. These diverse properties enable 2D nanostructures as promising nanoplatforms for applications in medicine, pharmaceutical industry, and artificial intelligence (AI) including anticancer drug delivery, photothermal therapy (PTT), biosensing, gene transportation, and photodynamic therapy (PDT), among others [1,2]. The morphology and atomic arrangements of these nanomaterials have a profound impact on their biomedical performances including cellular uptake, bioavailability, biodistribution, and pharmacokinetic properties [1,2]. Representatively, a number of biocompatible carbon-based 2D nanomaterials such as graphene oxide (GO), reduced graphene oxide (RGO), and other types of graphene analogues (such as graphitic carbon nitride); silicate clays; and black phosphorus have enriched the family of 2D nanostructures as appealing candidates in this vein [1,2].

The research on 2D nanostructures is still in its infancy, with the majority focusing on elucidating their unique material characteristics and a few reports focusing on the biomedical applications [1]. These nanostructures seem to have kindled intense research activities in the scientific community with the advent of new therapeutic approaches ranging from biosensors to bioimaging to personalized therapies with minimal invasive procedures [1,2]. An ambitious umbrella approach encompassing the entire value chain starting from basic research through engineering and technology innovation to industry and marketplace, the infrastructure required, societal implications, and market challenges is adopted for the complete transfer of technology from inception to the market [1,5]. In other words, amalgamation of numerous worldwide research and business initiatives is required to attain a prominent market position. The technology transfer to market domination process involves a detailed investigation including evaluating the potential of products, identifying market features and competitors, and pondering a promising market way in and share at different time horizons [5]. The pivotal role of science, technology, and innovation (STI)—the driving force for independent development and sustainable economic growth—has long been realized by the countries and the aggressive STI national policies have also been adopted worldwide [6]. Following a long research and development (R&D) incubation period, a handful of industrial segments have started emerging, and surprisingly, rapid market growth and mass market opportunities are envisaged as future potentials in this context. We today reside in an era where everything from innovations to technology development is looked through the prism of trade and commerce [5,6]. Magnetized by the potential applications of 2D nanostructures, the biopharmaceutical and biomedical companies have started to foresee partnerships between the biomedical enterprises and technology start-ups [7].

Although the investment community can expertly grasp and launch 2D-nanostructure-based viable products in the market, the products are limited by their corresponding biological effects in the human body. In other words, the clinical applicability of these nanostructures has been eclipsed by the poor understanding of their long-term in vivo safety, including potential short- and long-term toxicity, biopersistence, and hemo-, histo-, cyto-, and biocompatibility behaviors [8,9]. 2D nanostructures have increased surface areas that warrant consideration of these nanomaterials with respect to nanotoxicity [9]. The exposed specific surface area and prominent facet of the nanomaterials towards the exterior environment play a pivotal role in the determination of their biological reactivity and response of the living beings [10]. Additionally, the appropriate dose metric and the probable binding of these nanostructures to proteins in the biological fluids of living beings has also been debated in the context of nanotoxicologic studies [11]. The toxicologic responses of these nanostructures have attracted considerable attention from across government sectors and scientific communities [12,13]. However, these nanostructures a priori should not be regarded as dangerous, but scientific paradigm is needed to facilitate the standardization of bioassays for both in vitro and in vivo testing [14,15]. Overall, unprecedented challenges and opportunities exist in developing 2D nanostructures for next-generation biomedical technologies such as basic cell biology, medical diagnostics, regenerative medicine, and so on. In this milieu, attracting funds is the key, yet challenging, issue, and the investors have started visualizing the prospective marketplace of these nanostructures [16,17]. However, 2D-nanomaterial-based biomedical technology belongs to those budding sectors wherein business expertise is yet to establish till date. Momentum is gradually building up for the successful development of biomedical products based on 2D nanostructures, and a growing number of the products are in the pipeline to make it to the market.

As the chase for the commercialization of 2D nanostructures is at its pinnacle, the apparent question that arises at this juncture is can the researchers raise a toast to the dividends of 2D-nanomaterial-based products and flood the market with the same? In response to the aforementioned question, it becomes apparent to recollect the snapshot on the applicability of these 2D nanostructures in biomedical sciences, including therapeutic (PTT/PDT, chemotherapy, and synergistic therapy), diagnostic (fluorescent imaging/magnetic resonance imaging [MRI]/computed tomography [CT]/photoacoustic imaging), and theranostic (concurrent diagnostic imaging and therapy) applications [1,18]. In addition, the biosensing and bioengineering applications of these nanostructures are also scrutinized. Herein, we focus on the state-of-the-art biomedical applications of 2D nanostructures as well as recent developments and commercial viability that are shaping this emerging field.

9.2 The spectrum of biomedical applications: a snapshot of technologies and market opportunity

2D nanostructures exhibit diverse properties in terms of their mechanical, chemical, and optical attributes, apart from uniqueness in their size, aspect ratio, shape, biocompatibility, and biodegradability [4]. These varied properties afford a suitable platform for these nanostructures for myriad applications in the healthcare sector, including drug delivery, bioimaging, tissue engineering, biosensors, and so on [1,4,18]. However, their low dimensionality and exceptionally large aspect ratios (surface area-to-volume ratio) make these nanostructures invaluable for uses demanding high levels of surface interactions on a nanoscale regime [1,4]. To cite an example, these nanostructures provide large anchoring sites for adsorption of drug molecules [1,2,10]. This property is exploited in drug delivery systems, which afford advanced control over the release kinetics of a drug [1,2]. Additionally, their large aspect ratios and high modulus make them suitable nanofillers (at low concentrations) for enhancing the physicomechanical properties of biomedical nanocomposites [1,2]. The distinctive thinness of these nanostructures was found to be instrumental for advancements in biosensing and gene sequencing utilities [1,2]. Furthermore, the rapid response of these nanostructures to the external environment, such as light, and their fluorescence properties paved way for breakthroughs in optical therapies, including bioimaging, PTT, and PDT [1,2]. In view of the aforementioned properties, 2D nanostructures can be described as "multifunctional therapeutics." Still in its infancy, much of the works in the discipline of biomedical sciences involve R&D exercises and it is, therefore, prudent to foresee the 2D-nanostructure-based products in the pilot scale. In tune with this vision, the following subsections provide a summary of 2D-nanostructure-based technologies and their potent market share.

9.2.1 Therapeutic applications

The advent of 2D nanostructures marked several research breakthroughs in the realm of biomedical sciences in the past decade [1,2]. The superior stability, drug loading/carrier efficiency, viability of adsorption/absorption of hydrophilic/hydrophobic moieties, modulation of time- and dose-dependent pharmacokinetics, and ability to ensure compliance to prescribed therapy and compatibility with various routes of administration such as oral, parenteral, inhalation, intraarticular, and subcutaneous of these nanostructures merit their inclusion in therapeutic applications [1,2]. 2D nanostructures aim to enhance the therapeutic index, efficacy, and bioavailability of drugs apart from decreasing the dosing frequency and possibility of systemic side effects [1,19]. They can noncovalently adsorb drug molecules and act as drug delivery vehicles, which affords sustained release of the same from the matrix. In this context, Vallet-Regí et al. [20] tailored the surface properties of the silicates for active adsorption of drug molecules via noncovalent functionalization using alkoxysilanes. On a similar note, Rámila et al. [21] and Balas et al. [22] studied the pharmacokinetics of drug release upon oral administration. They explored the stability, release profile, dissolution, and bioavailability of a drug under different milieus of the gastrointestinal tract on oral intake [21,22]. Mellaerts et al. [23] made significant headway with regard to in vivo studies using silicate-loaded itraconazole drug. The presence of intricate porous structure of the silicates afforded high effective surface area for adsorption and carrier capability of the administered drug. The use of 2D porous matrix thus resulted in the enhancement of the bioavailability performance of the loaded drug molecule [23]. Furthermore, Warheit et al. [24] studied the toxicity behavior response of the commercially available 2D silica using cellular, biochemical, and autoradiographic techniques. The pulmonary cellular response of silica was delved into using a rat model. The study revealed no adverse effect in the inflammatory responses of the rat upon oral administration of 10 mg m^{-3} concentration of silica [24]. Radu et al. [25] and Slowing et al. [26] conducted toxicologic studies of cationic surfactant—modified nanoporous 2D silica MCM-41 nanomaterials. The study ascertained low cytotoxicity of these 2D hexagonal matrices toward the HeLa cell line [25,26]. The silicate matrixes were also explored as drug nanocarriers for antitumor therapy by Li et al. [27]. In vitro and long-term in vivo studies of doxorubicin-loaded silicate exhibited higher drug uptake by the tumor cells relative to that of the free drug molecules [27]. The enhanced efficacy of doxorubicin-loaded silicate underscores the importance of the enhanced permeability and retention effect [27]. This effect embodies high drug loading capacity and large effective surface area for conjugating target sites of silicates, preventing biodegradation, the ability of the silicate matrix to conjugate and passively target tumor cells, and promoting controlled release of the drug [27]. Furthermore, the reduced lymphatic drainage results in prolonged retention of doxorubicin inside the tumor vessels. The systemic administration of the free drug results in hematologic toxicities, whereas the use of drug loaded silicate in mice demonstrates normal blood parameters, thereby reducing systemic adverse effects and improving anatomy [27]. Gonçalves and his coworkers [28] exploited the use of nanosilicates in injectable hydrogels. The doxorubicin-loaded alginate-silicate—based hydrogels exhibited shear thinning rheologic behavior apart from sustained drug release kinetics, which makes them fit for injectable applications for cancer treatment [28]. The cohesive interaction among silicates, alginate, and doxorubicin resulted in enhanced drug loading capacity, reduced burst release, and a pH-dependent controlled release profile of the drug, as compared with alginate-drug hydrogels [28]. The silicate clays also offer the substantial advantage of superior hemostatic performance, which helps address the unmet need of hemostatic drugs in prehospital settings of ballistic and traumatic injuries or in battlefronts [29]. The inclusion of nanosilicates in collagen-based hydrogels not only engineered their mechanical properties and rheologic behavior, including shear thinning, but also imparted self-healing ability (activation of clot formation) to the hemostatic gels [30]. Of note, these gels exhibited clot time comparable to that of thrombin and profound biocompatibility. In this milieu, WoundStat, a silica-based commercially available hemostatic wound dressing gauge approved by the US Food and Drug Administration (FDA) for treating external hemorrhage, forms a promising leap ahead in hemostatic technology [31]. The high adsorption capacity of the hydrated aminosilicates amalgamated with chitosan and polyacrylic acid also ensured a prominent hemostatic effect [32].

The advantage of Mg/Al layered double hydroxides (LDHs) as a biocompatible nanodelivery agent was explored by Saifullah et al. [33] for the delivery of the antituberculosis drug isoniazid. This loaded LDH demonstrated sustained drug release kinetics in the normal human murine fibroblast and lung cells, as compared with the free drug molecules [33]. The favorable cytocompatibility, target-specific cell types, pH-dependent drug release, drug accumulation in the targeted cells, prolonged drug half-life in blood, and high surface charge density and anion exchange capacity rank LDH among the promising 2D nanomaterials for efficient drug delivery processes [33]. A study in this regard by Ma et al. [34] highlighted an 11-fold enhancement of the effectiveness of cisplatin upon loading onto LDH nanosheets, apart from decreasing the toxicity imparted to the normal cells. The increased efficacy of the anticancer drug molecule was attributed to endocytosis, which resulted in increment of the uptake of the drug by cancer cells on one hand with no significant apoptosis to healthy cells on the other hand [34]. LDHs were also utilized for the dual delivery of both anticancer drug (5-fluorouracil) and nucleic acid (AllStars Cell Death siRNA) to impart synergistic cytotoxicity to cancer cells [35]. The ability to exchange the surface anions of LDHs with the anionic drug and nucleic acid in the interlayer LDH spaces was exploited herein for the delivery of drugs and oligonucleotides [35]. Regarding the nutraceutical applications of LDHs, one key study demonstrated the potentiality of CaAl-LDH as calcium supplements exploiting the diversity in the exchange of Ca ions. As Ca^{2+} readily dissolves at pH ≤ 7, CaAl-LDH is supposed to get absorbed by the gastric and small intestinal mucosa [36]. Furthermore, the biocompatibility, biodegradability, and antacid and antipepsin properties of MgAl-LDH forward them for use in antacid formulations. It is pertinent to mention that hydrotalcite is a carbonate-type LDH containing Mg^{2+}, Al^{3+}, and CO_3^{2-} in the interlayer of the laminate host [37]. Of note, Juggat Pharma marketed the use of hydrotalcite (generic name) as "Talsil Forte" (500 mg or 50 mg5 mL^{-1}) as an antacid in dealing with dyspepsia.

Graphene is another material of choice among 2D nanostructures, which are envisaged to revolutionize the biomedical technologies owing to their wonder properties including high specific surface area, biocompatibility, unprecedented mechanical strength, ease of functionalization, and electric and thermal conductivities [1]. A study by Lin et al. [38] ascertained the feasibility of the use of 2D graphene and its derivatives for chemotherapeutic drug delivery. The high effective surface area, supramolecular $\pi-\pi$ stacking, and hydrophobicity of 2D graphene and its derivatives bestow strong adsorption affinity for anchoring anticancer drugs and empower them to be efficient gene carrier [38]. The graphitic analogues such as graphitic-phase carbon nitride (g-C_3N_4) have been reported to encapsulate, load, and deliver anticancer drugs such as doxorubicin (up to 18,200 mg g^{-1}) [38]. The doxorubicin-loaded nanomaterials exhibited comparable cytotoxicity as the unloaded drug molecules. The high specific surface area and presence of functional groups on GO account for their use as nanocarrier for drug delivery processes [38]. The nanocomposites of GO and poly(vinyl alcohol) were explored for the oral delivery of vitamin B$_{12}$ [39]. The high specific surface area of GO and self-assembling networks between GO and the polymer were instrumental in the sustained release of the entrapped acid-sensitive drug molecule [39]. Furthermore, the presence of aromatic moieties and pronounced bioactivity of GO forwarded graphene-polypyrrole nanocomposites as stimulus-responsive drug delivery agents [40,41]. These nanocomposites triggered the sustained diffusion (via electric stimulation) of the entrapped antiinflammatory steroid (dexamethasone) by twofold magnitude, as compared with the release profile by the pristine polymer [40,41].

The strong light absorbance of graphene and GO in the near-infrared (NIR) window (700−1300 nm) and their ability to efficiently transform light into heat energy make them suitable for PTT, a noninvasive substitute to conventional treatments such as surgery, chemotherapy, radiation therapy, and hormonal therapy [1,2,42]. Furthermore, the transparency of living tissues in the NIR window results in high photothermal efficacy within cancer cells [1,2,42]. Literature reports the use of graphene and GO as attractive candidates for photothermal ablation (both in vitro and in vivo assessments) of cancerous cells [1,2,42]. Towards this end, Calevia Inc. (private biotechnology company) in partnership with Grafoid Inc. and ProScan Rx Pharma Inc. designed an injectable tumor-specific photoresponsive MesoGraf Xide-PSC1700 bioconjugate to target prostate cancer cells [43]. This injectable anticancer therapeutic agent harnessed the photothermal property of graphene and targeted the adherence of prostate cancer antibody ligand (PSC1700) to prostate-specific membrane antigen proteins (overexpressed proteins on tumor cells) to molecularly target the cancerous cells [43]. The product is presently in the R&D phase involving malignant cells of animal models (mice xenograft and dog models) [43].

9.2.2 Diagnostic imaging

The unique optoelectronic properties including the intrinsic fluorescence property, photostability and tunable emission spectrum of 2D nanostructures have been explored for wide-sweeping applications including contrast-enhanced imaging processes [1,2]. The pronounced photoluminescence and absorbance in the NIR window brings transition metal dichalcogenides (TMDs) to the fore for bioimaging applications [1,2]. The broad photoresponsive WS_2 nanosheets were exploited as contrast agents in CT imaging owing to the high opacity of tungsten to absorb X-rays and consequently increase the image contrast of the area under observation [1,2]. The NIR absorption of WS_2 nanosheets also finds them use as contrast agents in photoacoustic tomographic imaging modality [44,45]. The intravenous or intratumor administration of WS_2 produced strong photoacoustic signals from tumor cells in response to excitation using laser light [45].

The intrinsic two-photon fluorescence of GO finds applications in fluorescence cell imaging [19]. To this end, the enhanced photoluminescence of six-arm branched polyethylene glycol (PEG)-grafted GO in the visible-NIR window was probed into as NIR fluorophore for selective lymphoma cell recognition and imaging [46]. Sun and his coworkers [46] covalently conjugated anti-CD20 antibody Rituxan to PEGylated GO to selectively anchor onto B cells. These lymphoma cells exhibited NIR photoluminescence and thus were imaged using 658-nm laser light [46]. The presence of lattice imperfections or structural defects in GO nanolayers aids in enhancing the reactivity and adsorptivity of other atoms, including radioactive nuclides, onto its defective edges [47]. To this end, ^{125}I-labeled PEGylated GO (on the defect edges) finds use in radionuclide-based imaging techniques (positron emission tomography [PET] or single-photon emission CT) [47,48]. PET imaging also demonstrated the conjugation of ^{64}Cu-labeled PEG-grafted GO with TRC105 antibody for in vivo targeting of murine breast tumor cells [47,49]. The ability of the oxygenated functional groups of GO to coordinate and chelate the toxic paramagnetic metal ions into its cavities (which mitigates the toxicity of metal ions) enables their use as T_1 contrast agents in MRI [47]. In this regard, Gizzatov et al. [50] employed carboxyphenylated graphene nanoribbons as the nanoplatform for sequestering Gd^{3+} for enhancing MRI relaxivity (relative to that of the clinically available Gd^{3+}-based T_1 agents). The enhancement of Raman peaks of GO (strong D and G peaks) upon decoration with metal nanoparticles was effectively utilized for producing surface-enhanced Raman scattering (SERS) imaging [51]. For instance, folic acid—conjugated Ag-decorated GO nanohybrids found use in selective labeling and SERS imaging of folate-receptor-positive A549 cells [52]. The Ag-decorated nanohybrid SERS label selectively targeted the folate-receptor-positive cells and demonstrated intense Raman peaks from the cytoplasmic membrane of the cells. On the contrary, the folate-receptor-negative cells exhibited no such peaks in SERS imaging [52].

Notably, the number of research works in this vein has provided a proof of concept and opened new vistas toward understanding the in vivo applications of 2D nanostructures as contrast agents in bioimaging [1,2,44−52]. However, the clinical application and commercialization of these technologies is yet to make advances in the market.

9.2.3 Theranostic applications

Anchoring drug molecules onto the surface of the 2D-nanostructure-based contrast agents amalgamates their synergistic therapeutic performance and photoresponsive property [1,2]. This combined diagnostic bioimaging and therapeutic outcome of 2D nanostructures is explored for theranostic applications [1,2]. In this milieu, 2D topological TMDs possess specific properties, such as photosensitizing and X-ray attenuation abilities, for simultaneous use in therapeutics and imaging-guided diagnosis (theranostics) [1,2]. Cheng et al. [45] conjoined the PTT functionality and bioimaging modalities (owing to the high atomic number of W in WS_2) of PEG-functionalized WS_2 nanosheets in a theranostic device. The intravenous injection of these biocompatible PTT-efficient nanosheets selectively concentrates in the tumor cells and provides an apparent contrast-enhanced CT imaging, as compared with iopromide (a clinically available CT contrast agent) [45]. Continuing in the same vein, high CT imaging performance (increase of Hounsfield units) was also observed for intratumoral administration of bovine serum albumin−modified WS_2 nanosheets to human cervical cancer cells [53]. Furthermore, the in vitro incubation of HeLa cells using these nanosheets showed efficient PDT selectivity and NIR-induced PTT. Thus the pronounced drug loading capacity of these nanosheets and X-ray attenuation coefficients of W were used as a multifunctional anticancer platform for concurrent therapeutics and bioimaging (X-ray CT contrast agent) [53]. Bi_2Se_3 nanoplates too possess twin properties of concurrent PTT and CT imaging for use in theranostics [2,54]. The occurrence of brighter CT signals within the cancer cells relative to those of the healthy cells after in vivo administration of these nanoplates demonstrated superior CT imaging performance of Bi_2Se_3, as compared with WS_2 [2,54]. In other words, low dosages of Bi_2Se_3 were capable of producing equivalent CT image contrasts in comparison to iopromide [2,54]. Furthermore, the intrinsic fluorescence, ease of modification of surface functionalities to optimize drug loading characteristics, and significant electric and mechanical properties of graphene and GO qualify them for theranostic applications [55]. In this regard, poly(amidoamine) dendrimer and Gd-functionalized GO nanocarriers efficiently delivered epirubicin (chemotherapeutic drug) and microRNA (gene-targeting agents) to human glioblastoma (U87) cells, apart from being used as an MRI contrast agent

[55]. These nonviral nanovectors identified the permeability of the blood-brain barrier and quantified the chemogene therapy required [56].

It is thus clear from the aforementioned reports that these engineered 2D nanostructures integrated many functions in a single theranostic system to afford modulations in therapeutics in addition to molecular imaging diagnosis for next-generation clinical applications.

9.2.4 Bioelectronic devices and biosensors

Bioelectronic devices and biosensors are found to play a paramount role in biomedical sciences [1,2]. These are employed as minimally invasive detection approaches to monitor vitals for early detection of diseases [1,2]. The unique 2D morphology with higher aspect ratio, surface functionalities (to immobilize large receptor molecules), significant electric properties, and carrier mobility features make 2D nanostructures a smart choice for bioelectronic and biosensing technologies [1,2]. Towards this end, the presence of large surface area and versatile surface functional groups of graphene and its analogues aid to immobilize receptor molecules (via physisorption) [1,2,43]. Graphene and its analogues then subsequently act as a transducer (of biosensors) for conversion of the interactions (owing to the high rate of electron transfer) between the receptor and target molecules to detectable measurements [1,57−60]. In view of this, the RGO-based field-effect transistor (FET) immunosensor was fabricated for the detection of prostate-specific antigen at low concentrations [1,57−59]. This biosensor was found to be rapid, ultrasensitive, highly specific, and label free [1,57−59]. The graphene-based biosensor also found uses in dentistry for the detection of bacterial binding on the enamel of tooth [1,60].

Moreover, the estimation of blood glucose levels plays a vital role in diabetes management. The graphene-based glucose oxidase biosensor demonstrates profound sensitivity, selectivity, and reproducibility [61]. The synergism of metals and graphene, for instance, CdS nanocrystal−decorated graphene, showed enhanced sensitivity, allowing glucose detection at lower concentrations (detection limit as low as 0.7 mM) [62]. Similar results were obtained upon incorporating Pt nanoparticles to chitosan-graphene nanocomposites, which showed low glucose detection limits of 0.6 μM [63]. Continuing in the same vein, graphene-based enzymatic biosensors were used for the detection of levels of cytochrome c, coenzyme (NADH), hemoglobin, cholesterol, catechol, and more [64]. The nonenzymatic graphene-electrode-based biosensors were employed for the detection of ascorbic acid, dopamine, and uric acid [64]. The graphene-based bioelectronic devices find use in the detection of biomolecules including nucleic acids, antigens, and proteins [64]. Furthermore, the electric and photoluminescence properties and intercalatable morphologies of TMDs offer biorecognition capabilities for their use as biosensors [65]. The changes in the photoluminescence potentials of MoS_2 nanosheets were used for cation (H^+, Li^+, Na^+, K^+) detection in biological systems [65]. Another biosensing study revealed glucose sensing by MoS_2 nanoflakes after conjugation with glucose oxidase [65,66]. These studies showed enhancement of photoluminescence potentials of TMDs with increasing glucose concentration [65,66]. The peroxidase-like activity of TMD (as a function of the surface charges) was effectively exploited for the design of a glucose testing kit [67]. WS_2 nanosheets in conjunction with tetramethylbenzidine and glucose oxidase find use in glucose-sensing applications (via monitoring the change in the color of tetramethylbenzidine with glucose concentration) [68]. The FET potential of MoS_2 was also explored for fast, inexpensive, robust, and sensitive FET biosensing of proteins and biotin, apart from pH detection [65,69,70].

Featured with profound physicochemical properties, the "wonder material" graphene has been found to revolutionize the field of bioelectronics [2,18,43,71]. The smart health watch "GF1" is one such graphene-based commercially available product launched by Wuxi Graphene Film (a subordinate of The Sixth Elements Materials). Chemical vapor deposited—graphene film acts as the touch screen conductive element in lieu of commonly used indium tin oxide glass [72]. GF1 is a smart health monitoring watch that provides a detailed health index and scrutinizes the dynamic electrocardiogram, heart rate, temperature, and blood pressure, apart from tracking the calories burned during physical workouts.

9.2.5 Miscellaneous applications

2D nanostructures provide wide-sweeping benefits ranging from tissue engineering to biomedicines [1,2]. The profound cyto- and biocompatibilities, osteoinductive characteristics, and in vivo biodegradation of silicates hold the potential for musculoskeletal tissue engineering avenues [73,74]. These silicates stimulated the adhesion, proliferation, and differentiation of stem cells without exogenous growth factors and promoted osteogenic pathways [73,74]. The significant mechanical strength, stiffness, and biocompatibility of graphene have shown to promote differentiation of human mesenchymal stem cells [75]. Furthermore, the electric conductivity of graphene was explored for designing a biocompatible graphene substrate for culturing of neural stem cells [76,77]. The cardiomyocytes seeded on the methacrylated gelatin-GO hydrogel surface exhibited higher proliferation rate for use as a cardiac patch [78]. It is thus clear from the aforementioned reports that these 2D nanostructures hold potential for serving as an efficient future bioengineered stratum for tissue regeneration.

The low cost, hydrophilicity, ease of surface functionalization, and biocompatibility of silicates have enabled their use as a versatile ingredient (abrasive, bulking, suspending, opacifying, and anticaking agents; viscosity enhancer; adsorbent; emulsifier; and more) in the compositions for topical dermatologic products [79,80]. In this context, sodium silicate is formulated into ointments or creams for the treatment of dry skin conditions including psoriasis, facial malar rash, dermatitis, acne rosacea, dermatophyte infections, cradle cap, and eczema [79,80]. Silicates are used in "Plexaderm Rapid Reduction Cream PLUS," a dermocosmetic product (marketed by PLEXADERM) that reduces the appearance of under-eye bags and wrinkles. "Dermocalm Lotion" marketed by GlaxoSmithKline and "Calamine Lotion BP" by Thornton and Ross Ltd. are antiskin irritants containing silicates in their formulations.

With regard to healthcare products, the high surface energy, hydrophilicity, biocompatibility, chemical versatility, cost-effectiveness, and ease of processing of silicates merit for their inclusion in the formulations of nutraceuticals [81,82]. FDA approved the usage of silicates as safe anticaking additives in dietary health supplementary foods and oral drugs. Swanson Health Products, USA has developed a dietary supplement "Alta," which contains silica fortified with horsetail extract (*Equisetum arvense*, bioflavonoids).

In keeping with the edge of digitalization, contemporary times have witnessed the benefits of AI in addressing healthcare issues [83]. AI holds great promises in democratizing healthcare and empowering healthcare consumers [83]. Clinical decision support systems constitute one of the most recent advances of AI for improving medical care [83]. Herein, the preeminent properties of graphene, such as profound electric conductivity, and physicomechanical properties, find use in the fabrication of neural networks, flexible electronics, and so on [71,84]. To this end, graphene-based "Nuance AI-Powered Virtual Assistants" provides smart solutions to healthcare establishments. The advanced features such as voice biometrics, text-to-speech technology, and so on aim to improve patient outcome, doctor-patient relationship, and care efficiency [71,85]. The global leader in biomedical technology, Royal Philips, also launched "Philips HealthSuite Insights" to provide AI-based smart solutions in the healthcare domain.

9.3 Challenges in the clinical translation of biomedical technology

Although the current armory of 2D nanostructures has been effectively demonstrated in laboratory scale (and to some extent as commercial products) in the field of biomedical sciences, the real challenge remains in bringing this translational research from "bench-to-bedside" [86,87]. In other words, the clinical translation of these 2D nanostructures has not progressed as rapidly as the multitude of preclinical research works available. The rapid advancement in 2D nanostructures wrestles with the obvious question of their physiologic interactions with biological moieties for evaluation of their biocompatibility (to judge their relevance in biomedical applications) [88]. The oxidative damage from free radicals (produced from 2D-nanostructure-induced immune reactions, the presence of contaminants in 2D nanostructures, or their biodegraded products) remains the prime concern of toxicity to healthy cells, apart from damages caused by apoptosis, thrombosis, or hemolysis [89]. The high aspect ratio of 2D nanostructures is envisaged to augment the ability to interact with living cells and protein adsorption and thus might increase the immunogenicity to these implanted nanostructures. The current trends and challenges for translational development and commercialization of 2D nanostructures in the healthcare industry constitutes the biocompatibility (in vivo fate) and biosafety issues, difficulties in large-scale manufacturing, intellectual property (IP) rights, government regulations, and cost-effectiveness with regard to existing therapies [88]. These key issues associated with the challenges in the clinical development of 2D nanostructures are discussed in the following.

9.3.1 Challenges in the nano-bio interface

The biological challenges to understand the influence of introduction of 2D nanostructures into the living body and their bioaccumulation, biodistribution, and retention form the major deterrent for the investment of healthcare industries in these nanomaterial-based products [84,90,91]. The preclinical experiments should broadly assess and optimize the therapeutic efficiency and pharmacokinetic data in multiple animal models against standard controls, apart from evaluating their biosafety and biodistribution to ascertain reproducibility of the results [84,90,91]. The dissimilarity of the anatomy and physiology of the animal model species relative to human bodies should be considered based on varied modes of their administration. Disappointingly, these detailed preclinical data seem incompletely documented in many published research works [8,9,25,26,33]. The missing results and the failure to address all the fronts in this regard in turn forms the bottleneck for clinical trials and applicability. Moreover, the optimized in vivo performances, dose schedule, and on-demand drug-releasing aspects add to the biological challenges in this milieu [88]. Thus a disease-driven approach to the understanding and development of 2D nanostructures through extensive experimentation to assemble the comprehensive preclinical data seems to be an answer to the clinical translation of these nanomaterials.

9.3.2 Challenges in manufacturing

One of the key requirements for large-scale manufacturing of 2D nanostructures is to have access to the preparative methods, which allows large scalable production of high quality having batch-to-batch reproducibility [88]. The commercialization of these nanostructures is pillared on expenditure and quality assessment [88]. However, the high input costs, poor quality control, product stability (both during storage and clinical administration), complexities in scalability and product purification processes, low production yield and reproducibility, lack of infrastructure and venture funds, instability issues, and so on are the stumbling blocks in realizing the commercialization of 2D nanostructures [88].

Collaboration among research institutions, clinicians, and pharma industries seems imperative to conduct clinical trials for translating the research findings into clinically approved therapeutic outcomes [88]. The process of transfer of the superior characteristics of these 2D nanostructures from virtual research models to real-world-based solutions has many limitations including regulatory risks, biosafety concerns, and human health impacts.

9.3.3 Regulatory risks, biosafety concerns, and human health impacts

The detailed toxicology data is of paramount importance for the commercialization of 2D nanostructures in the context of biosafety issues for use by mankind [88,91]. The rational plan starting with material selection phase, preclinical and clinical evaluations, nanopharmaceutical design, and optimization of production are important in accessing the clinical translation potential of these nanostructures [88]. The regulatory risks of 2D nanostructures calls for the need of standardized assays for in vitro, ex vivo, and in vivo experiments (both short- and long-term studies) to ascertain the potential toxicities (cyto-, immuno-, and genotoxicities) of 2D nanostructures before preclinical evaluations [88]. Furthermore, adequate biosafety assessment protocols are required to examine the pharmacokinetics, drug-release profile and delivery process, biodistribution and bioaccumulation at target sites, and therapeutic efficacy before clinical trials [88]. To this end, real-time imaging after in vivo administration of drug-loaded nanostructures suffices the need. Although there is limited availability of current testing approaches for complete biosafety evaluation [92], a number of alternate techniques including computational modeling (subtractive genomics, molecular docking analysis, etc.) and bioinformatics are under development.

Furthermore, the biosafety issues are intertwined with the problems concerning biomedical waste management, which is acknowledged as a grave environmental and public health concern on a global level [93]. The public outcry in this milieu called for drafting the legislative and regulatory aspects of biomedical wastes, apart from the execution of national waste management policies by the healthcare institutions and governments [93]. To this end, India implemented the Biomedical Waste (Management and Handling) Rules in 1998 in the purview of the Environment Protection Act (EPA), 1986 [93]. US Congress passed the Medical Waste Treatment Act (MWTA) to address the potential biomedical waste hazards. The concern over these hazardous wastes also galvanized Germany to respect the "Closed Substance Cycle Waste Management Act" that regulates proper prevention, recycling, and disposal of waste.

Combating this issue has thus entwined multiple humanitarian and environmental challenges and demands for an immediate cooperative responsiveness for our common world. However, there remain other issues and herculean tasks to confront, including legislative regulations and rights over ownership of biomedical technologies in the clinical translation from laboratory to marketplace.

9.3.4 Intellectual property rights and government regulations

The progress toward creating economic values from biomedical innovations of 2D nanostructures hinges on the successful exploitation and translation of basic science discoveries to produce commercially viable technologic innovations [88]. The new innovations reach out to the commercial entities and finally to public use from the bench of basic research by the process of "technology transfer" [88]. This process includes granting exclusive access (legal rights) to other for-profit commercial entities to capitalize on their discoveries (also known as IP rights) [88]. These ownership rights (be it either permission or license) also exclude other market players from using the invention. In its IP portfolio, the companies in the high-tech arena intend to secure their core technology particularly through patents, commercial licenses, trademarks, copyrights, and trade secrets [88]. In other words, one of the critical elements of product development begins with patent protection and ownership.

India's action plan on forming a regulatory body for nanotechnology to regulate nanotechnologic products, including biomedical products, was called for in 2010 by Prof. C.N.R. Rao (chairman of the Nano Mission Council), but the National Regulatory Framework road map is yet to be laid down. The Department of Science and Technology, India has made efforts toward constituting government regulatory systems [94]. There also exist a handful of Indian laws, such as the Drugs and Cosmetics Act, 1940, National Pharmacovigilance Protocol, and Medical Devices Regulation Bill, concerning the production and marketing of biomedical products [95]. However, the effort to streamline these regulations to include nanotechnology-related biomedical technologies hitherto to foresee a grand vision for India. Moreover, section 3(b) of the Indian Patents Act, 1970 mirrors the key challenges in patenting these nanostructure inventions [96]. The section 3(b) forms a barrier to nanobiotechnology-based patenting owing to the inherent nanotoxicity issues of such products [96].

Moreover, while the premarketing approval for biomedical devices and drugs is strenuous and requires an average time of 7–12 years, less than 1 in 10 putative drugs that survive preclinical trials head on through FDA approval [97]. FDA approval of the drugs or biomedical devices guarantees the manufacturers to recreate the same and make profits.

9.4 Commercialization of two-dimensional nanostructure innovations—how close we are?

Commercialization of 2D-nanostructure-based biomedical products, as of today, has been lagging its technologic development owing to multifaceted problems, not exclusively monetary or managerial but more of fundamental, including manufacturing at large scale (with), quality control (with precision in size and shape), and biosafety issues [98]. The holy grail in this sphere of business is mass production with reasonable resources, with precision in size and shape, and at a competitive cost utilizing existing machinery, which is intended to suffice consumers' demand and environmental safety concerns in addition to making high revenues [98]. Although patenting 2D-nanostructure-based biomedical technology seems an arduous task, the payoffs could be the freedom of operation and leapfrogging STI through open innovations and bridging innovative research and commerce in cutting-edge economies. Toward this goal, exploiting academe-industry-government collaborations appears more appeasing. Thus STI cooperation, academia-industry partnerships, socioeconomic implications, and allocation of funds play a pivotal role in the road map of commercialization of the same [98]. As of today, although the insecurity of spending billions of money in this commerce is high, the role of funding from angel investors, risk-takers, and mavericks taking significant strides (while keeping balanced with prudence) could be a fillip with regard to the old aphorism "pioneering ventures into little-known waters." Albeit a handful of 2D-nanostructure-based biomedical products are available in the market, commercialization of the advanced biomedical products has lagged behind. From the aforementioned sections, it is clear that R&D in this vein has matured enough, but a paradigm shift of biomedical technologies from laboratory to market calls for extensive public participation. Although these initiatives seek to complement the government's endeavor to rev up the commercialization phase and expand its workforce, the broader question to be pondered over is how "powerful beacons" are these 2D-nanostructure-based biomedical technologies to make their mark on global news cycle? The realization of these incredibly diverse applications standing at the pipeline would certainly epitomize new hopes for a new epoch.

9.5 Future landscape

Inspired by their unprecedented advantages and profound physiochemical features, unparalleled structural and compositional attributes, and superior performances, 2D nanostructures remain the material of choice in the realm of biomedical sciences. But the spectrum of probable damage and risks threatens to slow down the pace of commercialization of these technologic innovations. As goes the old gambler's adage "Scared money never wins," so thus the industrial profit-makers need to strive to delineate and execute booming approaches in these uncertain biotech business road maps. These biomedical technologies represent a strategic gamble typified by the convergence of various stakeholders including researchers, policy-makers, investors, medical fraternity, and public. The bold headship in this business is in no way an easy task, as its outcomes are never assured, but fear of not investing in fortune is worthwhile. The integration of research, training, and industry has made a significant march in the context of academia STI excellence initiatives, such as encouraging nationwide student start-ups (such as the Student Startup Madness event in Austin; the Student Startup Initiative and startup café at the University of Illinois, Science and Technology for harnessing innovation; and SATHI in India International Science Festival 2018) to train young scientists and the community at large, as business entrepreneurs. These "lab to start-up" concepts aim toward creating a nanoculture, which would focus on problems relevant to the societal healthcare and thereby strengthen the technology-based global healthcare sector. The worldwide interest in nanobiomedical technology is increasing at a fast pace with more and more nations entering the race for supremacy in the technologic front. Despite the proper establishment of nanoculture at research institutions, the culture still remains a visionary for decision-makers, funding agencies, and the general public at large. Although there exists risks in overestimating these technologies as the immediate solution for biomedical problems owing to their inherent toxicologic issues, the broader question that remains is "will these proof-of-concepts of biomedical engineers and scientists become the clinical tools of physicians one day?" Although the innovative and revolutionary breakthroughs place these 2D nanostructures at the pole position, the issue of commercialization of the same warrants bold and strong action with regard to translating these biomedical technologies into real healthcare products and devices. Such a mission directing the society to harvest the welfare of the investments in biomedical technologies is highly applaudable. Thus prophesy of the swift shift of these biomedical technologies from hype and hope to hat trick (propel the industrial revolution, make way from laboratory to customer, and realization of the millennium development goal of healthcare for all) aspires to conquer this far-reaching science.

References

[1] D. Chimene, D.L. Alge, A.K. Gaharwar, Two-dimensional nanomaterials for biomedical applications: emerging trends and future prospects, Advanced Materials 27 (45) (2015) 7261–7284.

[2] Y. Chen, C. Tan, H. Zhang, L. Wang, Two-dimensional graphene analogues for biomedical applications, Chemical Society Reviews 44 (9) (2015) 2681–2701.

[3] M. Auffan, J. Rose, J.Y. Bottero, G.V. Lowry, J.P. Jolivet, M.R. Wiesner, Towards a definition of inorganic nanoparticles from an environmental, health and safety perspective, Nature Nanotechnology 4 (10) (2009) 634.

[4] Y. Chen, Z. Fan, Z. Zhang, W. Niu, C. Li, N. Yang, B. Chen, H. Zhang, Two-dimensional metal nanomaterials: synthesis, properties, and applications, Chemical Reviews 118 (13) (2018) 6409–6455.

[5] V. Morigi, A. Tocchio, C. Bellavite Pellegrini, J.H. Sakamoto, M. Arnone, E. Tasciotti, Nanotechnology in medicine: from inception to market domination, Journal of drug delivery 2012 (2012).

[6] P. Aghion, P.A. David, D. Foray, Science, technology and innovation for economic growth: linking policy research and practice in 'STIG systems', Research Policy 38 (4) (2009) 681–693.

[7] N.S. Gray, Drug discovery through industry-academic partnerships, Nature Chemical Biology 2 (12) (2006) 649.

[8] Y. Chen, H. Chen, J. Shi, In vivo bio-safety evaluations and diagnostic/therapeutic applications of chemically designed mesoporous silica nanoparticles, Advanced Materials 25 (23) (2013) 3144–3176.

[9] A.B. Seabra, A.J. Paula, R. de Lima, O.L. Alves, N. Durán, Nanotoxicity of graphene and graphene oxide, Chemical Research in Toxicology 27 (2) (2014) 159–168.

[10] S. Prabha, G. Arya, R. Chandra, B. Ahmed, S. Nimesh, Effect of size on biological properties of nanoparticles employed in gene delivery, Artificial Cells, Nanomedicine, and Biotechnology 44 (1) (2016) 83–91.

[11]   Y.W. Huang, M. Cambre, H.J. Lee, The toxicity of nanoparticles depends on multiple molecular and physicochemical mechanisms, International Journal of Molecular Sciences 18 (12) (2017) 2702.

[12]   A.D. Maynard, R.J. Aitken, T. Butz, V. Colvin, K. Donaldson, G. Oberdörster, M.A. Philbert, J. Ryan, A. Seaton, V. Stone, S.S. Tinkle, L. Tran, N.J. Walker, S.S. Tinkle, Safe handling of nanotechnology, Nature 444 (7117) (2006) 267.

[13]   C. Wilkinson, S. Allan, A. Anderson, A. Petersen, From uncertainty to risk?: scientific and news media portrayals of nanoparticle safety, Health, Risk and Society 9 (2) (2007) 145−157.

[14]   C.F. Jones, D.W. Grainger, *In vitro* assessments of nanomaterial toxicity, Advanced Drug Delivery Reviews 61 (6) (2009) 438−456.

[15]   V. Kumar, N. Sharma, S.S. Maitra, In vitro and in vivo toxicity assessment of nanoparticles, International Nano Letters 7 (4) (2017) 243−256.

[16]   A.D. Romig Jr., A.B. Baker, J. Johannes, T. Zipperian, K. Eijkel, B. Kirchhoff, H.S. Mani, C.N.R. Rao, S. Walsh, An introduction to nanotechnology policy: opportunities and constraints for emerging and established economies, Technological Forecasting and Social Change 74 (9) (2007) 1634−1642.

[17]   M.C. Roco, Government nanotechnology funding: an international outlook, JOM 54 (9) (2002) 22−23.

[18]   M. Nurunnabi, J. McCarthy (Eds.), Biomedical Applications of Graphene and 2D Nanomaterials, Elsevier, 2019.

[19]   G. Yang, C. Zhu, D. Du, J. Zhu, Y. Lin, Graphene-like two-dimensional layered nanomaterials: applications in biosensors and nanomedicine, Nanoscale 7 (34) (2015) 14217−14231.

[20]   M. Vallet-Regí, F. Balas, D. Arcos, Mesoporous materials for drug delivery, Angewandte Chemie International Edition 46 (40) (2007) 7548−7558.

[21]   A. Rámila, S. Padilla, B. Muñoz, M. Vallet-Regí, A new hydroxyapatite/glass biphasic material: in vitro bioactivity, Chemistry of Materials 14 (6) (2002) 2439−2443.

[22]   F. Balas, M. Manzano, P. Horcajada, M. Vallet-Regí, Confinement and controlled release of bisphosphonates on ordered mesoporous silica-based materials, Journal of the American Chemical Society 128 (25) (2006) 8116−8117.

[23]   R. Mellaerts, R. Mols, J.A. Jammaer, C.A. Aerts, P. Annaert, J. Van Humbeeck, G. Van der Mooter, P. Augustijns, J.A. Martens, Increasing the oral bioavailability of the poorly water soluble drug itraconazole with ordered mesoporous silica, European Journal of Pharmaceutics and Biopharmaceutics 69 (1) (2008) 223−230.

[24]   D.B. Warheit, M.C. Carakostas, D.P. Kelly, M.A. Hartsky, Four-week inhalation toxicity study with ludox colloidal silica in rats: pulmonary cellular responses, Toxicological Sciences 16 (3) (1991) 590−601.

[25]   D.R. Radu, C.Y. Lai, K. Jeftinija, E.W. Rowe, S. Jeftinija, V.S.Y. Lin, A polyamidoamine dendrimer-capped mesoporous silica nanosphere-based gene transfection reagent, Journal of the American Chemical Society 126 (41) (2004) 13216−13217.

[26]   I.I. Slowing, B.G. Trewyn, V.S.Y. Lin, Mesoporous silica nanoparticles for intracellular delivery of membrane-impermeable proteins, Journal of the American Chemical Society 129 (28) (2007) 8845−8849.

[27]   K. Li, S. Wang, S. Wen, Y. Tang, J. Li, X. Shi, Q. Zhao, Enhanced *in vivo* antitumor efficacy of doxorubicin encapsulated within laponite nanodisks, ACS Applied Materials and Interfaces 6 (15) (2014) 12328−12334.

[28] M. Gonçalves, P. Figueira, D. Maciel, J. Rodrigues, X. Shi, H. Tomás, Y. Li, Antitumor efficacy of doxorubicin-loaded laponite/alginate hybrid hydrogels, Macromolecular Bioscience 14 (1) (2014) 110–120.

[29] Y.J. Zhang, B. Gao, X.W. Liu, Topical and effective hemostatic medicines in the battlefield, International Journal of Clinical and Experimental Medicine 8 (1) (2015) 10.

[30] A.K. Gaharwar, R.K. Avery, A. Assmann, A. Paul, G.H. McKinley, A. Khademhosseini, B.D. Olsen, Shear-thinning nanocomposite hydrogels for the treatment of hemorrhage, ACS Nano 8 (10) (2014) 9833–9842.

[31] B.S. Kheirabadi, J.E. Mace, I.B. Terrazas, C.G. Fedyk, J.S. Estep, M.A. Dubick, L.H. Blackbourne, Safety evaluation of new hemostatic agents, smectite granules, and kaolin-coated gauze in a vascular injury wound model in swine, Journal of Trauma and Acute Care Surgery 68 (2) (2010) 269–278.

[32] M. Ghadiri, W. Chrzanowski, R. Rohanizadeh, Biomedical applications of cationic clay minerals, RSC Advances 5 (37) (2015) 29467–29481.

[33] B. Saifullah, P. Arulselvan, M.E. El Zowalaty, S. Fakurazi, T.J. Webster, B.M. Geilich, M.Z. Hussein, Development of a biocompatible nanodelivery system for tuberculosis drugs based on isoniazid-Mg/Al layered double hydroxide, International Journal of Nanomedicine 9 (2014) 4749.

[34] R. Ma, Z. Wang, L. Yan, X. Chen, G. Zhu, Novel Pt-loaded layered double hydroxide nanoparticles for efficient and cancer-cell specific delivery of a cisplatin prodrug, Journal of Materials Chemistry B 2 (30) (2014) 4868–4875.

[35] L. Li, W. Gu, J. Chen, W. Chen, Z.P. Xu, Co-delivery of siRNAs and anti-cancer drugs using layered double hydroxide nanoparticles, Biomaterials 35 (10) (2014) 3331–3339.

[36] T.H. Kim, J.A. Lee, S.J. Choi, J.M. Oh, Polymer coated CaAl-layered double hydroxide nanomaterials for potential calcium supplement, International Journal of Molecular Sciences 15 (12) (2014) 22563–22579.

[37] X. Bi, H. Zhang, L. Dou, Layered double hydroxide-based nanocarriers for drug delivery, Pharmaceutics 6 (2) (2014) 298–332.

[38] L.S. Lin, Z.X. Cong, J. Li, K.M. Ke, S.S. Guo, H.H. Yang, G.N. Chen, Graphitic-phase C_3N_4 nanosheets as efficient photosensitizers and pH-responsive drug nanocarriers for cancer imaging and therapy, Journal of Materials Chemistry B 2 (8) (2014) 1031–1037.

[39] H. Bai, C. Li, X. Wang, G. Shi, A pH-sensitive graphene oxide composite hydrogel, Chemical Communications 46 (14) (2010) 2376–2378.

[40] J. Liu, L. Cui, D. Losic, Graphene and graphene oxide as new nanocarriers for drug delivery applications, Acta Biomaterialia 9 (12) (2013) 9243–9257.

[41] C.L. Weaver, J.M. LaRosa, X. Luo, X.T. Cui, Electrically controlled drug delivery from graphene oxide nanocomposite films, ACS Nano 8 (2) (2014) 1834–1843.

[42] S.C. Patel, S. Lee, G. Lalwani, C. Suhrland, S.M. Chowdhury, B. Sitharaman, Graphene-based platforms for cancer therapeutics, Therapeutic Delivery 7 (2) (2016) 101–116.

[43] H. Shinohara, A. Tiwari, Graphene: An Introduction to the Fundamentals and Industrial Applications, John Wiley & Sons, 2015.

[44] S. Park, U. Jung, S. Lee, D. Lee, C. Kim, Contrast-enhanced dual mode imaging: photoacoustic imaging plus more, Biomedical Engineering Letters 7 (2) (2017) 121–133.

[45] L. Cheng, J. Liu, X. Gu, H. Gong, X. Shi, T. Liu, C. Wang, X. Wang, G. Liu, H. Xing, W. Bu, B. Sun, Z. Liu, Imaging: PEGylated WS2 nanosheets as a multifunctional theranostic agent for in vivo dual-modal CT/photoacoustic imaging guided photothermal therapy, Advanced Materials 26 (12) (2014) 1794.

[46] X. Sun, Z. Liu, K. Welsher, J.T. Robinson, A. Goodwin, S. Zaric, H. Dai, Nanographene oxide for cellular imaging and drug delivery, Nano Research 1 (3) (2008) 203–212.

[47] J. Lin, Y. Huang, P. Huang, Graphene-based nanomaterials in bioimaging, in: Biomedical Applications of Functionalized Nanomaterials, Elsevier, 2018, pp. 247–287.

[48] K. Yang, S. Zhang, G. Zhang, X. Sun, S.T. Lee, Z. Liu, Graphene in mice: ultrahigh in vivo tumor uptake and efficient photothermal therapy, Nano Letters 10 (9) (2010) 3318–3323.

[49] H. Hong, K. Yang, Y. Zhang, J.W. Engle, L. Feng, Y. Yang, T.R. Nayak, S. Goel, J. Bean, C.P. Theuer, T.E. Barnhart, Z. Liu, W. Cai, *In vivo* targeting and imaging of tumor vasculature with radiolabeled, antibody-conjugated nanographene, ACS Nano 6 (3) (2012) 2361–2370.

[50] A. Gizzatov, V. Keshishian, A. Guven, A.M. Dimiev, F. Qu, R. Muthupillai, P. Decuzzi, R.G. Bryant, J.M. Tour, L.J. Wilson, Enhanced MRI relaxivity of aquated Gd^{3+} ions by carboxyphenylated water-dispersed graphene nanoribbons, Nanoscale 6 (6) (2014) 3059–3063.

[51] S.S. Sukumaran, C.R. Rekha, A.N. Resmi, K.B. Jinesh, K.G. Gopchandran, Raman and scanning tunneling spectroscopic investigations on graphene-silver nanocomposites, Journal of Science: Advanced Materials and Devices 3 (3) (2018) 353–358.

[52] Z. Liu, Z. Guo, H. Zhong, X. Qin, M. Wan, B. Yang, Graphene oxide based surface-enhanced Raman scattering probes for cancer cell imaging, Physical Chemistry Chemical Physics 15 (8) (2013) 2961–2966.

[53] Y. Yong, L. Zhou, Z. Gu, L. Yan, G. Tian, X. Zheng, X. Liu, X. Zhang, J. Shi, W. Cong, W. Yin, Y. Zhao, WS_2 nanosheet as a new photosensitizer carrier for combined photodynamic and photothermal therapy of cancer cells, Nanoscale 6 (17) (2014) 10394–10403.

[54] J. Li, F. Jiang, B. Yang, X.R. Song, Y. Liu, H.H. Yang, D.R. Cao, W.R. Shi, G.N. Chen, Topological insulator bismuth selenide as a theranostic platform for simultaneous cancer imaging and therapy, Scientific Reports 3 (2013) 1998.

[55] S. Roy, A. Jaiswal, Graphene-based nanomaterials for theranostic applications, Reports in Advances of Physical Sciences 1 (04) (2017) 1750011.

[56] H.W. Yang, C.Y. Huang, C.W. Lin, H.L. Liu, C.W. Huang, S.S. Liao, P.Y. Chen, Y.J. Lu, K.C. Wei, C.C.M. Ma, Gadolinium-functionalized nanographene oxide for combined drug and microRNA delivery and magnetic resonance imaging, Biomaterials 35 (24) (2014) 6534–6542.

[57] B. Cai, S. Wang, L. Huang, Y. Ning, Z. Zhang, G.J. Zhang, Ultrasensitive label-free detection of PNA–DNA hybridization by reduced graphene oxide field-effect transistor biosensor, ACS Nano 8 (3) (2014) 2632–2638.

[58] D.J. Kim, I.Y. Sohn, J.H. Jung, O.J. Yoon, N.E. Lee, J.S. Park, Reduced graphene oxide field-effect transistor for label-free femtomolar protein detection, Biosensors and Bioelectronics 41 (2013) 621–626.

[59] B. Zhan, C. Li, J. Yang, G. Jenkins, W. Huang, X. Dong, Graphene field-effect transistor and its application for electronic sensing, Small 10 (20) (2014) 4042–4065.

[60] M.S. Mannoor, H. Tao, J.D. Clayton, A. Sengupta, D.L. Kaplan, R.R. Naik, N. Verma, F.G. Omenetto, M.C. McAlpine, Graphene-based wireless bacteria detection on tooth enamel, Nature Communications 3 (2012) 763.

[61] F. Wang, L. Liu, W.J. Li, Graphene-based glucose sensors: a brief review, IEEE Transactions on Nanobioscience 14 (8) (2015) 818–834.

[62] K. Wang, Q. Liu, Q.M. Guan, J. Wu, H.N. Li, J.J. Yan, Enhanced direct electrochemistry of glucose oxidase and biosensing for glucose via synergy effect of graphene and CdS nanocrystals, Biosensors and Bioelectronics 26 (5) (2011) 2252–2257.

[63] H. Wu, J. Wang, X. Kang, C. Wang, D. Wang, J. Liu, I.K. Aksay, Y. Lin, Glucose biosensor based on immobilization of glucose oxidase in platinum nanoparticles/graphene/chitosan nanocomposite film, Talanta 80 (1) (2009) 403–406.

[64] S.K. Krishnan, E. Singh, P. Singh, M. Meyyappan, H.S. Nalwa, A review on graphene-based nanocomposites for electrochemical and fluorescent biosensors, RSC Advances 9 (16) (2019) 8778–8881.

[65] S. Barua, H.S. Dutta, S. Gogoi, R. Devi, R. Khan, Nanostructured MoS$_2$-based advanced biosensors: a review, ACS Applied Nano Materials 1 (1) (2017) 2–25.

[66] J. Yu, D. Ma, L. Mei, Q. Gao, W. Yin, X. Zhang, L. Yan, Z. Gu, X. Ma, Y. Zhao, Peroxidase-like activity of MoS$_2$ nanoflakes with different modifications and their application for H$_2$O$_2$ and glucose detection, Journal of Materials Chemistry B 6 (3) (2018) 487–498.

[67] T. Lin, L. Zhong, L. Guo, F. Fu, G. Chen, Seeing diabetes: visual detection of glucose based on the intrinsic peroxidase-like activity of MoS$_2$ nanosheets, Nanoscale 6 (20) (2014) 11856–11862.

[68] T. Lin, L. Zhong, Z. Song, L. Guo, H. Wu, Q. Guo, Y. Chen, F. Fu, G. Chen, Visual detection of blood glucose based on peroxidase-like activity of WS$_2$ nanosheets, Biosensors and Bioelectronics 62 (2014) 302–307.

[69] J. Mei, Y.T. Li, H. Zhang, M.M. Xiao, Y. Ning, Z.Y. Zhang, G.J. Zhang, Molybdenum disulfide field-effect transistor biosensor for ultrasensitive detection of DNA by employing morpholino as probe, Biosensors and Bioelectronics 110 (2018) 71–77.

[70] J. Lee, P. Dak, Y. Lee, H. Park, W. Choi, M.A. Alam, S. Kim, Two-dimensional layered MoS$_2$ biosensors enable highly sensitive detection of biomolecules, Scientific Reports 4 (2014) 7352.

[71] H. Kim, J.H. Ahn, Graphene for flexible and wearable device applications, Carbon 120 (2017) 244–257.

[72] J.K. Wassei, R.B. Kaner, Graphene, a promising transparent conductor, Materials Today 13 (3) (2010) 52–59.

[73] X. Zhou, N. Zhang, S. Mankoci, N. Sahai, Silicates in orthopedics and bone tissue engineering materials, Journal of Biomedical Materials Research Part A 105 (7) (2017) 2090–2102.

[74] K.M. Pawelec, J. Shepherd, R. Jugdaohsingh, S.M. Best, R.E. Cameron, R.A. Brooks, Collagen scaffolds as a tool for understanding the biological effect of silicates, Materials Letters 157 (2015) 176–179.

[75] I. Lasocka, E. Jastrzebska, L. Szulc-Dąbrowska, M. Skibniewski, I. Pasternak, M.H. Kalbacova, E.M. Skibniewska, The effects of graphene and mesenchymal stem cells in cutaneous wound healing and their putative action mechanism, International Journal of Nanomedicine 14 (2019) 2281.

[76] M. Bramini, G. Alberini, E. Colombo, M. Chiacchiaretta, M.L. DiFrancesco, J.F. Maya-Vetencourt, L. Maragliano, F. Benfenati, F. Cesca, Interfacing graphene-based materials with neural cells, Frontiers in Systems Neuroscience 12 (2018) 12.

[77] L.W. Kenry, K.P. Loh, C.T. Lim, When stem cells meet graphene: opportunities and challenges in regenerative medicine, Biomaterials 155 (2018) 236–250.

[78] B.J. Klotz, D. Gawlitta, A.J. Rosenberg, J. Malda, F.P. Melchels, Gelatin-methacryloyl hydrogels: towards biofabrication-based tissue repair, Trends in Biotechnology 34 (5) (2016) 394—407.

[79] P. Bakker, N. Wieringa, V. Gooskens, H. Van Doorne, Dermatological Preparations for the Tropics, University of Groningen, 1990.

[80] A. Sivaraman, A.K. Banga, Quality by design approaches for topical dermatological dosage forms, Research and Reports in Transdermal Drug Delivery 4 (2015) 9—21.

[81] N. Sozer, J.L. Kokini, Nanotechnology and its applications in the food sector, Trends in Biotechnology 27 (2) (2009) 82—89.

[82] L.N. Bell, Nutraceutical stability concerns and shelf life testing, in, REC Wildman. Handbook of Nutraceuticals and Functional Foods, pp. 467-483.

[83] C.C. Bennett, K. Hauser, Artificial intelligence framework for simulating clinical decision-making: a Markov decision process approach, Artificial Intelligence in Medicine 57 (1) (2013) 9—19.

[84] A.F. Girão, M.C. Serrano, A. Completo, P.A. Marques, Do biomedical engineers dream of graphene sheets? Biomaterials Science 7 (4) (2019) 1228—1239.

[85] H.C. Koydemir, A. Ozcan, Wearable and implantable sensors for biomedical applications, Annual Review of Analytical Chemistry 11 (2018) 127—146.

[86] B. Chawla, Bench to bedside: translational research demystified, The Official Scientific Journal of Delhi Ophthalmological Society 29 (1) (2018) 4—5.

[87] A.L. Van der Laan, M. Boenink, Beyond bench and bedside: disentangling the concept of translational research, Health Care Analysis 23 (1) (2015) 32—49.

[88] S. Hua, M.B. De Matos, J.M. Metselaar, G. Storm, Current trends and challenges in the clinical translation of nanoparticulate nanomedicines: pathways for translational development and commercialization, Frontiers in Pharmacology 9 (2018).

[89] J. Ai, E. Biazar, M. Jafarpour, M. Montazeri, A. Majdi, S. Aminifard, M. Zafari, H.R. Akbari, H.G. Rad, Nanotoxicology and nanoparticle safety in biomedical designs, International Journal of Nanomedicine 6 (2011) 1117.

[90] C. Martín, K. Kostarelos, M. Prato, A. Bianco, Biocompatibility and biodegradability of 2D materials: graphene and beyond, Chemical Communications 55 (39) (2019) 5540—5546.

[91] R. Gupta, H. Xie, Nanoparticles in daily life: applications, toxicity and regulations, Journal of Environmental Pathology, Toxicology and Oncology: Official Organ of the International Society for Environmental Toxicology and Cancer 37 (3) (2018) 209.

[92] K. Hund-Rinke, M. Herrchen, K. Schlich, K. Schwirn, D. Völker, Test strategy for assessing the risks of nanomaterials in the environment considering general regulatory procedures, Environmental Sciences Europe 27 (1) (2015) 24.

[93] P. Datta, G.K. Mohi, J. Chander, Biomedical waste management in India: critical appraisal, Journal of Laboratory Physicians 10 (1) (2018) 6.

[94] K. Beumer, S. Bhattacharya, Emerging technologies in India: developments, debates and silences about nanotechnology, Science and Public Policy 40 (5) (2013) 628—643.

[95] M. Imran, A.K. Najmi, M.F. Rashid, S. Tabrez, M.A. Shah, Clinical research regulation in India-history, development, initiatives, challenges and controversies: still long way to go, Journal of Pharmacy and Bioallied Sciences 5 (1) (2013) 2.

[96] P. Rastogi, Patenting of nanotechnology invention: issues and challenges in India, Nanotechnology Law and Business 10 (2013) 139.

[97] G.A. Van Norman, Drugs, devices, and the FDA: Part 1: an overview of approval processes for drugs, Journal of the American College of Cardiology: Basic to Translational Science 1 (3) (2016) 170–179.

[98] M.H. Nayfeh, Fundamentals and Applications of Nano Silicon in Plasmonics and Fullerines: Current and Future Trends, Elsevier, 2018.

Index